# New Forms of Urban Agriculture: An Urban Ecology Perspective

Jessica Ann Diehl • Harpreet Kaur
Editors

# New Forms of Urban Agriculture: An Urban Ecology Perspective

Springer

*Editors*
Jessica Ann Diehl
Department of Architecture
National University of Singapore
Singapore, Singapore

Harpreet Kaur
Agroparistech
Paris, Paris, France

ISBN 978-981-16-3737-7 ISBN 978-981-16-3738-4 (eBook)
https://doi.org/10.1007/978-981-16-3738-4



This Springer imprint is published by the registered company Springer Nature Singapore Pte Ltd.
The registered company address is: 152 Beach Road, #21-01/04 Gateway East, Singapore 189721, Singapore

# Acknowledgements

Jessica A. Diehl would like to thank the financial support from the Singapore Ministry of Education (R-295-000-141-133). A special thanks to Ruiee Dhuri, who helped in the layout, organization, and editing of this book. She is currently a graduate student at the National University of Singapore in the Master of Landscape Architecture Programme.

# Contents

# Editors and Contributors

## About the Editors

**Jessica Ann Diehl** is Assistant Professor in the Department of Architecture at the National University of Singapore, where she teaches geodesign and a community design studio addressing issues of urban food security in the Master of Landscape Architecture Program. Research interests include place-based investigation of social networks, health equity, and alternative food systems. She holds a PhD in Health and Behavioral Sciences from the University of Colorado Denver, USA, where she was a National Science Foundation (NSF) IGERT PhD Fellow in Sustainable Urban Infrastructure Systems and a 2013–2014 Fulbright-Nehru Research Fellow affiliated with the School of Human Ecology at Ambedkar University Delhi, India. She holds a BLA/MLA in Landscape Architecture from The Pennsylvania State University, and a BA in English with a concentration in creative writing from West Virginia Wesleyan College.

**Harpreet Kaur** Harpreet Kaur's research interest focuses on taking an ecological approach towards understanding functions and drivers within agroecosystems and applying this knowledge towards sustainable management of farming systems. At Agroparistech, she investigated the effects of agriculture on biogeochemical cycles at local and national scales. By quantifying how nutrients flow across various components of agricultural production, she seeks to evaluate the factors that lead to nutrient surpluses or losses. Her previous research work mainly focused on high altitude agriculture in the Indian Cold Desert. She holds an MPhil in Environmental Sciences from the Jawarharlal Nehru University, New Delhi, India, where she was a Senior Research Fellow funded by the University Grants Commission, India.

## Contributors

**Sumer Alali** Department of Environmental Science and Policy, Università degli Studi di Milano, Milan, Italy

**Federico Andreotti** Laboratory of Geo-Information Science and Remote Sensing, Wageningen University & Research, Wageningen, The Netherlands

CIRAD, UPR GREEN, Montpellier, France
GREEN, University of Montpellier, CIRAD, Montpellier, France

**Stefano Bocchi** Department of Environmental Science and Policy, Università degli Studi di Milano, Milan, Italy

**Jessica Ann Diehl** Department of Architecture, National University of Singapore, Singapore, Singapore

**Molly Doane** Department of Anthropology, University of Illinois at Chicago, Chicago, IL, USA
The Field Museum, Chicago, IL, USA

**Stephen Fisher** Resilient Sustainable Infrastructure, Tetra Tech, Inc., Denver, CO, USA

**Elisha Goodman** Massachusetts Institute of Technology, Boston, MA, USA

**S. Gopakumar** Directorate of Education, Kerala Agricultural University, Thrissur, Kerala, India

**Madhavi Jain** School of Environmental Sciences, Jawaharlal Nehru University, New Delhi, India

**Sarah W. James** Australian National University, Canberra, ACT, Australia

**Harpreet Kaur** Agroparistech, Paris, France

**Bindu Gauri Kottaram** Agri Business School for Women, Krishi Vigyan Kendra, Indian Council for Agricultural Research, Coimbatore, India

**Rajesh S. Kumar** Indian Forest Service (IFS), New Delhi, India

**Shilpi Kundu** Department of Agricultural Extension & Information Systems, Sher-e-Bangla Agricultural University, Dhaka, Bangladesh

**M. G. Leech** IERM, Sandton, South Africa
SACLAP, Gauteng, South Africa

**Haley Lerand** Department of Anthropology, University of Illinois at Chicago, Chicago, IL, USA

**David C. Love** Johns Hopkins University, Baltimore, MD, USA

**Leslie Miller-Robbie** University of Colorado Denver, Denver, CO, USA

**Emily S. Minor** Department of Biological Sciences, University of Illinois at Chicago, Chicago, IL, USA
Institute for Environmental Science and Policy, University of Illinois at Chicago, Chicago, IL, USA

**Francesca Orlando** Department of Molecular and Translational Medicine (DMMT), Università degli Studi di Brescia, Brescia, Italy

**Kate Oviatt** University of Colorado Denver, Denver, CO, USA

**David W. Ow** Plant Gene Engineering Center, Chinese Academy of Sciences Key Laboratory of South China Agricultural Plant Molecular Analysis and Genetic Improvement, South China Botanical Garden, Chinese Academy of Sciences, Guangzhou, China

**Anu Ramaswami** Princeton University, Princeton, NJ, USA

**Michael Roberts** University of Illinois at Chicago, Chicago, IL, USA

**Ching Sian Sia** Department of Architecture, National University of Singapore, Singapore, Singapore

**Anindya Sinha** Frontier Elephant Programme and Animal Behaviour and Cognition Programme, National Institute of Advanced Studies, Bangalore, India
Indian Institute for Scientific Education and Research Kolkata, Mohanpur, India
Centre for Wildlife Studies, Bangalore, India

**Alexis D. Smith** University of Illinois at Chicago, Chicago, IL, USA

**Nishant Srinivasaiah** Frontier Elephant Programme, Foundation for Ecological Research, Advocacy and Learning, Pondicherry, India
Frontier Elephant Programme and Animal Behaviour and Cognition Programme, National Institute of Advanced Studies, Bangalore, India
SOAS University of London, London, UK

**Yun-Lin Su** Plant Gene Engineering Center, Chinese Academy of Sciences Key Laboratory of South China Agricultural Plant Molecular Analysis and Genetic Improvement, South China Botanical Garden, Chinese Academy of Sciences, Guangzhou, China

**Raman Sukumar** Asian Nature Conservation Foundation, Bangalore, India
Centre for Ecological Sciences, Indian Institute of Science, Bangalore, India

**Valentina Vaglia** Department of Environmental Science and Policy, Università degli Studi di Milano, Milan, Italy

**Srinivas Vaidyanathan** Frontier Elephant Programme, Foundation for Ecological Research, Advocacy and Learning, Pondicherry, India

**Saleema Vellani** The World Bank, Washington, DC, USA

**Dorte Verner** The World Bank, Washington, DC, USA

**Ivonne Weichold** Department of Geography and Spatial Planning (DGEO), Programme in Architecture, European Urbanization and Globalization, University of Luxembourg, Esch-sur-Alzette, Luxembourg

**Ambrogio Zanzi** Department of Environmental Science and Policy, Università degli Studi di Milano, Milan, Italy

# Introduction: New Forms of Urban Agriculture Embedded in Urban Resources—Where Is the Evidence? 1

Jessica Ann Diehl and Harpreet Kaur

## 1.1 Introduction

UN-Habitat estimates that by 2030 urban populations will account for 60% of the world's population, putting immense pressure on urban resources. One of the most urgent requirements of any urban agglomeration is the question of urban food security. Urban populations depend on the reliable and stable availability of food products, as well as affordable and convenient to access, that does not compromise their health and well-being. Acknowledging that cities are reliant on food imports, Bloem and de Pee (2017) suggest enhancing rural-urban linkages to improve urban food systems. But, as city populations continue to grow, rural land moves farther away and foodsheds—the area of land needed to grow enough to support a particular population (Blay-Palmer et al. 2018)—overlap, causing competition for the same rural resources.

Urban Agriculture (UA) has increasingly being considered a remedy to inadequate and often expensive access to food in cities. UA appears in various forms—community farms, rooftop gardens, urban orchards—and can be formal or informal. Defined as is an industry located within (intra-urban) or on the fringe (peri-urban) of a town, a city, or a metropolis, which grows or raises, processes, and distributes a diversity of food and non-food products (Mougeot 2000), however, UA is not just about where it occurs, rather, the distinguishing characteristic of urban agriculture is "that it is an integral part of the urban economic, social and ecological system: urban agriculture uses urban resources (land, labor, urban organic wastes, water), produces for urban citizens, is strongly influenced by urban conditions (policies, competition

J. A. Diehl (✉)
Department of Architecture, National University of Singapore, Singapore, Singapore
e-mail: akijac@nus.edu.sg

H. Kaur
Agroparistech, Paris, France





J. A. Diehl, H. Kaur (eds.), *New Forms of Urban Agriculture: An Urban Ecology Perspective*, https://doi.org/10.1007/978-981-16-3738-4_1

for land, urban markets and prices) and impacts the urban system (effects on urban food security and poverty, ecological and health impacts)" (van Veenhizen 2006, p. 2).

There is an increasingly expanding body of literature expounding the potential of UA as a strategy for addressing diverse social, economic, and environmental issues facing cities: food security, unemployment and economic opportunity, education, habitat creation, stormwater mitigation, and other ecosystem services (France 2016; Smith 2013). Bridging empirical studies, theoretical approaches, and policy analysis, UA is hailed to fulfill a variety of needs across multiple scales. It has the potential to meet both short-term population needs by increasing food security and providing livelihood opportunities, and long-term system needs by helping to mitigate global climate change (Beddington et al. 2012; Lesher 2006). But, at the same time, agriculture in, around, or outside the city can be polluting, exclusionary, marginalizing, and fail to be productive (Agrawal et al. 2003; Cole et al. 2008; Horst et al. 2017; van Veenhizen 2006). UA is threatened by land scarcity and insecurity due to land use competition, and is at risk for contamination by pollution emitted by vehicles, heavy metals in soils, organic chemicals, and urban sanitation (Armanda et al. 2019). Furthermore, although UA contributes to urban consumption requirements, practitioners and academics contest growing food in the city as a reliable and plausible strategy for feeding city residents (Morgan 2009; Raja et al. 2008; Siegner et al. 2018).

The explicit framing of UA as embedded in the urban "system" challenges cities to address agriculture not as a discrete activity, but as an important component of economic, social, and ecological systems. Coupled with this, land competition and rising land values in and around cities have intensified a focus on increasing the food supply without increasing agricultural area (Diehl et al. 2020). Intensification of agriculture is not a new idea, as witnessed in the Green Revolution. But, increased inputs of mineral fertilizer, agro-chemicals, and other external inputs are unsustainable and cannot be the future model. Options are emerging with new technology adoption that improves food production efficiency (Armanda et al. 2019; Smith 2013). Many of these, including hydroponic and aquaponic systems, are well-suited to combine with other land uses. With diverse strategies and systems thinking, there is opportunity for cities to leverage services in addition to food production that urban agriculture can provide such as stormwater mitigation, building energy reduction, and soil remediation. Integration of UA in multi-functional forms could be a critical adaptation for the sustainability of future cities. However, incorporating UA as part of the city system requires a diverse range of technical, scientific, planning, horticultural, and social expertise to be feasible and sustainable (Weidner et al. 2019).

The conception of UA as being *necessarily* integrated in urban systems has sparked attention from researchers in this post-productivist era in which there is broad evidence and awareness of our finite resources (La Rosa et al. 2014; McFarland 2015). Forms of UA that provide benefits in addition to food production—ecosystem services such as those already mentioned—have been termed New Forms of Urban Agriculture (NFUA). Growing evidence suggests that incorporating NFUA into the urban environment will greatly improve the sustainability of cities, taking advantage of the multiple benefits and services it can provide. For example, as highly managed plant communities, productive spaces can exhibit high levels of biodiversity, often exceeding that of other green space areas within the city. Additionally, it is likely that variation in vegetation cover, diversity, and structure influence not only the biodiversity in UA, but also the quantity and quality of ecosystem services supported by such systems. While NFUA have the potential to provide diverse benefits to humans, we shift the focus to investigate the potential impacts of urban resources on NFUA. Using an urban ecology lens, we wanted to know more about how urban resources of land, water/waste, labor, and biodiversity impact NFUA. Simply defined, urban ecology is the scientific study of the relation of living organisms with each other and their surroundings in the context of the urban environment (Forman 2014). As a concept, it is complex and dynamic, comprised of energy flows and feedbacks among the components. For example, how park vegetation impacts quality and quantity of stormwater runoff from an adjacent road, provides respite to nearby office workers during their lunch break, and provides birds a stepping stone in their migratory route. Boundaries are not fixed. And, investigation requires diverse disciplines, expertise, and geographic contexts—which was precisely our starting point as editors.

## 1.2 Origin of the Book

We met as a PhD candidate and a master's student: Jessica conducting fieldwork in Delhi on social networks of marginalized urban farmers and Harpreet finishing an M. Phil in Environmental Sciences at JNU-Delhi. Jessica hired Harpreet as an interpreter, but both soon found deep overlapping research interests in complementary aspects of urban agriculture: why were farmers using so much chemical fertilizer to grow crops on the floodplain of an already highly polluted river? As a health and behavioral scientist (and landscape architect), Jessica asked Harpreet about the ecological impacts. As an environmental scientist, Harpreet asked Jessica about the social drivers. Ten years later, we have continued to ask each other such relevant questions, tapping into our different expertise and experiences. Harpreet going on to study nutrient flows of rural farmers in Lahaul in north India due to rural-urban tourism and then on to Agroparistech to study effects of agriculture on biogeochemical cycles, and Jessica continuing to explore social and spatial social networks of urban farmers and the impacts on agricultural practices in major cities in southeast Asia and Australia. But, as we have come to understand the intrinsic relationships and dependency of urban agriculture in its many manifestations on basic resources of land and water as well as humans and biota, we have become keenly aware of a gap in the empirical research. NFUA are increasingly discussed and promoted by academics, practitioners, and governments and the social and policy research is trending up; however, where is the empirical data on urban water use and pollution? Where is the data on pollinators and biodiversity? What about land versus landless systems? And what about the cultivators? There is important research being conducted in all these areas, but it is like fragmented forest patches requiring corridors to establish linkages. We hope this book offers that.

## 1.3 Purpose of the Book

Thus, the purpose of this book is to address what we see as the absence of reliable empirical data on the scale and impact of urban resources on NFUA, which is a serious gap in our understanding of the viability and sustainability of UA and may also explain urban planner's reluctance to embrace the concept. There is thus an evident need for a systematic approach to gather and integrate available data on UA across varied typologies. The current volume intends to discuss the critical perspectives related to the actual and potential role of urban and peri-urban agriculture in the developing and the developed world, where forms, adaptations, and debates around NFUA vary distinctively. Using an urban ecology lens, the chapters are organized into four basic components of growing in the city: land, water/waste, labor, and biodiversity.

## 1.4 Structure of the Book

### 1.4.1 Land

The first section of this book explores the requirement for land. Growing food requires land for cultivation—or does it? Beginning with a study on land management through the development and application of a land use suitability model, Weichold argues that planners can evaluate ecological qualities of the land to make better informed decisions on where to develop land. The Analytic Hierarchical Process (AHP) method utilized in her Luxembourg study comprised five criteria of soil water holding capacity, erosion, slope, evaluation, and aspect, which were weighted based on expert feedback. More than identifying the most suitable land for agricultural use, it can show potential land use conflicts where planners could decide to preserve the land for agriculture or define a mixed-use zone that integrates both development and agriculture. A suitability model can be modified to fit different urban contexts, thereby giving planners and other stakeholders a basis for negotiation. Chapter 3 looks deeper at the issues and impacts of land use and land cover (LULC) changes in Delhi, India. Jain begins with an overview of urban versus rural areas, specifically highlighting the city's need for economic generation, which usually target natural and agricultural land in pursuit of urbanization—resulting in urban heat islands and heat stress. She proposes opportunities for NFUA on idle land and land unfit for development to mitigate climate risks and describes how agriculture can be combined with other economic activities. Chapter 4 investigates the potential for rooftop farming. Su and Ow address a gap in research on the economic feasibility of rooftop farming in the city of Guangzhou, China. Using empirical data, they test multiple scenarios including which vegetable types are suitable for cultivation, vegetable quality, production capacity, costs, yields, market price, and consumer preferences. In Chap. 5, Kumar et al. apply a discourse analysis of the current state of UA in India to summarize case examples of different models of UA that address constraints on land availability, major supply side factors that facilitate such models, and what institutional arrangements are driving UA across urban areas in India. The case examples summarized in this chapter reveal diverse and innovative ways that land is made available for cultivation, who participates, and enabling stakeholders and resources. The last chapter (Chap. 6) in this section moves us to the capital city of Port Vila in the Republic of Vanuatu in the Pacific Islands. Threatened by natural disasters, isolated geography, and heavy reliance on imported food, it is both a unique context and one that lends deeper insight into land use complexities that go beyond issues of physical availability, to understand economic and political impacts on land accessibility and utilization for agricultural uses.

### 1.4.2 Water/Waste

The second section of this book focuses on the essential resource of water in the functioning and sustainability of UA. Water availability and water quality are considered as an input and an output of NUFA, across different scales, and as a vector of nutrients and waste flows. In Chap. 7, Fisher takes an engineering perspective, first describing characteristics of water in global cities including water stress, water resources, and wastewater as a resource. Applying a Life Cycle Assessment (LCA), he summarizes water requirements for urban vegetable production. He urges planners to ask whether the city should feed itself and how water should be allocated for agriculture. Chapter 8 links water, energy, and health in a case study of urban agriculture in Hyderabad, India. Miller-Robbie and Ramaswami evaluate the impacts of three qualities of wastewater on water use, energy use and greenhouse gas emissions, nutrient uptake, and crop pathogen quality. Verner et al. link water, livelihoods, and health in Chap. 9. Focusing on the Middle East and North Africa (MENA), they show that Frontier Agriculture (climate-smart, water-saving technologies, e.g., hydroponics, aquaponics) can contribute to improve the well-being and nutrition status of refugee populations. In Chap. 10, Leech provides a case study of wastewater use by poor urban agriculturalists in and around Durban, South Africa due to the issue of water scarcity. He compares policy and legislation with actual agricultural practices related to stormwater capture and describes ways to bridge top-down and bottom-up approaches in order to implement UA support and protect human health.

### 1.4.3 Labor

The third section of this book turns attention to labor. As a managed ecosystem, UA requires someone to cultivate it. Beginning with Chap. 11, Oviatt acknowledges that while UA has many benefits, they are not experienced uniformly among participants.

She presents a case study on the AGRUPAR urban agriculture program in Quito, Ecuador to explore how the practice and benefits of urban agriculture differ among producers based on three primary characteristics: migration history, age, and gender. While Oviatt focuses on small-scale, mostly subsistence UA, Chap. 12 shifts to commercial-scale UA. Sia and Diehl describe a trend away from traditional toward high-tech farming in Singapore, specifically reflecting on the changing demographic profile of the urban farmer. They conducted interviews with farmers to understand their agricultural background (knowledge, experience, skills), educational level and wages, motivation, and community involvement. Digging deeper into the entrepreneurial aspects of NFUA, Zanzi et al. investigate the OpenAgri project in Milan, Italy, a start-up project focused on food production and agroecological land restoration. This research focuses on the quantification and evaluation of strategies for enhancing ecosystem services and investigating their link with job opportunities. In Chap. 14, Diehl conceptualizes labor as a resource and a social network as part of a larger goal of creating sustainable farming livelihoods. And, describes four case studies with diverse farmer-labor social networks in Delhi, India; Jakarta, Indonesia; Singapore; and Sydney, Australia.

### 1.4.4 Biodiversity

The final section of this book looks at biodiversity and NFUA. Chapter 15 focuses on pollinators—a critical resource for the production of many edible crops. Smith et al. conducted a study on honey bees and wild bees in community gardens in Chicago, USA. They explored questions about urban apiculturists' perceptions and knowledge of wild bees, as well as the impact of urban apiculture on wild bees in community gardens. NFUA can attract beneficial insects, but also other animals foraging for food. In Chap. 16, Srinivasaiah et al. track the impact of land use change on Asian elephants in peri-urban southern India. While the land area under agriculture has only slightly increased, there have been significant changes in the shift from single-cropping to double- or multiple cropping every year and an overall increase in built-up areas. Although elephants generally prefer forested habitats, the increased availability of nutritious crops and forest cover in the form of agroforestry plantations outside protected forests has led them to move extensively across peri-urban areas and successfully adapt to this novel anthropogenic ecological regime—with increased potential for human–animal conflict. In the final chapter in this section, Chap. 17, Kaur summarizes the current state of the research on NFUA and biodiversity. She reminds us that loss of biodiversity and ecosystem function are among the top five threats to humankind in the next decade. First, she focuses on why biodiversity in an urban area should be conserved and common constraints for its conservation. Examples are described from various cities on the new approaches undertaken to maintain and develop green spaces for improving floral and faunal biodiversity. Then, she summarizes traditional and new forms of urban agriculture, range from cultivated open spaces to organoponics, and their role in conservation of agri- and related biodiversity.

While the chapters in this edited book cover diverse topics, scales, geographies, disciplines, and research methodologies they are by no means comprehensive or conclusive. Rather, they provide snapshots of evidence of the complexity and dynamic system of urban ecology as it interacts with NFUA. Our intent is that by bringing together these "patches" of research, we can start to see a larger system at work and identify important gaps for further investigation. Before NFUA can make a meaningful and sustainable contribution to urban economic, social, and ecological systems, it must first be sustained by urban resources—and we need evidence of how that can happen.

## References


Agrawal M, Singh B, Rajput M, Marshall F, Bell JNB (2003) Effect of air pollution on peri-urban agriculture: a case study. Environ Pollut 126:323–329

Armanda DT, Guinée JB, Tukker A (2019) The second green revolution: innovative urban agriculture's contribution to food security and sustainability–a review. Glob Food Sec 22:13–24

Beddington J, Asaduzzaman M, Clark M, Fernández A, Guillou M, Jahn M et al (2012) Achieving food security in the face of climate change: final report from the Commission on Sustainable Agriculture and Climate Change. Commission on Sustainable Agriculture and Climate Change, Copenhagen, Denmark

Blay-Palmer A, Santini G, Dubbeling M, Renting H, Taguchi M, Giordano T (2018) Validating the city region food system approach: enacting inclusive, transformational city region food systems. Sustainability 10(5). https://doi.org/10.3390/su10051680

Bloem S, de Pee S (2017) Developing approaches to achieve adequate nutrition among urban populations requires an understanding of urban development. Glob Food Sec 12:80–88. https://doi.org/10.1016/j.gfs.2016.09.001

Cole D, Grace D, Diamond M (2008) Researchers' approaches to evidence on urban agriculture and human health. In: Cole D, Lee-Smith D, Nasinyama G (eds) Healthy city harvests: generating evidence to guide policy on urban agriculture. CIP/Urban Harvest and Makerere University Press, Lima, Peru

Diehl JA, Sweeney E, Wong B, Sia CS, Yao H, Prabhudesai M (2020) Feeding cities: Singapore's approach to land use planning for urban agriculture. Glob Food Sec 26:100377

Forman RT (2014) Urban ecology: science of cities. Cambridge University Press, Cambridge

France R (2016) Integrated urban agriculture, precedents, practices, prospects. Green Frigate Books, Oxfordshire

Horst M, McClintock N, Hoey L (2017) The intersection of planning, urban agriculture, and food justice: a review of the literature. J Am Plan Assoc 83(3):277–295

La Rosa D, Barbarossa L, Privitera R, Martinico F (2014) Agriculture and the city: a method for sustainable planning of new forms of agriculture in urban contexts. Land Use Policy 41:290–303. https://doi.org/10.1016/j.landusepol.2014.06.014

Lesher CW (2006) Urban agriculture: a literature review. Alternative Farming Systems Information Center, National Agricultural Library, Beltsville, MD

McFarland P (2015) The peri-urban land-use planning tangle: an Australian perspective. Int Plan Stud 20(3):161–179. https://doi.org/10.1080/13563475.2014.965250

Morgan K (2009) Feeding the city: the challenge of urban food planning. Routledge, Milton Park

Mougeot LJA (2000) Urban agriculture: definition, presence, potentials and risks. IDRC, Ottawa, Canada

Raja S, Born BM, Russell JK (2008) Planners guide to community and regional food planning. American Planning Association, Chicago

Siegner A, Sowerwine J, Acey C (2018) Does urban agriculture improve food security? Examining the nexus of food access and distribution of urban produced foods in the United States: a systematic review. Sustainability 10(9):2988

Smith P (2013) Delivering food security without increasing pressure on land. Glob Food Sec 2(1): 18–23

van Veenhizen R (2006) Cities farming for the future: urban agriculture for green and productive cities. IIRR, RUAF, IDRC, Leusden, The Netherlands

Weidner T, Yang A, Hamm MW (2019) Consolidating the current knowledge on urban agriculture in productive urban food systems: learnings, gaps and outlook. J Clean Prod 209:1637–1655

# Part I

# Land

# Managing Land: Protecting, Integrating, and Allocating Agriculture in Urban Design and Planning—The Case of Luxembourg

2

Ivonne Weichold

**Abstract**

The aim of this study was to highlight the potentials and challenges of managing land-use planning through a contemporary case study addressing the future of spatial development in the Luxembourg region. The "Analytic Hierarchy Process (AHP)" method, commonly used in Agricultural land-use suitability analysis, was utilised in this study. The application comprises the five criteria of soil water holding capacity, erosion, slope, elevation, and aspect. In determining the weights of the parameters, experts were consulted for their opinions in order to generate an agricultural land suitability map. At the end of the assessment, it was estimated that 27.9% of the study area was of high to highest suitability for agriculture. From this initial investigation, the distribution of suitable areas within the existing perimeters of the built-up area was further estimated, based on a selection of municipalities.

The main contribution of this study is the combination of the parameters used in terms of agricultural production and the detailed explanation of the scoring approach, which has not previously been applied to the case of Luxembourg. The study adds valuable insights into current planning discussions in Luxembourg by providing spatial planning guidelines—at an urban and regional scale—on where to develop land while respecting the ecological qualities of its land, in this case the soil quality.

**Keywords**

Multi-criteria decision analysis · Analytic hierarchy process · Agricultural land-use suitability · Urban design and planning · Luxembourg

I. Weichold (✉)
Department of Geography and Spatial Planning (DGEO), Programme in Architecture, European Urbanization and Globalization, University of Luxembourg, Esch-sur-Alzette, Luxembourg
e-mail: ivonne.weichold@uni.lu





J. A. Diehl, H. Kaur (eds.), *New Forms of Urban Agriculture: An Urban Ecology Perspective*, https://doi.org/10.1007/978-981-16-3738-4_2

## 2.1 Rethinking Urban Design and Planning with Suitability Measures

> Land development and management should be based on the study of the structure and history of the local landscape, on meticulous soil capability classification, on the top-soil map, the geological map, and so on. (Mumford 1971, p. 102)

Throughout the history of western urbanisation, for example, in the modernist manifesto of the Athens Charter from 1933 (CIAM/Le Corbusier), which classified different spatial urban functions in the city plan, agricultural or food production as a function was marginalised,[1] nor has land development according to the soil capability been considered (Mumford 2002).

Today, more than ever, urban development pressure, the decline of arable land, the rise in climate uncertainty, and the current global pandemic COVID-19 make us aware of the shortcomings of current planning practices in food security and the sustainable development of regional agriculture. However, the emergence of a sustained interest in alternative farming methods in and around cities (Rio Summit in 1992, Local Agenda 21) allowed urban food planning to develop from a "stranger in the planning field" (Pothukuchi and Kaufman 1999, 2000; Morgan 2009) into a conceptual niche for research and practice in post-industrial cities (Ilieva 2016) at the end of the twentieth century. Since then, many scholars and practitioners have written about the significance of urban farming for agricultural production, public policy, and food as a cultural element (AESOP Sustainable Food Planning Group; American Planning Association). This has resulted in a gradual rethinking of planning practices in terms of vulnerability, climate change, and disruption of global supply lines.

With more than two-thirds of the global population forecast to live in urban areas by 2050—up from 56% today (UN 2019)—planners also have to balance the competing demands of agriculture and urban development for land. In particular, the problem of soil fertility plays a sensitive and crucial role and is increasingly under threat: besides erosion and pollution, artificial covering has increased dramatically in recent decades in many areas of Europe, increasingly impacting global warming (UN 2017).

Research concerning alternative development strategies, including the ecological quality of agricultural zones and their soil fertility, calls for a careful rethinking of urban design and planning. Valuable insights on how to develop land without losing fertile, productive land can be found in Artur Glikson and Lewis Mumford's early consideration of the importance of ecology as a basis in planning (Mumford 1971) and the land-use analysis method of Ian McHarg (1969). Glikson and Mumford's reflections on land development and management included the history of local landscapes and their soil capability as a significant landscape-shaping factor. Ian

[1] In fact, in Le Corbusier's "Trois établissements humains" from 1945, agriculture was a key element. However, his proposals remained theoretical and were never realised, due to low acceptance in government and the private sector (Le Corbusier 1979; Arredondo-Garrido 2016).

McHarg, in his consideration of the environment in land-use decisions, supported such an argument with an overlay analysis method. Given today's development pressure, decision-support tools such as those of land-use analysis—which so far, have been used to choose between buildable or agricultural land—can also help to determine which agricultural land can be transformed in an agri-urban area.

Against this background, this chapter aims to explore the potentials and challenges of managing land-use planning. It examines the integration and protection of productive agricultural land in urban development plans using one of the most pervasive methodologies of landscape analysis and design: landscape suitability analysis through composite mapping. Land-use suitability analysis is an urban and regional planning tool that helps to identify the use that is most suitable for the land (Wang and Hofe 2007). In particular, the Analytical Hierarchy Process (AHP) method (Saaty 1990, 2000; Saaty and Vargas 2012) will be used for exploring the agricultural land-use suitability analysis in this research. The Grand Duchy of Luxembourg, a country with a rapid economic and demographic growth rate and an ambitious agricultural agenda, has been chosen as an example for examining this complex area. In this chapter, the author presents preliminary results of her on-going PhD research, which is currently being conducted at the University of Luxembourg. In the next section, the methodological approach for the construction of a land-use suitability for the case study of Luxembourg is illustrated. This section explores the future development of Luxembourg through a composite "envelope", tracing optimum suitability of productive agricultural land. In subsequent sections, the main findings from further research are tackled. The chapter ends with a discussion around the contribution and implementation of such suitability measures in urban design and planning.

## 2.2 Methodological Approach

### 2.2.1 The Luxembourg Context: Agricultural and Urban Development

The Grand Duchy of Luxembourg (GDL) is a country of 2586 km$^2$, sharing borders with France in the south, Germany in the east, and Belgium in the west (Fig. 2.1). Compared to other European countries, Luxembourg is one of the smallest member states of the European Union, and in recent decades it has seen above-average favourable economic transformation and demographic growth. In essence, the annual economic growth rate is in the range of 2–5%, with a cumulative increase in the population of more than 40% and growth of more than 250% in the number of trans-border commuters since the 2000s (STATEC 2020a, b, c). Although these trends impose enormous pressures upon both the capital city and the rural landscape, a third of the territory is covered by forests (35%), and half of it is used for farming and wine-growing purposes (51% agricultural land use), with the rest covered by settlements (10%) and infrastructure links (4%) (ibid. 2020) (Fig. 2.2).

**Fig. 2.1** Geographical positioning of Luxembourg in the Greater Region

Demographic growth and urbanisation tend to occur equally in urban and rural areas in Luxembourg. This development pattern explains the country's landscape, an extensive assemblage of small and medium-sized towns, which has a relatively low building density for residential and commercial areas (Table 2.1). Land development is rather diffuse and spread out, and mainly occurs in new areas, on agricultural land, instead of existing built surfaces. This high degree of sprawl has led to excessive land consumption, soil overuse, and a very high degree of landscape fragmentation, which is critical for biodiversity, landscape, and recreation. In fact, Luxembourg has the lowest degree of densification in Europe apart from Cyprus.[2]

There is a high demand for planning in Luxembourg, but urban planning practice in the country is rather young and far from being well established and commonly accepted (Hesse 2015). Planning sovereignty lies with the 102 municipalities. In addition, there is an administrative gap between the state and the municipality, which might explain why governmental studies are missing a broader appreciation of the

[2]The indicator of the degree of landscape fragmentation is developed by the European Environment Agency (EEA 2012a) in order to measure the degree to which new development is taking place over land already developed (in terms of construction).

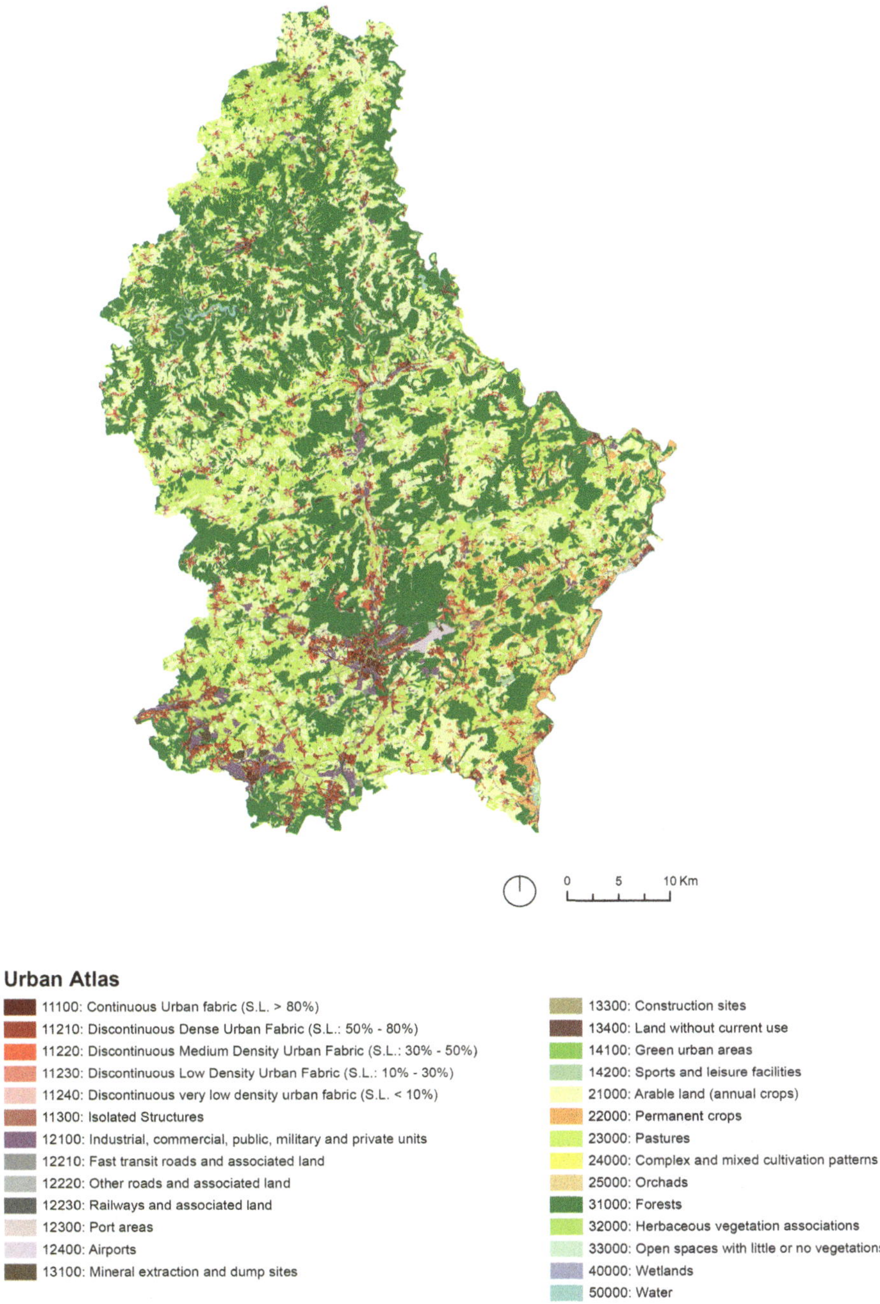


**Fig. 2.2** Urban Atlas 2012—Land-use map

**Table 2.1** Urban Atlas Categories—land development (CLMS 2016)

| Area | 2006 | 2012 | Change | Change | Change of total |
|---|---|---|---|---|---|
| Urban Atlas Categories | Hectares (ha) | ha | ha | % | % |
| Agricultural, semi-natural areas, wetlands | 133,868.3 | 132,815.8 | −1052.5 | −0.8 | −93.6 |
| Construction sites | 366.3 | 406.3 | 40.1 | 10.9 | 3.6 |
| Continuous urban fabric (S.L.[a]: >80%) | 817.7 | 862.7 | 45.0 | 5.5 | 4.0 |
| Discontinuous dense urban fabric (S.L.: 50–80%) | 6825.2 | 6965.2 | 140.0 | 2.1 | 12.5 |
| Discontinuous low-density urban fabric (S.L.: 10–30%) | 271.4 | 365.7 | 94.2 | 34.7 | 8.4 |
| Discontinuous medium density urban fabric (S.L.: 30–50%) | 4059.7 | 4097.4 | 37.7 | 0.9 | 3.4 |
| **Discontinuous very low-density urban fabric (S.L.: <10%)** | **0.7** | **153.8** | **153.1** | **21,597.1** | **13.6** |
| Fast transit roads and associated land | 644.6 | 653.1 | 8.4 | 1.3 | 0.8 |
| Forests | 94,822.1 | 94,784.7 | −37.5 | 0.0 | −3.3 |
| Green urban areas | 755.0 | 755.6 | 0.6 | 0.1 | 0.0 |
| Industrial, commercial, public, military, and private units | 6164.9 | 6667.1 | 502.1 | 8.1 | 44.7 |
| Isolated structures | 369.1 | 450.1 | 81.0 | 21.9 | 7.2 |
| Land without current use | 171.8 | 146.7 | −25.1 | −14.6 | −2.2 |
| Mineral extraction and dump sites | 593.9 | 584.5 | −9.3 | −1.6 | −0.8 |
| Other roads and associated land | 6339.5 | 6348.0 | 8.5 | 0.1 | 0.8 |
| Sports and leisure facilities | 1268.6 | 1277.5 | 8.8 | 0.7 | 0.8 |
| Water | 1242.4 | 1247.2 | 4.8 | 0.4 | 0.4 |

[a]*S.L.* sealing layer

qualities of the built fabric and socio-spatial relationships (such as accessibility, mixed use, density). The development plans (PAG) of the municipalities are revealed as extremely bottom-up, with local communities having reliable agency in defining development trajectories. Municipalities are mostly small, and private players, such as homeowners and real-estate developers, are very powerful. In fact, less than 10% of land ownership in Luxembourg is publicly owned, which in turn has far-reaching implications for architecture and urban development (Hertweck 2020). Finally, the liberal tradition of the government, the different agendas of local municipalities, and Luxembourg's current planning strategy, which is based on three Agglomeration development clusters (Agglomeration North, Luxembourg and South) and currently under construction (MECDD 2016), are challenging for spatial planning in Luxembourg.

**Table 2.2** Agricultural land use in Luxembourg (SER 2016)

| Agricultural area (2015) | Area<br>Hectares (ha) | Area<br>% |
|---|---|---|
| Utilised agricultural area | 131,384 | 100 |
| Permanent grassland | 66,923 | 51 |
| Arable land (annual crops) | 62,798 | 48 |
| Other agricultural land | 1663 | 1 |

Further, peculiar issues such as land speculation on undeveloped but buildable land are prevalent and have increased the pressure on undevelopable land in recent decades (see agricultural land prices in SER 2018). In turn, in addition to farmers and municipalities, real-estate developers also represent potential buyers for whom land is a profitable property. Therefore, municipalities and the state have recently increased their buying of large properties in order to retain more land in public hands. This positive act is enhanced by the fact that sellers do not have to pay value-added tax if they sell their property to public bodies.

Yet, even with the legislation in place, to protect farmland, for example, transforming permanent meadows and pastures into land for arable crops has been prohibited by law since 1996 in the "Loi Protection Nature" and "PEEN—Prime entretien espace naturel" (MECDD and MA 2016; MECDD and FI 2018). This does not prevent repurposing of good quality farmland by the government (municipalities) and other private stakeholders. Luxembourg's agricultural sector consists mainly of peri-urban agriculture. The utilised agricultural area is predominantly defined by arable land and permanent grassland (permanent grassland and pasture) (Table 2.2). Around 63,000 ha of the approximately 131,384 ha of agricultural land are cultivated; the rest of the area is covered by permanent grassland, with 67,000 ha (see Box 2.1: Crop production).

**Box 2.1 Crop Production**

The main component of Luxembourg's arable farming is the cultivation of cereals, with almost 29,288 ha (22% of the agricultural area). Of these, the main crops are wheat, with more than 14,494 ha, followed by barley, with around 7713 ha. Other important cereal crops are triticale (4604 ha) as well as oats and oat-weight mixed cereals (1381 ha). Forage plants (maize, forage or arable grass, and fodder legumes) cover around 26,091 ha (20% of the agricultural area), which can be explained by the fact that agriculture is characterised by livestock farming, especially dairy. With around 5199 ha, industrial plants (mainly rapeseed) comprise only 4% of the agricultural area, and potatoes are cultivated on only around 570 ha, most of which is used to produce seedlings, in the north of the country. Horticulture (102 ha), vineyards (1296 ha), and other permanent crops account for a limited proportion. Noticeably, the vineyards have a significant positive impact on the farms and the cultural landscape in the east of the country along the Moselle River (MA 2016; SER 2016).

Standing in contrast to the dominant presence of agricultural land is a relatively low percentage of local food production produced for local consumption: an estimated 90% of food is imported, mainly from the European Union. Local production is primarily meat, milk products, and grains (STATEC 2020a, b, c; Eurostat 2012). Farming consists of conventional farms that account for 95% (MECDD 2019) and organic agriculture at 5% (ibid. 2019), but has diversified in recent decades. Concerns over food security and sustainable agricultural development have given rise to a rethinking of Luxembourg's food systems.

In 2019, Luxembourg's government conducted a "National strategy for urban agriculture", aiming to increase local organic agricultural production at the national level 25% by 2020 and 100% by 2050 (ibid. 2019), thereby being in line with the European Green Deal *Farm to Fork*[3] policy. The results of this study are a series of recommendations to facilitate the development and implementation of urban farming in the GDL. The application of the recommendations is still vague, and it will take several years to assess their real impact on the development of agricultural land. So far, no fixed measures concerning the way food and land are currently produced and consumed have been integrated, nor have they been integrated into the land use and development plan of Luxembourg or supplementary documents and urban plan. The lack of a cross-sector strategy further reflects Luxembourg's weakness in planning, which does not manage to anchor all of the different demands of other fields in one strategy. Nevertheless, the governmental strategy to go 100% organic is a positive act. This ecological approach is the right step towards sustainable development for Luxembourg, whether those measures are feasible or not.

All in all, this brief overview of urban and agricultural developments illustrates the fact that Luxembourg faces specific challenges, to which future spatial and regional planning will have to react.

### 2.2.2 Agricultural Land-Use Suitability Using GIS and the AHP Technique

Against this background, the question of **how Luxembourg can develop land without abandoning its fertile, productive agricultural land** was investigated. With an increased awareness of the need to protect the vital ecosystem of farmland, identification of suitable agricultural land can be helped by decision-support tools such as land-use suitability analysis, which can be integrated with the decision-making processes of urban and regional planning. This section provides a snapshot of such an application tool.

[3]The European Green Deal is a set of policy initiatives by the European Commission with the overarching aim of making Europe climate-neutral in 2050. One goal is to reach the objective of having at least 25% of the EU's agricultural land under organic farming by 2030 and achieving a significant increase in organic aquaculture (EC 2020).

#### 2.2.2.1 The Analytical Hierarchy Process Method

The most widely known multi-criteria decision-analysis approach, the "Analytic Hierarchy Process" (AHP) method, is used in this study to determine the distribution and areas of land suitable for agriculture in GDL. The AHP method has been developed by Saaty (2000), Saaty and Vargas (2012) and is commonly used in agricultural land-use suitability analysis (Akıncı et al. 2013; Malczewski 2004; Pramanik 2016).

It is a useful tool for dealing with the problem of designing alternatives which optimise objectives. It allows decision-making to take account of complex sustainability issues and can help to recognise and define a problem in detail. It is widely used to deconstruct a decision-making problem into its constituent parts, which are then structured hierarchically (Saaty and Vargas 2012).

The hierarchical model consists of objectives, criteria, and alternatives used for every issue (Saaty 1990). Once the problem has been set in a hierarchical structure, the weights of the criteria maps forming the hierarchy are calculated (Akıncı et al. 2013). Saaty (2000), Saaty and Vargas (2012) suggested that the criteria are then evaluated by comparison with other criteria in a hierarchy level. This involves the construction of a matrix where each criterion is compared with other criteria, relative to its importance, on a scale of 1–9. Scoring is undertaken by using the preference "fundamental scale" (Table 2.3), and a pairwise comparison matrix is created. The pairwise comparison matrix consists of $n\ (n - 1)/2$ comparisons for $n$ number of criteria (Akıncı et al. 2013; Malczewski 2004).

Weights or priorities are determined by normalising the pairwise comparison matrix. A "normalised pairwise comparison matrix" is obtained by calculating the column and row elements as shown in Table 2.4: by dividing the column elements of the matrix by the sum of each column; by totalling the row elements in the matrix that has been obtained; and by dividing the total value by the number of elements in

**Table 2.3** Fundamental scale for pairwise comparison—AHP method (Saaty and Vargas 2012)

| Intensity of importance | Definition | Explanation |
|---|---|---|
| 1 | Equal importance | Two activities contribute equally to the objective |
| 3 | Weak importance of one over another | Experience and judgement slightly favour one activity over another |
| 5 | Essential or strong importance | Experience and judgement strongly favour one activity over another |
| 7 | Demonstrated importance | An activity is strongly favoured and its dominance demonstrated in practice |
| 9 | Absolute importance | The evidence favouring one activity over another is of the highest possible order of affirmation |
| 2, 4, 6, 8 | Intermediate values between the two adjacent judgements | Intermediate values between the two adjacent judgements |
| Reciprocals | If activity $i$ has one of the above numbers assigned to it when compared with activity $j$, then $j$ has the reciprocal value when compared with $i$. | |

**Table 2.4** Pairwise comparison—based on expert opinion

| Criteria | Soil capacity | Erosion | Slope | Aspect | Temperature | Weights[a] % | Weights 0–1 |
|---|---|---|---|---|---|---|---|
| Soil capacity | 1 | 7 | 5 | 9 | 9 | 59.5 | 0.595 |
| Erosion | 1/7 | 1 | 1 | 3 | 5 | 13.3 | 0.133 |
| Slope | 1/5 | 1 | 1 | 7 | 7 | 18.9 | 0.189 |
| Aspect | 1/9 | 1/3 | 1/7 | 1 | 3 | 5.2 | 0.052 |
| Temperature | 1/9 | 1/5 | 1/7 | 1/3 | 1 | 3.1 | 0.031 |

[a]Relative weights: Max. eigenvalue ($W_{max}$) = 5.413; $n = 5$; Consistency index (CI) = ($W_{max} - n$)/($n - 1$) = 0.10325; Random index (RI) = 1.12; Consistency ratio (CR) = CI/RI = 0.092

the row. In this way, a priority relative vector or weight vector is generated. The weights produced in this procedure are within the range of 0–1, and their sum is equal to 1 (ibid. 2013; ibid. 2004).

Further, the AHP method also has the capability to identify inconsistencies in judgement, when performing pairwise comparisons of criteria. Saaty (2000), Saaty and Vargas (2012) therefore proposed a consistency ratio (CR) of the pairwise comparison, to measure inconsistencies in judgement. The consistency index (CI) of a matrix of comparisons is given by CI = ($W_{max} - n$)/$n$ − 1. The consistency ratio (CR) is obtained with the consistency index (CI) and the random index (RI) by using the formula CR = CI/RI.[4] According to Saaty (2000), Saaty and Vargas (2012), the upper limit of the CR value is 0.10. If the CR is below 0.10, it is considered that the judgements $n$ exhibit a sufficient degree of consistency and that the assessment can be continued. If the CR is above 0.10, then the judgements are considered untrustworthy because they are too close to randomness. In this case, the quality of the judgements needs to be improved and repeated.

#### 2.2.2.2 Criteria, Data Sets, and Methodology Used in the Land Suitability Analysis

This study is novel for two important reasons. First, the agricultural land suitability analysis in this study was not carried out for a particular type of crop production. This characteristic is one of the most important differences between this study and others and generates significant results for development pressure in Luxembourg. The second reason is the combination of the main parameters used in terms of agricultural production and the detailed explanation of the scoring approach, which has not previously been applied to the case of Luxembourg.

In order to determine the land in the study area that is suitable for agriculture, the criteria of soil water holding capacity, erosion, slope, elevation/temperature, and

[4]The formulas of the calculations are the following:

Max. eigenvalue: $W_{max} = M/(n + M - 1)$. Consistency Index (CI) = ($W_{max} - n$)/($n - 1$). Random Index (RI). Consistency ratio (CR) = CI/RI aspect were used (see Box 2.2). There were three main reasons for using these criteria in this study. The first was the fact that the criteria were adequate for determining the areas where vegetative production can be carried out. The second reason was that because the study was not conducted for a particular type of agricultural crop, the use of parameters for precipitation, humidity, natural efficiency, pH, salinity, and organic substance content was not needed. Third, the geographical data concerning the specified criteria had already been generated and made available by the related institutions Administration des Services Techniques de l'Agriculture (ASTA) and the Luxembourgish data platform. The framework for a suitability analysis (Fig. 2.3) depends heavily on data available, which also explains why five criteria could be assessed.

**Box 2.2 Criteria (See Fig. 2.4a–e and Table 2.5)**

The criteria used within the scope of the study are explained in detail in the following text. A more detailed overview of the compilation and equation of each dataset can be found in the references provided by MA/ASTA (Ministère de l'Agriculture, de la Viticulture et de la Protection des consommateurs, Administration des services techniques de l'agriculture) (Marx 2019; Steffen and Marx 2019).

**Soil water holding capacity**: Soil quality is defined very broadly as the capacity of a soil to function within ecosystem and land-use boundaries to sustain biological productivity, maintain environmental quality, and promote plant and animal health (Doran and Parkin 1994, 1997). Water storage is one of the most important parameters used to characterise the capacity of the soil to provide water for plant growth and influence the efficiency of rainfall used by crops. For the calculation of the soil water holding capacity, the spatial estimation of soil texture, volumetric stone content, bulk density, organic carbon content, and potential soil depth were integrated (Stevens et al. 2014; Steffen and Marx 2019). An overview of this criterion is presented in Map Soil Capacity. A very dominant soil water capacity of 15.14% is situated in the Gutland region, in the centre of Luxembourg, and in the northern part of the country. A medium capacity of 22.43% is evenly distributed all over the country, but mainly in the south and the wine-growing area. Very low capacities of 7.27% can be found across Luxembourg city and in the wine-growing area.

**Erosion**: According to the European Environment Agency (EEA 2012a, b), erosion is one of the main factors affecting the many functions of soil which are linked to climate, topography, and vegetation. Floods, mudslides, and the loss of fertility in agricultural soils are the consequences of too much run-off and accelerated erosion. The erosion dataset represents mean annual soil loss rates by sheet and rill erosion. The cover-management factor takes into account the 3-year crop rotation (2013–2014–2015), reduced soil tillage as

(continued)

**Box 2.2** (continued)
applied by farmers, and the presence of cover crop. An overview of the erosion criterion is presented in the Erosion Map. In 69.16% of the study area, tolerable erosion is observed. In 6.34% of the study area, moderate and low erosion is seen, and in 2.55% of the area, very severe and highly severe erosion is observed.[5]

**Slope**: The normal development of soils is closely related to the topography of the area with geomorphological properties. For one thing, slope indirectly limits agricultural production by affecting soil properties negatively; for another, slope negatively affects agricultural production directly by restricting the possibility of using machines and management applications such as soil tillage, irrigation, and drainage. The Slope Map shows that approximately 67.44% of the land in Luxembourg has a slope of 1–12%. Only 1.46% has a slope above 45%.

Generally, land areas with a slope up to 15% consist of large continuous areas of arable land. Land areas with a slope of more than 25% are only dominated by pastures and permanent meadows.

**Aspect**: To maintain their physiological activities, plants need sun exposure at specific intervals. The duration of this need varies according to the species of plant. However, in general, most cultigens exhibit optimum growth in the southern and western aspects, which receive sunlight for a substantial portion of the day. For this reason, aspect is taken into consideration as an assessment criterion for selecting the land to be used for agriculture.

The Aspects Map shows that 40% of the study area receives sufficient sunlight; only 13.3% has insufficient sunlight.

**Elevation/Temperature**: The elevation criterion is an essential factor that plays a part in the variation of plant cover by causing temperature changes. In the study area, the altitude varies between the northern and south-eastern parts: the highest elevation of 560 m (Wilwerdange) is found in the northern part, in the Oesling area. Luxembourg city has a medium elevation of 300 m. The lowest elevation, of 130 m (Wasserbillig), is in the Moselle valley, in the south-eastern part where the Moselle River flows. The average temperature for the period 1971–2000 was used for this parameter (Stevens et al. 2014). The annual mean temperature has a decreasing south-east/north-west gradient, with >10 °C on the Moselle and <7 °C in the extreme north. The annual rainfall is between 950 and 1000 mm in the north and west of the country. Only 700 mm rainfall occurs in the centre to the east of the country.

[5] Around 21.94% of the study area could not be assessed in this dataset due to a limitation of exact soil data on the scale of 1/100,000. Since these areas are mainly developed and flat land, the susceptibility to erosion was assessed with unsuitable with a value of 1.

**Fig. 2.3** Procedure followed in generating agricultural land-use suitability map

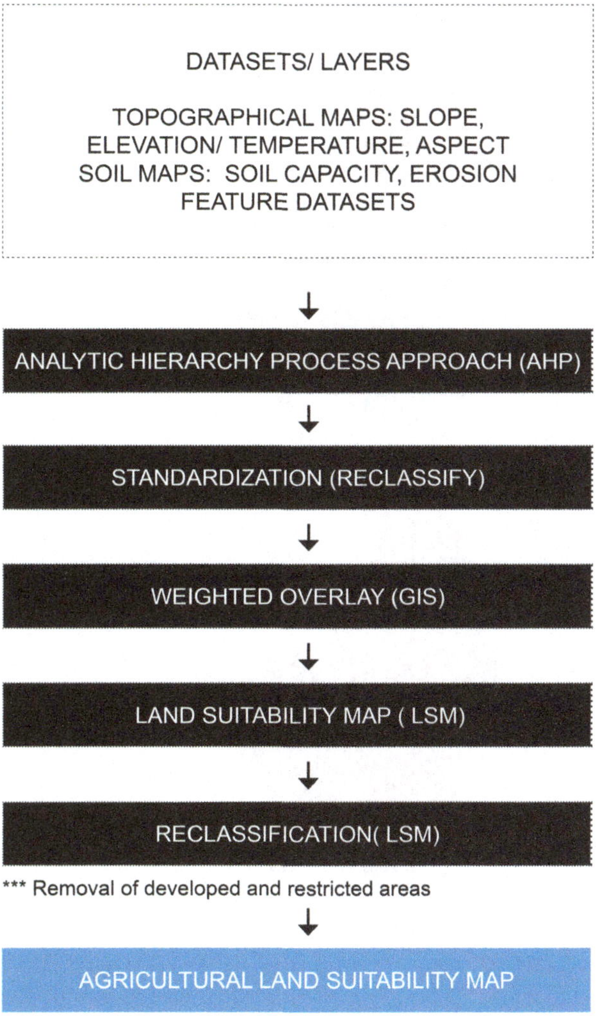


The datasets concerning the topographical parameters used in this study (slope, aspect, elevation) were obtained from a standard topographical map, Digital Elevation Model (DEM). The datasets regarding erosion, soil water holding capacity, and temperature by elevation were obtained from ASTA (MA 2016; Stevens et al. 2014; Marx 2019; Steffen and Marx 2019). The soil water holding capacity dataset was a synthesis of the soil map (1:25,000), the soil map from Vermeire (1:50,000), and the general soil map (1:100,000) (Steffen and Marx 2019). For the erosion dataset, the "Erosion_effective_avec_MAE_appliquees" map was used (Marx 2019; Steffen and Marx 2019).

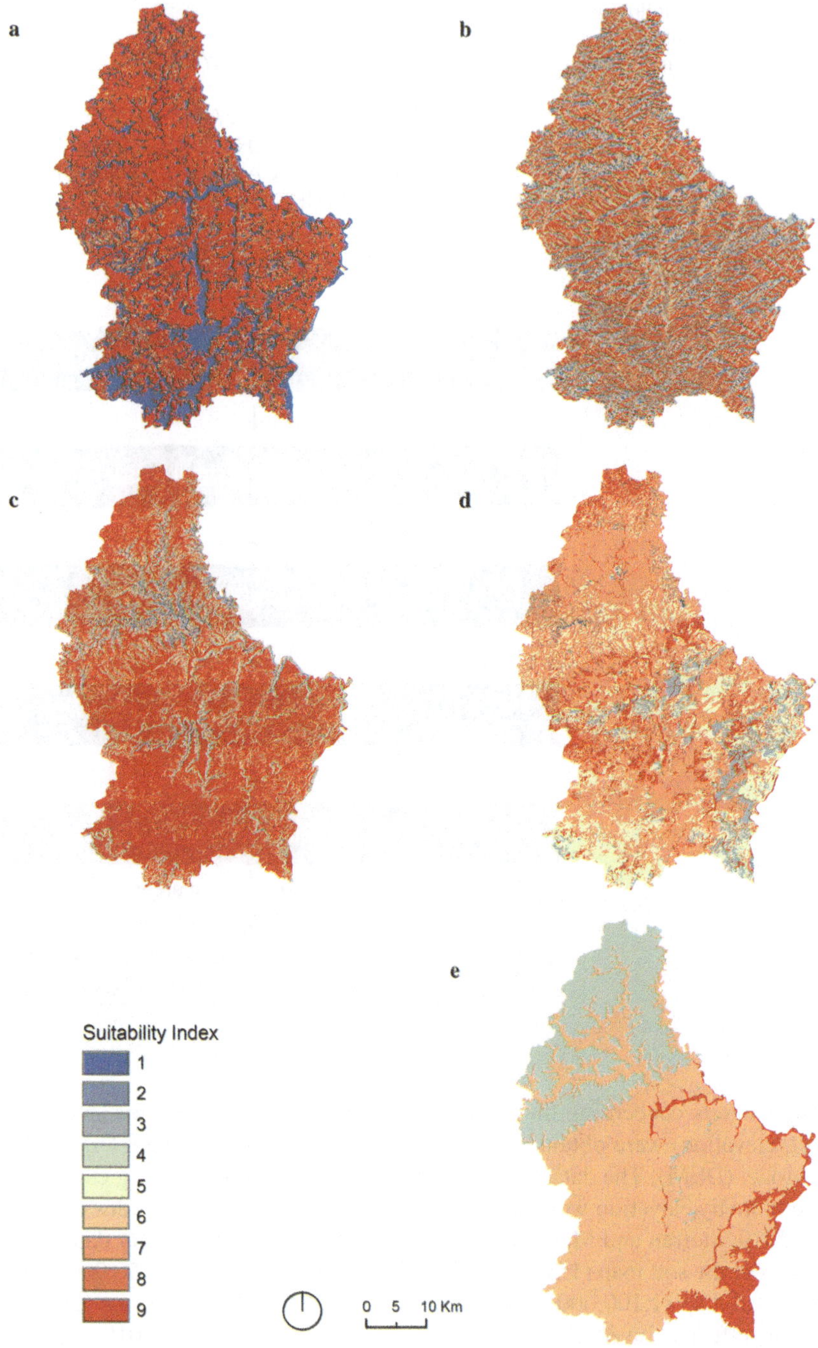


**Fig. 2.4** (**a**–**e**) Criteria maps used in study. (**a**) Erosion map; (**b**) Aspect map; (**c**) Slope map; (**d**) Soil water holding capacity map; (**e**) Elevation/temperature map

**Table 2.5** Ranking and weight distribution of main and sub-criterion parameters in Luxembourg

| Main criterion | Sub-criterion | Ranking 1–9 | Weight 0–1 |
|---|---|---|---|
| Soil capacity[a] | | | 0.595 |
| | Very low (cell number 0–50) | 1 | |
| | Low (cell number >50–80) | 3 | |
| | Medium (cell number >80–105) | 6 | |
| | High (cell number >105–140) | 8 | |
| | Very high (cell number >140) | 9 | |
| Erosion degree | | | 0.133 |
| | Tolerable erosion (<6 t/ha/year) | 9 | |
| | Low erosion (6–11 t/ha/year) | 7 | |
| | Moderate erosion (11–22 t/ha/year) | 4 | |
| | High erosion (22–33 t/ha/year) | 2 | |
| | Severe erosion (>33 t/ha/year) | 1 | |
| Slope | | | 0.189 |
| | <1–12% | 9 | |
| | 12–15% | 8 | |
| | 15–25% | 5 | |
| | 25–45% | 3 | |
| | >45% | 1 | |
| Aspect | | | 0.052 |
| | S, SW, SE | 9 | |
| | N | 2 | |
| | NW, NE | 5 | |
| | W, E | 7 | |
| Temperature[b] | | | 0.031 |
| | 7 °C *(44.6 °F)* | 4 | |
| | 8 °C *(46.4 °F)* | 5 | |
| | 9 °C *(48.2 °F)* | 6 | |
| | 10 °C *(50 °F)* | 7 | |

[a]Soil water holding capacity
[b]Average temperature by elevation

Another major data source included the datasets on a cadastral parcel level (2018) retrieved from the Luxembourgish data platform (Luxembourgish data platform, Plan cadastral numérisé 2020). The Copernicus services of the European Environmental Agency and the European Commission (CLMS 2015), especially the Urban Atlas dataset for Luxembourg, were the source of land-use and land-cover data (last updated for 2015).

The use of computer-based geographic information systems (GIS) provided a means for integrating and displaying the various data. While many of the datasets were available, several elements were either not accessible at the time of this study or limited in their scope (e.g. the precipitation dataset was aggregated in the soil water capacity dataset; the erosion degree of whole Luxembourg could not be assessed[6]).

#### 2.2.2.3 Standardisation of Criteria Maps

The selected criteria maps were initially in different units. Maps required conversion into a similar scale through standardisation technique into the format with 10 × 10 m cell size. For standardisation, all the criteria vector maps were converted to raster data formats. The raster maps were then reclassified into a range of 1–9. Once all the criteria maps are standardised, the weights of each criterion map can be calculated using AHP.

#### 2.2.2.4 Calculation of Weight for Criteria Maps

The land suitability analysis was created with a pairwise comparison matrix to determine the weights of parameters according to the AHP method (Table 2.4). The judgements in the pairwise comparison matrix (relative levels of importance of the parameters) were determined by consulting a team of experts, including from ASTA and faculty members working in the Department of Geography and Spatial Planning at the University of Luxembourg. The consistency ratio of the pairwise comparison judgements was calculated as 0.10. Criteria were scored within the range of 1–9, again according to experts' opinions. High point scores were given to criteria that positively affect agricultural land use, whereas lower point scores were given to those that restrict agricultural land use.

To provide an example of the analysis process, when evaluating the soil water holding capacity, the highest scores were given to areas with high capability of holding water. The lowest scores were given to areas with a low capability of holding water. When evaluating the study area in terms of degree of erosion, severe erosion (>33 t/ha/year) areas with a high level of slope were given 1 point. Areas with tolerable erosion (<6 t/ha/year) were scored with 9 points. Slope directly affects the diversity of plant species and the productivity of agricultural land. A 15% slope does not affect agricultural production, while between 15% and 25% there are limits to mechanisation. For slopes of 25% and 35%, mechanical production is limited (Marx 2019). Areas with a slope of less than 15%, therefore, scored 9 points. Only 1 point was given to areas with a slope above 45%. It was considered that the areas with southern, south-western, and south-eastern aspects receive more light, which allows cultivation of more products due to the duration of sunlight. Such areas scored 9 points. In contrast, areas with a northern aspect, which receive inadequate daylight, scored 2 points. The temperature by elevation criteria was evaluated according to the highest degree of temperature. Generally, it can be determined that vegetation starts to grow at 5 °C in Luxembourg. Thus, areas with 10 °C scored 7 points and areas with 7 °C scored 4 points (Table 2.5).

[6] A small proportion in the map could not be assessed in this dataset due to a limitation of exact soil data on the scale of 1/100,000. Since these are only flat areas, the susceptibility to erosion could be assessed as very low (ASTA).

**Table 2.6** Distribution of agricultural land suitability analysis

| Suitability degree | | Total area classified by the suitability analysis (parcel area) | |
|---|---|---|---|
| | | Hectares (ha) | % |
| 1 | Unsuitable | 9790.29 | 3.79 |
| 2 | Most unsuitable | 83.33 | 0.03 |
| 3 | Very low | 633.62 | 0.25 |
| 4 | Low | 3825.67 | 1.48 |
| 5 | Moderately low | 15,784.23 | 6.10 |
| 6 | Moderate | 24,328.55 | 9.41 |
| 7 | High | 83,183.17 | **32.17** |
| 8 | Very high | 104,185.82 | **40.29** |
| 9 | Highest | 16,785.32 | **6.49** |
| | Total | 258,600.00 | 100 |

After the criteria weights were assigned, raster maps were combined through a weighted overlay with their different respective weights along the 9-point scale (Table 2.6). An additional research step combined the final raster map with the actual parcel file (see cadastral parcel file). Further, to the criteria inputs for agricultural land suitability assessment, data on land cover (permanent grassland/pastures, forest—retrieved from the Urban Atlas) and already developed areas (existing building (footprint) and infrastructure—retrieved from cadastral parcel file) were collected for the overlay analysis to serve as constraint layers on the final suitability map. The result was an agricultural suitability indexation on a parcel level of all land in Luxembourg, which is presented in the following section.

## 2.3 Results and Discussion

### 2.3.1 The Agricultural Land-Use Suitability Map

#### 2.3.1.1 Distribution of Agricultural Land Suitability Analysis

The previous section briefly demonstrated the application of land-use suitability analysis for the case study of GDL. The results of this investigation are presented in the following section, showing an envelope of optimum areas for agricultural production.

The agricultural land-use suitability map, which was carried out using the criteria layers with their respective weights, was ranked from 1 to 9 (unsuitable to highly suitable, respectively). According to this map, it was determined that (Fig. 2.5):

- Approximately 78.95% (204,154.31 ha) of the total land area is suitable to highly suitable (classes 7–9).
- Approximately 15.51% (40,112.78 ha) has moderately low and very low suitability (classes 5–6) for agriculture.

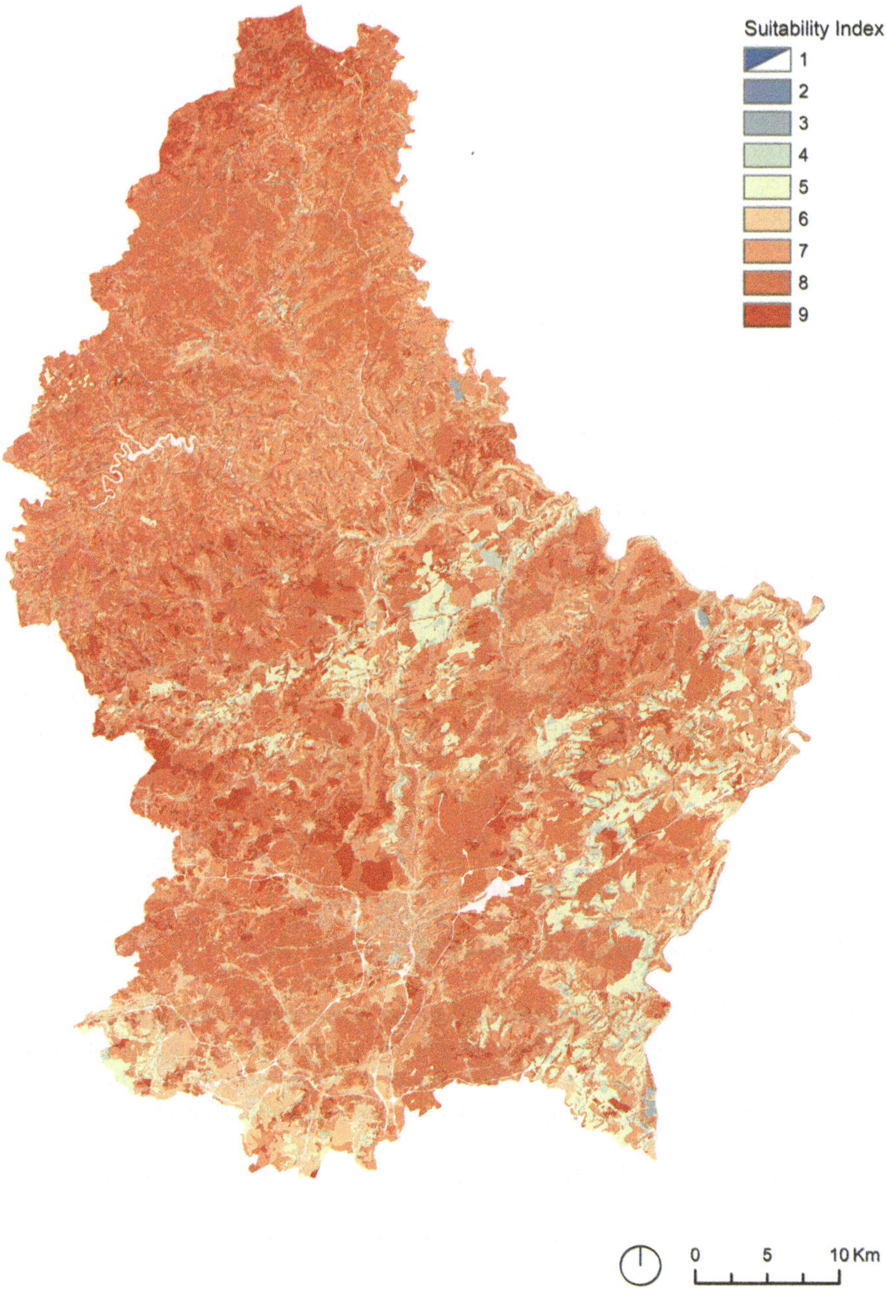


**Fig. 2.5** Agricultural land-use suitability map of Luxembourg

- Approximately 5.55% (14,332.91 ha) is unsuitable or of low suitability (classes 1–4).

Further, the constraint layers were superimposed on top of the computed suitability map. It was assumed that forest and permanent grassland/pasture areas were unavailable (and hence "restricted") for agriculture. A forest may, for instance, be deforested for large scale agriculture and thus change its land suitability. Another argument is that land-use change into arable land cannot be carried out on permanent grassland/pasture due to a protection regulation imposed by the Luxembourg government (see PEEN, 1996; Loi Protection Nature, 2018) and in favour of other ecological services (biodiversity conservation). The already developed areas, such as existing building (footprint) and infrastructure, were reclassified with an additional class 0 (and hence "developed").

At the end, the constraint layers, which covered approximately 58% of the study area (restricted) and approximately 5.28% (developed), were superimposed on the map to determine the final suitability map (Fig. 2.6).

The reclassified, final agricultural map shows (Table 2.7):

- Approximately 27.9% (72,101 ha) of the total land area is suitable to highly suitable (classes 7–9); the most highly suitable land, with 2.3% (5873.58 ha), is found in the Gutland region, above Luxembourg city, and the northern part of the Oesling region, in the north of Luxembourg.
- Approximately 7.8% (20,292 ha) has moderately low and very low suitability (classes 5–6) for agriculture.
- Approximately 1% (2434 ha) is most unsuitable or of very low suitability (classes 2–4).
- Approximately 58% (150,115 ha) is defined as restricted areas (class 1), representing forest and permanent grassland/pastures areas.
- Approximately 5.3% (13,656 ha) is defined as developed areas (class 0), corresponding to existing building footprint and infrastructure.

Overall, it was determined that 36.67% of the land in Luxembourg is available for agricultural production. This result was obtained from the geomorphological properties of the study area. The high degrees of slope and erosion, and the low soil water holding capacity, affect the suitability for agricultural use. However, 27.9% of land in the study area was found to have high to highest suitability for agricultural production.

#### 2.3.1.2 Agro-Urban Potentials

An additional exercise in this research identified potential agricultural productivity areas, called "agro-urban potentials", within each municipality. For this exercise, the available PAG development plans of 40 municipalities (Luxembourgish data platform, PAG 2020) were overlaid with the agricultural land-use suitability map. This zooming in on an urban, local scale would allow us to identify a series of agro-urban potentials relating to suitable areas for agricultural production in the urban fabric.

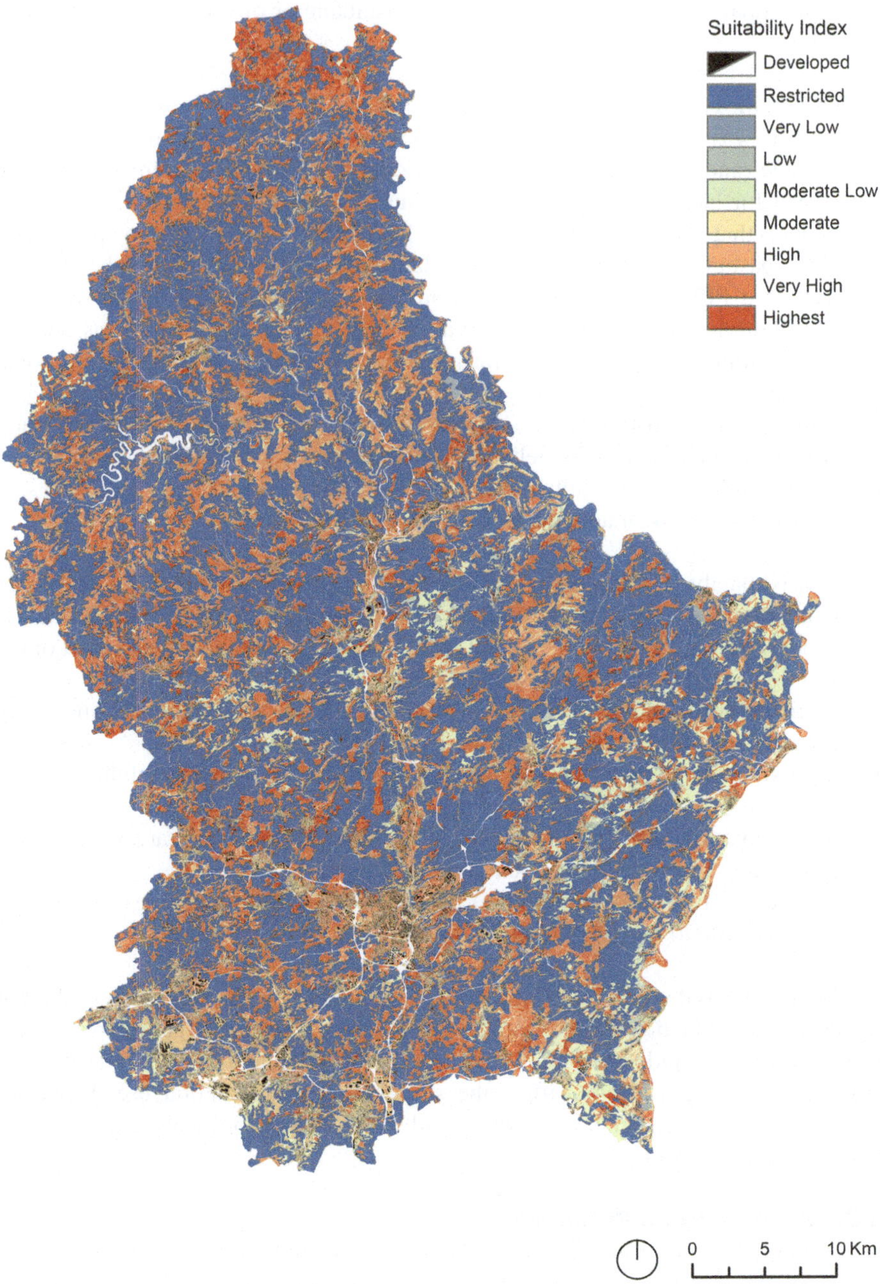


**Fig. 2.6** Agricultural land-use suitability map of Luxembourg—reclassified

**Table 2.7** Distribution of agricultural land suitability analysis—reclassified

| Suitability degree | | Total area reclassified by the suitability analysis (parcel area) | | Suitable for agricultural production (excl. restricted and developed areas) | |
|---|---|---|---|---|---|
| | | Hectares (ha) | % | ha | % |
| 0 | Developed | 13,656.00 | **5.28** | / | / |
| 1 | Restricted | 150,115.39 | **58.05** | / | / |
| 2 | Most unsuitable | 17.10 | 0.01 | 17.10 | 0.01 |
| 3 | Very low | 401.13 | 0.16 | 401.13 | 0.16 |
| 4 | Low | 2016.32 | 0.78 | 2016.32 | 0.78 |
| 5 | Moderately low | 7143.39 | 2.76 | 7143.39 | 2.76 |
| 6 | Moderate | 13,149.10 | 5.08 | 13,149.10 | 5.08 |
| 7 | High | 31,929.41 | 12.35 | 31,929.41 | **12.35** |
| 8 | Very high | 34,298.58 | 13.26 | 34,298.58 | **13.26** |
| 9 | Highest | 5873.58 | 2.27 | 5873.58 | **2.27** |
| | Total | 258,600.00 | 100 | **94,828.61** | **36.67** |

**Table 2.8** Agro-Urban potentials in Luxembourg

| Municipality | | Erpeldange-sur-Sûre | | Luxembourg | | Kayl | |
|---|---|---|---|---|---|---|---|
| Suitability degree | | ha | % | ha | % | ha | % |
| 0 | Developed | 149.66 | **8.35** | 1011.64 | **19.58** | 330.03 | **22.21** |
| 1 | Restricted | 1293.35 | **72.17** | 1734.75 | **33.58** | 929.43 | **62.56** |
| 2 | Most unsuitable | 0 | 0 | 0 | 0 | 0 | 0 |
| 3 | Very low | 0 | 0 | 2.24 | 0.04 | 1.61 | 0.11 |
| 4 | Low | 1.38 | 0.08 | 39.24 | 0.76 | 6.89 | 0.46 |
| 5 | Moderately low | 12.35 | 0.69 | 44.23 | 0.86 | 40.45 | 2.72 |
| 6 | Moderate | 56.54 | 3.15 | 197.39 | 3.82 | 71.43 | 4.81 |
| 7 | High | 184.56 | **10.30** | 1692.14 | **32.76** | 35.58 | **2.39** |
| 8 | Very high | 88.81 | **4.96** | 410.57 | **7.95** | 57.28 | **3.86** |
| 9 | Highest | 5.52 | **0.31** | 33.30 | **0.64** | 12.99 | **0.87** |
| | Total | 1792.17 | 100 | 5165.50 | 100 | 1485.69 | 100 |

A limitation of this study was that investigation could only be undertaken for 40 of a total 102 PAGs of Luxembourg's municipalities. The rest of the PAGs are still under revision and therefore not publicly accessibly. This chapter, therefore, only presents, by way of an example, one municipality from each development cluster of the North, Luxembourg, and South Agglomerations. A first assumption calculated the available suitable areas for each municipality (Table 2.8):

- Approximately 15.6% (279 ha) of areas with high and to highest agricultural suitability (classes 7–9) lie in the municipality Erpeldange-sur-Sûre, Agglomeration North.
- Approximately 41.35% (2136 ha) of areas with high to highest agricultural suitability (classes 7–9) lie in the municipality Luxembourg, Agglomeration

Luxembourg (Fig. 2.7). This result is particular interesting, since the development pressure in the capital city is extremely high.

- Approximately 7.12% (106 ha) of areas with high to highest agricultural suitability (classes 7–9) lie in the municipality Kayl, Agglomeration South.

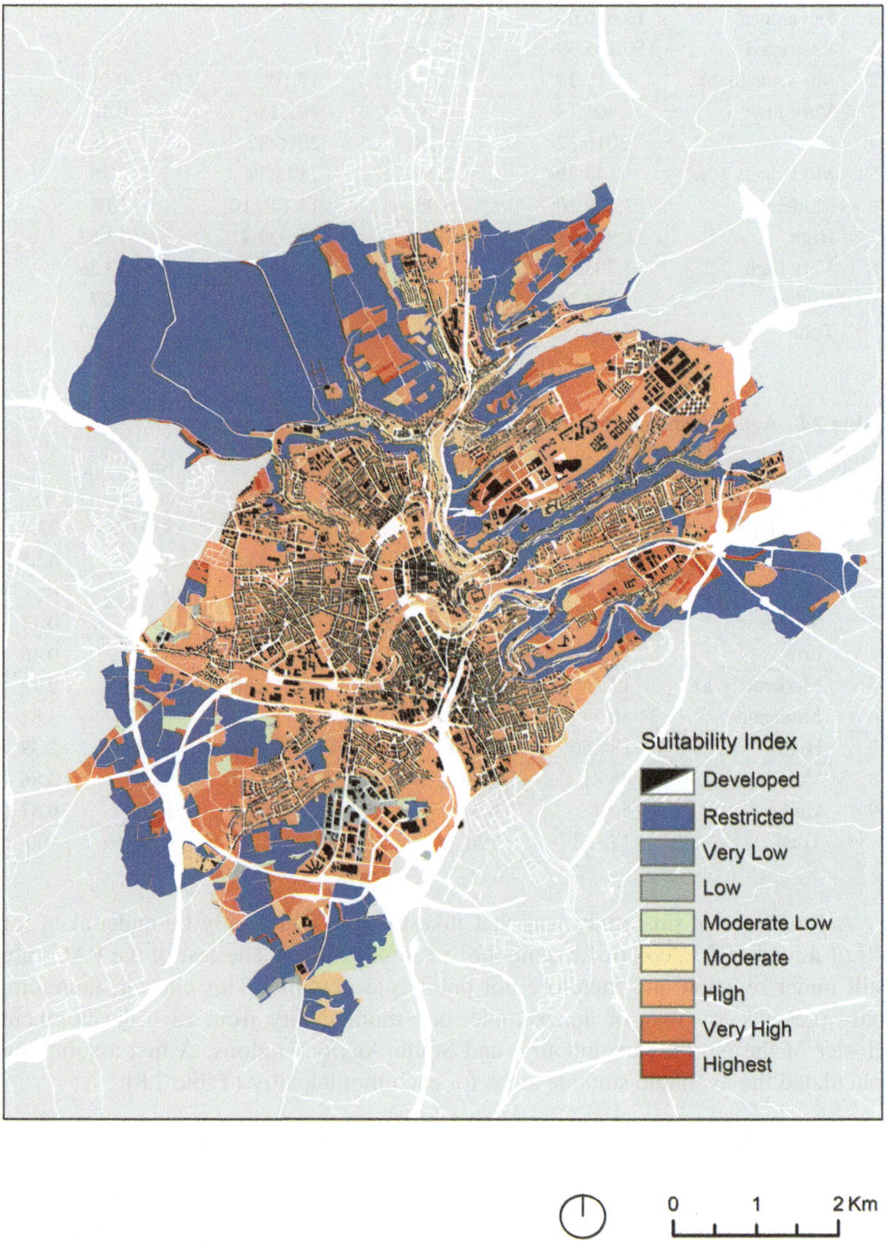


**Fig. 2.7** Agricultural land-use suitability map of the municipality Luxembourg

In spite of the limitation, the study adds to the understanding of the importance of land-use suitability analysis for urban design and planning processes. This section showed, using the examples of three municipalities, the potential productive agriculture areas within an urban perimeter. Those results can create a new basis for negotiation for the development of urban areas.

### 2.3.2 How Can Land Suitability Contribute to Urban Design and Planning Processes?

The discussion in this section is organised around two questions: How can land-use suitability contribute to urban design and planning processes? And, is this tool adequate to maintain sustainable agriculture, particularly in and around major urban and metropolitan centres?

Land-use suitability, as demonstrated in the previous sections, can be used to define potential productive agriculture areas. The integration of such a tool in the design decision process can help to direct growth, maintain sustainable agriculture, and show land-use conflicts and potential agro-urban areas in the planning trajectory. Such a design decision tool can be valuable for urban agriculture practices, particularly in and around major urban and metropolitan centres. Agro-urban potentials, as described in Sect. 2.3.1, could be made visible through this process, and it is then necessary to respond to them on a planning level. One possibility would be to rethink and adapt existing planning instruments for land use and urban development. One could consider the introduction of a mixed zone, for instance. An agri-urban zone that both protects and allows development on agricultural land could be one option. The implementation of such an alternative planning instrument could help to limit urban sprawl or allow the development of land to a certain degree by supporting the protection of good farmland and healthy agricultural production. Further, politics should stimulate public discourse on land-use conflicts caused by the withdrawal of good quality farmland by various stakeholders.

Overall, by using such a design decision tool, Luxembourg's future land development could be more in line with the ecological qualities of its agricultural land. Consequently, this could yield guidelines on which areas can be reserved for protection, allocation, or integration in future land use and urban development. Land-use suitability maps can create a basis for negotiations on where to urbanise land without abandoning fertile, productive agricultural land.

However, while the multi-criteria AHP method helps to inform the analysis of the land in question, it also has some limitations: it is highly dependent on the availability of accurate datasets. Besides comprising the physical properties (topographical properties, soil and geological characteristics), the study also needs to include for instance socioeconomic criteria for agricultural production. In the AHP process, the pairwise comparison method is based on expert opinions which are mostly subjective in nature. Therefore, any inaccurate judgement on any selected parameters could adversely affect the score assignment and weighting designation.

Nevertheless, the AHP technique is a useful tool which enables planners and local decision-makers to analyse interactions in various ways. Such a design decision tool can help elected officials and land managers make decisions and establish policies regarding the use of particular areas of land. Further research could usefully explore the implementation of such outcomes on various levels by including different stakeholders. It could explore different architectural and urban typologies which combine both growth and protection of the land. The AHP methods certainly indicate a direction for more detailed analysis.

## 2.4 Conclusion

Efficient and thoughtful use of land is an essential step in managing and developing any area, especially the vast developing landscape of Luxembourg. In order to determine the optimum direction for future development, in this study the suitability for agricultural land was explored to direct growth to the most appropriate places. The study was conducted using five criteria reflecting the topographical properties, soil, and geological characteristics of the area. At the end of the assessment, it was estimated that 27.9% of the study area had high to highest suitability for agriculture. From this initial investigation, the distribution of suitable areas within the existing perimeters of the built-up area was estimated, using the example of a selection of municipalities.

The continuation of an existing method, the Analytical Hierarchy Process Method (AHP), applied to the Luxembourg context opens a different view and discussion about land development. The study adds valuable insights into current planning discussions in Luxembourg, by providing spatial planning guidelines, on an urban but also a regional scale, regarding where to develop land while respecting the ecological qualities of its land—in this case, by respecting the soil quality for agricultural production.

The main contribution of this study is the combination of the main parameters used in terms of agricultural production and the detailed explanation of the scoring approach, which has not previously been applied to the case of Luxembourg. Further, it adds value to Luxembourg's ambitious agricultural agenda by defining suitable areas for agricultural production, particularly in and around urban centres. It can help to identify priority areas for potential management and/or policy interventions. Eventually, it may be possible for the findings from this research to be integrated into land use and development and the urban development plan. In this way, the goal of Luxembourg's agricultural agenda of increasing local productivity by 2050 could be directed towards a more sustainable land development.

Evidently, land suitability alone is not sufficient in itself to guarantee the maintenance of a good quality of land for sustainable agriculture. Rather, farmers must connect agriculture more to protection of biodiversity and the environment in order to maintain sustainable agriculture. This approach also ideally requires the integration of various stakeholders such as farmers and their families as actors and participants, as well as the appropriate local and regional municipalities, who must be involved in the strategic development planning process.

**Acknowledgements** This chapter presents some preliminary results obtained in the author's on-going PhD research. Sections 2.2 and 3.1 are the result of work supported by ASTA (Ministère de l'Agriculture, de la Viticulture et de la Protection des consommateurs, Administration des services techniques de l'agriculture, Luxembourg), while Sect. 2.1 contains some research output (see Table 2.1) from work carried out in collaboration for the project "Composite Landscapes in Luxembourg—The Eco Century Project", which is currently being conducted at the University of Luxembourg, funded by the Braillard Architects Foundation in Geneva. Finally, I wish to thank Simone Marx and Matthieu Steffen from ASTA and my colleague Nikolaos Katsikis for their technical support through the process of this work.

## References

Akıncı H, Özalp AY, Turgut B (2013) Agricultural land use suitability analysis using GIS and AHP technique. Comput Electron Agric 97:71–82

Arredondo-Garrido D (2016) References in Le Corbusier's reorganization of rural habitat. J Archit Urban 40(2):99–109

CLMS - Copernicus Land Monitoring Service, Urban Atlas 2012 (2015) European Environment Agency (EEA) and the Joint Research Centre (JRC). https://land.copernicus.eu/local/urban-atlas/urban-atlas-2012. Accessed 18 Jun 2020

CLMS - Copernicus Land Monitoring Service, Urban Atlas Change 2006–2012 (2016) European Environment Agency (EEA) and the Joint Research Centre (JRC). https://land.copernicus.eu/local/urban-atlas/change-2006-2009?tab=metadata. Accessed 18 Jun 2020

Doran JW, Parkin TB (1994) Defining and assessing soil quality, Defining soil quality for a sustainable environment, vol 35. Social Science Society of America, Madison, pp 1–21

Doran JW, Parkin TB (1997) Quantitative indicators of soil quality: a minimum data set, Methods for assessing soil quality, vol 49. Social Science Soceity of America, Madison, pp 25–37

EC - European Commission (2020) A European green deal. European Commission, Brussels

EEA - European Environment Agency (2012a) Land recycling and densification European Union. https://www.eea.europa.eu/data-and-maps/indicators/land-recycling-and-densification/assessment-1. Accessed 18 Jun 2020

EEA - European Environment Agency (2012b) Soil erosion. European Union. https://www.eea.europa.eu/data-and-maps/indicators/soil-erosion-by-water-1/assessment. Accessed 18 Jun 2020

Eurostat - Statistical Office of the European Communities (2012) Agricultural census in Luxembourg. European Commission. https://ec.europa.eu/eurostat/statistics-explained/index.php?title=Archive:Agricultural_census_in_Luxembourg&oldid=379557. Accessed 18 Jun 2020

Hertweck F (ed) (2020) Architecture on common ground. The question of land: positions and models. Lars Müller Publishers and University of Luxembourg, Zurich

Hesse M (2015) Luxembourg. disP Plan Rev 51(1):54–55

Ilieva RT (2016) Urban food planning: seeds of transition in the global north. Routledge, London

Le Corbusier (1979) The three human establishments. National government publication edn. Punjab Govt., Dept. of Town & Country Planning, Chandigarh

Luxembourgish data platform, PAG (2020) Luxembourgish data platform. Ministère de l'Intérieur. https://data.public.lu/fr/search/?q=PAG. Accessed 18 Jun 2020

Luxembourgish data platform, Plan cadastral numérisé (2020) Luxembourgish data platform. Administration du cadastre et de la topographie. https://data.public.lu/fr/datasets/plan-cadastral-numerise-pcn/. Accessed 18 Jun 2020

MA - Ministère de l'Agriculture de la Viticulture & du Développement rural (2016) Convention de recherche relative à la revision des cartes thématiques sur les zones à risque d'érosion et de ruissellement au Grand-Duché de Luxembourg. Administration des Services Techniques de l'Agriculture (ASTA), Gembloux Agro-Bio Tech (University Liege), Ettelbruck, Gembloux

Malczewski J (2004) GIS-based land-use suitability analysis: a critical overview. Prog Plan 62(1): 3–65

Marx S (2019) Bodengüteklassen, Luxembourg. Ministère de l'Agriculture, de la Viticulture et de la Protection des consommateurs (MA), Administration des Services Techniques de l'Agriculture (ASTA), Ettelbrück

McHarg IL (1969) Design with nature. Published for the American Museum of Natural History [by] the Natural History Press, Garden City, NY

MECDD - Ministre de l'Environnement du Climat et du Développement durable (2016) Wéi e qualitative Wuesstem fir eist Land? Ministre de l'Environnement du Climat et du Développement durable, Département d'Amenagement du Territoire, Luxembourg

MECDD - Ministre de l'Environnement du Climat et du Développement durable, FI - Ministre des Finances (2018) Loi du 18 juillet 2018 concernant la protection de la nature et des ressources naturelles et modifiant. Article 60 – (4), Article 63 – (4): Article 67. The Government of the Grand Duchy of Luxembourg, Luxembourg

MECDD - Ministre de l'Environnement du Climat et du Développement durable, MA - Ministère de l'Agriculture de la Viticulture & du Développement rural (2016) Prime à l'entretien du paysage et de l'espace naturel. Règlement grand-ducal du 24 août 2016 instituant une prime à l'entretien du paysage et de l'espace naturel et à l'encouragement d'une agriculture respectueuse de l'environnement. Article 4-(3). The Government of the Grand Duchy of Luxembourg, Luxembourg

MECDD - Ministre de l'Environnement du Climat et du Développement durable, Neobuild, CDEC - Conseil pour le Développement Economique de la Construction (2019) Stratégie Nationale urban farming Luxembourg. The Government of the Grand Duchy of Luxembourg, Luxembourg

Morgan K (2009) Feeding the city: the challenge of urban food planning. Int Plan Stud 14(4): 341–348

Mumford L (ed) (1971) Ecological basis of planning. Springer, Dordrecht

Mumford E (2002) The CIAM discourse on urbanism, 1928–1960. MIT press, Cambridge

Pothukuchi K, Kaufman JL (1999) Placing the food system on the urban agenda: the role of municipal institutions in food systems planning. Agric Hum Values 16(2):213–224

Pothukuchi K, Kaufman JL (2000) The food system: a stranger to the planning field. J Am Plan Assoc 66(2):113–124

Pramanik MK (2016) Site suitability analysis for agricultural land use of Darjeeling district using AHP and GIS techniques. Model Earth Syst Environ 2(2):56

Saaty TL (1990) An exposition of the AHP in reply to the paper "remarks on the analytic hierarchy process". Manag Sci 36(3):259–268

Saaty TL (2000) Fundamentals of decision making and priority theory with the analytic hierarchy process, vol 6. RWS Publications, Pittsburgh, PA

Saaty TL, Vargas LG (2012) Models, methods, concepts & applications of the analytic hierarchy process, International series in operations research & management science. Springer, New York

SER - Service d'économie rurale (2016) Luxembourgish agriculture facts and figures. Ministère de l'Agriculture de la Viticulture & du Développement rural, Luxembourg

SER - Service d'économie rurale (2018) Die Luxemburgische Landwirtschaft im Wandel. Ministère de l'Agriculture de la Viticulture & du Développement rural, Luxembourg

STATEC – Institut national de la statistique et des études économiques du Grand-Duché de Luxembourg (2020a) Agriculture. https://statistiques.public.lu/stat/ReportFolders/ReportFolder.aspx?IF_Language=eng&MainTheme=4&FldrName=2&RFPath=7276. Accessed 18 Jun 2020

STATEC – Institut national de la statistique et des études économiques du Grand-Duché de Luxembourg (2020b) Land use (in %) 1972–2019. https://statistiques.public.lu/stat/TableViewer/tableView.aspx?ReportId=12695&IF_Language=eng&MainTheme=1&FldrName=1. Accessed 18 Jun 2020
STATEC – Institut national de la statistique et des études économiques du Grand-Duché de Luxembourg (2020c) Population and employment. https://statistiques.public.lu/en/population-employment/index.html. Accessed 18 Jun 2020
Steffen M, Marx S (2019) Larsim: Erforderliche Parameter für RoGeR Ansatz-Bodendaten für Belgien und Luxemburg. Ministère de l'Agriculture, de la Viticulture et de la Protection des consommateurs (MA), Administration des Services Techniques de l'Agriculture (ASTA), Ettelbruck
Stevens A, van Wesemael B, Marx S, Leydet L (2014) Mapping topsoil organic carbon in Grand-Duchy of Luxembourg. Ministère de l'Agriculture, de la Viticulture et de la Protection des consommateurs (MA), Ettelbruck, Louvain-la-Neuve
UN - United Nations, Department of Economic and Social Affairs, Population Division (2019) World urbanization prospects: the 2018 revision (ST/ESA/SER.A/420). New York
UNCCD - United Nations Convention to Combat Desertification (2017) The global land outlook. UNCCD, Bonn
Wang X, Hofe R (2007) Research methods in urban and regional planning. Springer, Berlin

**Ivonne Weichold** is an Architect and Urbanist working in the intersection of Architecture, Urban and Territorial Design. Her research breaks open the duality of thinking in urban and rural environments to investigate the role of agricultural land, production, and planning in wealthy economies. She is currently conducting PhD research at the University of Luxembourg, where she is also working as a lecturer in the Program on Architecture, European Urbanization and Globalization. In her professional life as an Architect and Urbanist, she has worked on different scales of projects in Belgium and Germany for over a decade.

# Mitigation of Urbanization Ill-Effects Through Urban Agriculture Inclusion in Cities

3

Madhavi Jain

**Abstract**

Rapid urbanization has left cities with shrinking vegetative and green cover. This study first outlines the role of urbanization in creating heat stressed cities including urban heat islands and health effects. Urban agriculture (UA) practices are an undervalued and untapped counter resource for such ill-effects. Various food and non-food UA systems help engage (1) marginal workers and unemployed, (2) youth and elderly, and (3) business or hobby seekers for family sustenance, economy growth, environmental sustainability, and increased happiness. UA integration has the potential for increasing building efficiency and thermal comfort, enhancing aesthetic value and urban biodiversity, and providing fresh food.

This chapter highlights past and projected land use land cover (LULC) changes in Delhi (India). Urban area has expanded from 7.7% (1977) to 39.3% (2014), and is projected to cover 53.8% in 2030. Likewise, rapid decrease in agriculture, allied activities, and green cover is noted. The current status of UA in the capital city is discussed. It is encouraging that the government is supporting farmers to switch to profitable food based UA systems, e.g. vegetable and flower farming. Businesses utilizing soil-less farming techniques, zero-acreage farming, etc. are on the rise. Vertical gardens, green walls, and other non-food UA strategies are being promoted. However, the city lacks impact based studies of UA systems. Further research is needed to maximize UA benefits to the society.

**Keywords**

Urban agriculture (UA) · Land use land cover (LULC) · UA systems · Delhi

---

M. Jain (✉)
School of Environmental Sciences, Jawaharlal Nehru University, New Delhi, India



J. A. Diehl, H. Kaur (eds.), *New Forms of Urban Agriculture: An Urban Ecology Perspective*, https://doi.org/10.1007/978-981-16-3738-4_3

## 3.1 Introduction

Historically speaking, the concept of a city with all its functional elements can be traced back to 4000 B.C. in Mesopotamia (Oates et al. 2007). In modern times, the Industrial Revolution by far spearheaded the rapid segregation of urban and rural land uses. Rural areas generate gross domestic product (GDP) mainly through agriculture and other allied activities. Where agriculture in developing countries still relies on abundant workforce and intensive manual labour, a high degree of mechanization and technological advancement is present in the developed nations. Urban areas on the other hand are primary food consumption zones, generating GDP through secondary and tertiary economy sectors.

Our society is, therefore, inexorably moving towards urbanism and any return to predominant rurality seems unlikely. Such has been the unprecedented rise of cities that almost two-thirds global population was rural in the mid-twentieth century, but by the mid-twenty-first century, two-thirds is projected to be urban (United Nations 2014). Along with the whirlwind expansion of urban populations, urban lands are projected to triple by 2030 (d'Amour et al. 2017). This estimate considers the year 2000 as the baseline. Moreover, by 2030, the number of megacities (cities having >10 million inhabitants) is expected to rise to 43 from the current total of 33, and incipient megacities (5–10 million inhabitants) will rise to 66 (United Nations 2018). Furthermore, most of these emerging cities, which are often haphazardly planned, are clustered in African and Asian nations, including India, which presently has five megacities: Delhi, Mumbai, Kolkata, Bengaluru, and Chennai. While cities occupy only ~3% of the Earth's surface area, they consume ~75% of its resources (Bechtel et al. 2015), contribute to ~80% of the global GDP (World Bank 2019), and produce ~75–80% greenhouse gas (GHG) emissions (Satterthwaite 2008). More importantly, 80% of cities are also highly vulnerable to natural disaster related losses (United Nations 2016). The expanding urban population demands greater food supply and also leads to over-exploitation of natural resources. This calls for new measures to ensure food security and reduction in urban footprint (Thomaier et al. 2015).

In the context of changing climate and increasing number of related disasters (Rosenzweig et al. 2018), creating urban sustainability, tackling urban issues (including GHG emissions, pollution, heat stress), and moving closer to achieving the sustainable development goals (SGDs) demand significant urgency (United Nations 2018). To address these challenges, several authors emphasize reintegration of food systems into the city itself (Pothukuchi and Kaufman 1999; Eigenbrod and Gruda 2015). Increasing awareness and desire to include urban agriculture (UA) into the city has tremendously gained momentum in the past decade or so, with many sophisticated projects other than traditional farming on the rise. In view of this, the aim of the present contribution is to:

1. Outline the role of by-products of urbanization, viz. land use land cover (LULC) change and urban heat islands (UHIs) in creating heat stressed cities.
2. Discuss the potential of UA integration into urban areas as a mitigation strategy.

3. Study the diminishing agriculture (1977–2030) in rapidly urbanizing Delhi and the scope for food and non-food based UA systems.

## 3.2 Urban vs. Rural Areas

Before jumping to tackle any urban issues, one needs to first understand the concept of a city. It is rather perplexing to realize the multi-functionality and the multi-layered complexities found in present day cities (Varzi 2019). For most cities the real outward expansion (urban sprawl rates) are on the decline; in the case of developed nations rapid urbanization has virtually ended (Zhang 2016). However, vertical urban (building heights) growth is tremendously rising. Monocentric cities are rapidly evolving into multi-nuclei cities. With constant evolution, the earlier separation of zones by their function is not necessarily found in present day cities (Tian et al. 2010). The functional specialization of satellite cities is often more complex than the city's central business district (CBD), sometimes generating far more services.

Aside from this, urban areas at the core are very different in structure and function to rural areas. In this regard, LULC proves to be an important distinguishing parameter. Rural areas are composed predominantly of natural land cover classes such as bare rock/soil, grasses and vegetation, forests, and waterbodies. Urban areas, however, include elements not necessarily found in the natural environment. Use of materials such as concrete, glass, and metals is widespread, for construction of buildings including residential complexes, industries, institutions, theatres, arenas, and community centres. Since all material surfaces have different intrinsic properties, e.g. albedo, heat capacity, surface percolation, any changes in the surface are reflected as a change in the overlying atmosphere (Grimmond and Oke 1999). Such land–air interactions are important in the present context. Urban sprawl usually happens on all unconstructed land whose existence has no justifiable purpose, i.e. rural elements and LULC classes. Since the basic idea of a city is economy generation thorough non-agricultural activities, natural ecosystems and agro-systems are usually targeted when more land for urbanization is needed. Further, the ecological and socio-economic functions are tightly interconnected within the spatial structure of a city and thus city morphology has important bearings on urban sustainability. Cities also face numerous issues due to the by-products of urbanization. One such pressing issue is UHI and is detailed in the following section.

## 3.3 Urban Heat Island (UHI) and Heat Stress

A phenomenon typically exclusive to cities and a common by-product of urbanization is the UHI effect. Witnessed across most cities of the world (Guo et al. 2015; Debbage and Shepherd 2015), UHI refers to higher average temperatures—by several degrees, even up to 10 °C in some cases—in comparison to rural counterparts (Kishtawal et al. 2010). In addition, the energy and water balance of cities are strongly influenced by urbanization-led LULC change, city-specific urban morphology, and the intrinsic properties of urban materials used (Grimmond and Oke 1999). Bowen ratio, an important meteorological parameter, the ratio of surface heat flux and latent heat flux, is indirectly linked to both surface and air temperature (Bowen 1926). When the percentage of impervious area (including both areal and vertical expansion) increases, it results in surface heat flux changes. In almost all cases, intrinsic properties of these urban materials cause an increase in the surface heat flux. Simultaneously, the latent heat flux decreases with the loss of vegetated green cover. As a direct consequence, urbanization elevates Bowen ratio and elevates the air temperature. Such effects can be noted even at a spatial scale of several meters.

In the larger context, the Intergovernmental Panel on Climate Change (IPCC) Fifth Assessment Report (AR5) released in 2014 clearly states "the warming of the climate system is unequivocal, and since the 1950s, many of the observed changes are unprecedented over decades to millennia". It further states that the recent time period from 1983 to 2012 was arguably the warmest 30 years recorded in the northern hemisphere (IPCC 2014). Even with the strictest policies, slowing of expected warming trends under various representative concentration pathways (RCPs) seems highly unlikely. Most forecasts predict that cities will face higher risks of heat stress, precipitation extremes (flooding and drought), water scarcity, storms, and coastal inundation under climate change (IPCC 2014). Reports also indicate heat waves are becoming increasingly frequent and more intense due to urbanization (Meehl and Tebaldi 2004). Cities in the developing nations have high population densities, lower technological adaptability, few warning systems in place, and slow dissemination of information; all factors severely increasing the inherent risk multi-folds. The implications and extent are further exacerbated in cities due to UHI formation (Lundgren et al. 2013), culminating into pressing human health and well-being issues such as heat stress.

For people working in warmer urban set-ups (during the day), occupational heat stress poses a major threat to health and greatly reduces the work productivity. UHIs in the cities also cause heat exhaustion, dehydration, strokes, cardiovascular issues, and high mortality among vulnerable age groups (Kovats and Hajat 2008). To further add to the misery, the development of night UHIs severely limits any relief from heat exposure faced during the day (Hajat and Kosatky 2010). Affected sectors include agriculture, industries, tourism, enterprises, and workshops without air-conditioning, construction, and insurance/finance (Lundgren et al. 2013). Despite some degree of adaption by the human body to heat (particularly in tropical cities), the effects of UHI exacerbated heat stress is faced by all rich and poor cities alike; some studies even suggesting 1–3% increased mortality per 1 °C rise in temperature (Hajat and Kosatky 2010). Examples include the infamous 2003 heat wave across Europe that led to estimated 22,000–35,000 premature deaths (Schär and Jendritzky 2004).

Along with heat stress, issues of urban poor are also on the rise, leaving planners and government agencies in search of immediate and long term solutions (Zhang 2016). UHI mitigation strategies include two main approaches (1) increasing green cover and (2) use of surface materials having high albedo. These approaches have the potential to reduce impacts of air pollution as well as lower Bowen ratio, urban temperatures, and heat stress (Gill et al. 2007). The present work focuses on green (and vegetated) cover LULC modifications through UA systems, which have the potential to mitigate or in some instances reverse UHI.

## 3.4 The Growing Importance of UA and Its Challenges

### 3.4.1 What Is UA?

The majority of the foreseeable future urban expansion is likely to occur over rural croplands in the peri-urban areas. UA is by far a profoundly distinct yet complementary practice to rural agriculture (Mougeot 2006) and highly adaptive to the preferences or needs of city residents (Lovell 2010). By definition UA is "the growing, processing, and distribution of food and non-food plant and tree crops and the raising of livestock, directly for the urban market, both within and on the fringe of an urban area" (Mougeot 2006, p. 4). Interestingly, an increase in UA practices has been noted in recent times worldwide, aligning well with the goal of building sustainable cities. In a global study using the year 2000 as a baseline, d'Amour et al. (2017) estimated that around 36% urban areas contributed to UA. Many cities in the USA are actively including UA into their LULC planning (Mukherji and Morales 2010), and more than 30% increase in UA has been noted in three decades (Alig et al. 2004).

The system of UA exclusively uses natural and human resources in the city, produces for its own population, and is greatly impacted by urban stressors, e.g. high land prices, urban markets, job opportunities, government policies (Lovell 2010). UA therefore is an umbrella term which includes management practices such as small-sized agricultural fields, community farms and farmer markets, urban orchards, edible landscaping, backyard kitchen gardens, rooftop gardening, and vertical wall gardening (Lin et al. 2015). These practices are not limited towards the goal of food production (vegetables, fruits, milk, eggs, meat, etc.), but also include cultivation of flowers, medicinal plants, and ornamental fauna. The diverse range and type of services offered makes it "highly heterogeneous in size, form and function" (Lin et al. 2015, p. 1).

### 3.4.2 Benefits of UA: Food Production, Ecosystem Services, and Thermal Comfort

Growing evidence suggests that integration of UA into cities has numerous socio-economic, cultural, and environmental benefits besides moving towards urban sustainability (Thomaier et al. 2015; Ferreira et al. 2018). For many cities, availability of fresh produce is limited due to long transportation journeys from rural food production farms. There is an increasing demand for healthier, organic and fresh food products motivating urban planners to consciously include agriculture within and around the urban areas (Lin et al. 2015). UA can provide city residents with an increased vegetable diversity through cultivation of crops not native or typical to the region. Baker (2004) surveyed urban gardens in Toronto (Canada) and found high priority for selective Asian vegetables not native to the country. Moreover, UA converts idle spaces or lands unfit for building large complexes (this includes LULCs such as barren lands and/or small marginal lands) into generating higher ecosystem and economic services (Kaufman and Bailkey 2000; Beniston and Lal 2012). Inclusion of UA into the city morphology helps provide essential ecosystem services such as rich flora and fauna, increased pollination, bio-control of pests, nutrient and organic waste recycling, increased thermal comfort, reduction in air pollution and dust, improved drainage, and recharge of ground water table (Cook et al. 2015). However, certain trade-offs are also associated with UA practices. These include eutrophication, water and soil pollution of urban system by excessive pesticide and chemical use, as well as increased breeding of mosquito and other parasites (Lin et al. 2015).

Where urbanism often disregards the essential importance of green cover (parks, ridges and forests, etc.) in cities, the concept of UA brings in renewed sense of relevance and purpose (Pothukuchi and Kaufman 2000). Green space is often a refuge for native biodiversity, along with providing vegetative and ecological function across various fragmented city habitats (Lin and Fuller 2013). Another benefit of green landscaping is the beauty and aesthetics (e.g. vertical gardens, rooftop gardens) it provides and a soothing "feel-good" emotion it evokes. UA inclusion can provide thermal comfort for heat stressed urban areas and also lower cities' energy demand. In a study covering urban trees, peak power, and energy savings through cooling, Akbari et al. (1997) note that gardens and street trees adjacent to buildings contribute 27–30% cooling energy savings. The cooling potential however depends on the tree species, maturity, and size, as well as placement in the urban fabric (Shashua-Bar and Hoffman 2000). Many such urban tree patches diversely placed across the city can multiply the local-scale cooling effects to a wider urban environment.

### 3.4.3 Inclusion of UA into City LULC

Considering the new generation of environmentally conscious urbanites, and the values of urbanity, there is a need to holistically include UA practices into the daily urban machinery. Thus arises a strong need to preserve agricultural or natural LULC classes in the city and at the same time gain from the socio-economic, ecosystem, environmental, and health services their co-existence in the urban fabric offers. However, the prime constraint in imbibing UA practices is land availability along with unaffordable land prices. Money, time, and human labour are also crucial constraints. In essence, the combined impact of above constraints is felt in both developing and developed world cities, however to different degrees. Since developed and developing world cities are governed by different LULC dynamics and policies, it affects which UA practices are most commonly adopted and how they are managed. In short, every city in its own right and judgement can selectively adopt UA forms and practices befitting to its resource availability and specific needs. A few innovative and resource saving UA strategies are described.

Zero-acreage farming presents a fresh outlook at food production in dense urban areas where growing societal needs must be met on scarce land availability (Thomaier et al. 2015). Such farms improve building energy efficiency, are small space oriented (e.g. on rooftops), mostly use soil-less farming techniques, harvest rainwater or at times practice hydroponics (Thomaier et al. 2015). For the same yield, use of hydroponics against conventional vegetable cultivation practices can reduce water consumption up to 75% (Astee and Kishnani 2010). Excess runoff and greywater from urban areas can also be utilized for green cover maintenance by the local municipality. A new concept of green architecture has gained popularity. Expanding on the concept of green rooftops, vertical gardening landscapes the walls of private residences, commercial building, flyovers, and other urban structures with numerous small pots of ornamental plants. Apart from aesthetics, it functions as energy saving. Similar to the cooling effect of tree plantations, as discussed in the previous sub-section, vertical gardening also reduces air temperatures sufficiently (Bass and Baskaran 2003). If such strategies are adopted on a city scale, significant reduction in issues related to UHI and heat stress could be realized. A case study of UA opportunities in Delhi, a rapidly urbanizing Indian city having gained the notorious position as one of most polluted and populated places, is presented in the following section.

## 3.5 UA Scenario in the National Capital Territory (NCT) of Delhi

### 3.5.1 Urbanization, Population Explosion, and Shrinking Agricultural Lands

Delhi, the capital of India is an important city for its role in governance, legislature, judiciary, education, healthcare, economy, and livelihood. At the time of India's independence in 1947, the NCT of Delhi (total administrative area—1483 km$^2$) was in a nascent stage of urbanization and the dominant LULC class was agriculture. However, a massive influx of migrant population was witnessed in the 1950s, spurring need for urbanization. In the following three decades, urbanization was led by a dramatic shift in economic focus, meaning a rapid reduction in agricultural and other fallow lands. Immigration rates in turn increased with high urbanization and abundant job opportunities. Eventually, job opportunities declined and further urbanization was required to meet the societal demands. Census of India (2011) highlights an exponential population growth in the NCT of Delhi over the last century—from just 0.4 million in 1901 to 16.8 million in 2011. By 2030, the population of NCT of Delhi and its satellite towns is expected to cross 39 million, and it is poised to become the world's most populous city (United Nations 2016, 2018). Population density is another important urbanization statistic (Jain et al. 2016a). In 2011, the average population density of the city was 11,320 persons/

km$^2$, opposed to a national average of 382 persons/km$^2$ (Census of India 2011). Such a disparity in this statistic shows high rural-urban migration and the trend is likely to continue in the foreseeable future. Of this migration, a considerable part consists of farmers, marginal workers, and skilled/unskilled labour. It is not surprising that even in the urban set-up, these migrants engage in agricultural activities if the opportunity exists (Diehl et al. 2019).

High resolution, multi-temporal satellite information offers valuable insights in understanding LULC based relationship between urbanization and agriculture, especially in data deprived regions. Jain et al. (2016b) exhaustively assessed the past and contemporary LULC change dynamics in the NCT of Delhi from 1977 to 2014 using satellite images (Landsat and Indian Remote Sensing Satellites; data available from USGS and ISRO web portals, respectively) and remote sensing tools. For each selected year, they created false colour composites primarily utilizing the near-infra red, red, and green wavelength bands of the satellite images. Supervised and unsupervised classification techniques were then employed to create the LULC database, ensuring ~90% accuracy was attained. Further extension of the study for future LULC change assessments (year 2030) was performed (Jain 2019). This was based on use of cellular automata and artificial neural networks on the past and contemporary LULC dataset created till 2014. Both studies considered a total of nine LULC classes. Of them, five different classes (dense forest, open forest, scrubs/degraded forest, plantations, and cultivable area) were used to describe the vegetative and green cover. Impervious surface area was used to estimate the built-up area class and along with road/rail network class was used as a proxy for physical extent of urbanization. They also considered two other LULC classes: wasteland and river/waterbodies. Figure 3.1 shows the past, contemporary, and future LULC estimations for the city. Since the present chapter focuses on UA and urbanizations, the figure is restricted to LULC classes pertaining to vegetative cover, green cover, and built-up area.

A major decline in cultivable areas is noted from 44.7% in 1977 to 21.9% in 2014 (Jain et al. 2016b). Future LULC scenario projection shows that area under cultivation in the megacity is likely to be reduced greatly to only 13.3% in 2030 (Jain 2019). Figure 3.1 shows four remaining green cover classes amounting to 38.2% LULC share in the past (1977) and estimated to be 22.3% in the future (2030). Built-up areas on the other hand show significant increase from 7.7% in 1977 to 39.3% in 2014, and are projected to cover 53.8% of Delhi's administrative area by 2030 (Jain et al. 2016b; Jain 2019). Both studies of LULC in Delhi conclude with high statistical confidence, that over the years, urbanization has taken place over lands previously dedicated to agriculture and those treated as wasteland (or barren lands). A similar pattern is likely to continue till 2030. However, in order to realize a part of SDGs, the state government plans to increase the green cover of the city back to 30% (Institute for Human Development 2019).

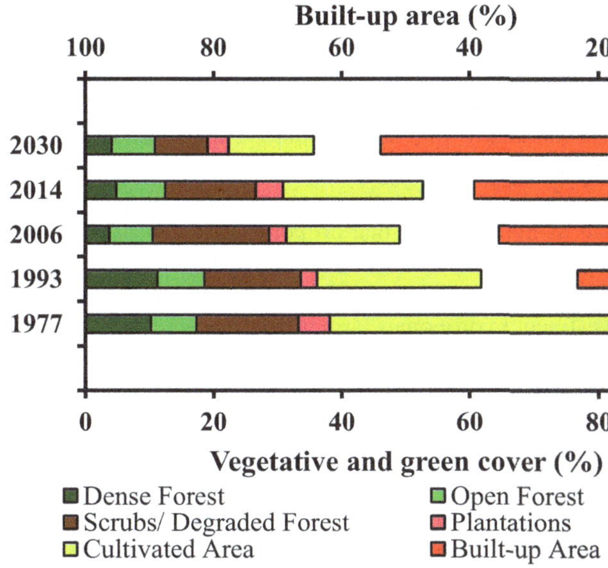


**Fig. 3.1** Past, contemporary, and future LULC change in NCT of Delhi, sourced from Jain et al. 2016b and Jain 2019. Here vegetative and green cover (%; left axis) is represented by five LULC sub-classes. In comparison, the right axis shows the growth of built-up area (including road/rail network) over time. Two LULC classes (wasteland and river/waterbodies) constitute the remaining variation (%; Jain et al. 2016b; Jain 2019) and are excluded in the current figure

### 3.5.2 Current Status of UA Opportunities and Future Outlook

Currently almost the entire city can be considered urban. In 2011, more than 97% of NCT of Delhi's population resided in urban areas compared to 53% in 1901 (Economic Survey of Delhi 2019). Also, the number of rural villages within the NCT has decreased from 300 (in 1961) to just 112 (in 2011) (Census of India 2011). Over the past 50 years it has been noted that approximately one-third of the state population constitutes as the workforce and only about 0.7% working population is currently engaged in agriculture and allied sector (Economic survey of Delhi 2019). This percentage includes both full-time and marginal workers. The survey report finds increase (by 0.17% per annum) in number of individual land holdings (operational) from 2010–2011 to 2015–2016 for agricultural purposes. However, a faster decrease (0.46% per annum) in the operational agricultural area is also noted. Urbanization driven LULC changes are undoubtedly responsible for continued decline of agriculture's contribution in state GDP of Delhi (NCT).

A field observation study conducted by Diehl et al. (2019) finds that farming practices in the city are mostly carried out on small plots (0.4–1.0 ha) along the fertile floodplains of river Yamuna. Despite shrinking cultivable area and less attractive returns from traditional cereal crop farming the value of cash crops, horticulture, and other commercial activities remains high. Their survey finds high priced vegetables such as eggplant, okra, and spinach and herbs such as cilantro and mint are increasingly preferred over wheat, millet, sorghum, and other traditional crops. All-season vegetable cultivation assures food security and earnings for migrant farmers and marginal workers to survive in the city. Having wholesale vegetable and produce markets within short distances, transportation costs are cut down immensely for the poor and marginal urban farmers (Diehl et al. 2019). Meanwhile, new-age urban farmers in the city are opting for advanced indoor farming practices such as hydroponics, that do not require use of soil and/or very less use of water. Hydroponics uses coconut fibre instead of soil, and liquid nutrients are provided to the plants in a controlled environment (Touliatos et al. 2016). One such large capital investment hydroponics farm, with an operational cost of $30,000 USD per year, operates near the outskirts of Delhi cultivating vegetables and herbs (The Better India 2017). The yield is high as compared to traditional agricultural methods, is free of pesticides and fertilizers, and the farming technique makes optimum use of urban space. A sharp rise in floriculture is also noted. A few business companies in the city also use hydroponics to grow rose and jasmine flowers for essential oils production (The Economic Times 2019). Over a span of 10 years (2005–2015) flower cultivations showed a 130% increase, and the state government is further encouraging farmers to take up floriculture, vegetable production, and mushroom cultivation (Economic Survey of Delhi 2019). Animal husbandry practices including livestock and fisheries, however, are on a continued decline since 2003 mainly due to conversion of agricultural lands to built-up areas.

Apart from the above information that was gathered from scientific research papers and economic surveys reports, newspaper articles, the author independently visited some of the food and non-food UA systems in NCT of Delhi. It should be noted that most of the urban farms practicing food UA systems in Delhi were found along the northern, south western, and southern outskirts. Out of these the author visited a cluster of farms in the southern peri-urban edge. Informal talks with practicing urban farmers in Chhatarpur area of south Delhi were then conducted and photographic documentation was carried out. The farms in this area were small in size (~0.5–5 acres), typical to city farming constraints. Vegetable farming was favoured over cereals due to better returns. Animal husbandry, a practice requiring higher investments was not preferred by low income farmers. Non-food UA systems were found to exist in both public and privately owned lands and, unlike the food UA systems, were not restricted to the peri-urban outskirts. These systems were also photographed by the author so as to document their scenario. The current scenario of food and non-food based UA systems is presented in Sects. 3.5.2.1 and 3.5.2.2, respectively.

#### 3.5.2.1 Food Based UA Systems

The author, in an independent field visit, surveyed an urban farm cluster in Chhatarpur area of South Delhi in May 2019. No formal interviews were conducted as the farmers were found reluctant in such a setting. However, they agreed to provide an informal tour of their farm and verbally gave permission to photograph. It was one of the very few privately owned farms (~5 acre plot) where animal husbandry is still in practice. The farm had 40–50 buffaloes (Fig. 3.2a), which contributed as its major income source. No further details about their monthly income were provided. The informal interview revealed that milk and other milk products were sold in nearby dairies and also home delivered to nearby households on a daily basis. Apart from buffaloes, five chickens (Fig. 3.2b) provided ample eggs for the family sustenance. As per the farmer, 7–8 fish used to be present in a small water tank but only one was surviving at the time of field survey. Additionally, a part of the land (Fig. 3.2c) was used to grow millet (sown in early May), and mustard (harvested by May). However, the remaining land in the farm was used for vegetable cultivation and provided maximum returns in a short period of time.

A variety of year round and seasonal vegetables such as potato, bitter gourd, pumpkin, eggplant, tomatoes, okra, sponge gourd, beetroot as well as garlic and onion (Fig. 3.2d–i) were found growing during the visit. Dung collected from animal husbandry practices (Fig. 3.2j) was used as manure in the agricultural fields and no additional pesticides were sprayed. Such organic vegetables were of high quality and high yield. Thus, the agricultural practices in this particular urban farm took care of the family's food needs for the entire year. All excess milk (and other milk products), vegetables, and food grains were sold in the open market, generating substantial monetary income. Other households in middle and high income residential areas instead opt for kitchen gardens or terrace gardens. Plants commonly used in food preparation such as curry tree and chili peppers along with herbs such as mint, cilantro, and basil are grown widely in such households.

#### 3.5.2.2 Non-food Based UA Systems

As discussed in previous sections in this chapter, non-food UA systems are necessary for providing thermal comfort in cities. In Delhi, the maximum temperatures normally exceed 30 °C, often reaching 40–45 °C during pre-monsoons causing huge thermal discomfort (India Meteorological Department 2019). Added pollution from vehicular emissions and 131 tonnes/day dust load (DNA 2018) currently make Delhi the world's most polluted city (Greenpeace 2018). Presence of green cover including densely forested ridges, roadside trees, plantation pockets, etc. in between the dense urban areas of NCT of Delhi has greatly mitigated the daytime UHI formation (Jain 2019). Such areas are noted to be 0.5–3 °C cooler than average, more so during the pre- and post-monsoon seasons. However, land for further expansion of forests is severely limited in the city and therefore the role of other non-food UA systems is indeed relevant. Inclusion of such systems on a large scale is relatively very recent in NCT of Delhi and is detailed in the following paragraph.

Following the Singapore model, the Delhi government (collaborative partnership of Delhi Development Authority, Public Works Department, and Municipal Corporation of Delhi—North (NDMC) and South (SDMC)) adopted a plan in April 2017 to install vertical gardens with integrated water irrigation systems across the entire city. While forest cover expansion projects in Delhi's urban areas are often met with lack of free space (land), vertical gardens can simply be installed over existing urban structures. They have proven to be an effective management strategy in land-crunched cities (Despommier 2013). The approximately two million dollar (USD)

**Fig. 3.2** Animal husbandry and agricultural practices found during field survey to a small privately owned UA farm in Chhatarpur area of South Delhi. The farm boasted (**a**) buffaloes and calves, (**b**) chicken and cultivation of crops, (**c**) millet (recently sown) and mustard (recently harvested) as well as vegetables such as (**d**) bitter gourd, (**e**) pumpkin, (**f**) eggplant, (**g**) sponge gourd, (**h**) beetroot, (**i**) garlic, (**j**) dung from animal husbandry provided manure for the fields. (Source: author)

**Fig. 3.2** (continued)

project aims to green 207 primary schools, 60 roads, and 26 civic markets under NDMC and SDMC jurisdiction in order to ameliorate the ill-effects to urbanization (DNA 2018). Such gardens trap roadside dust and smog, regulate carbon emissions, and have a cooling effect on the concrete structures on which they are erected. Above ground Delhi Metro rail pillars and vehicular traffic flyovers spanning the city are rapidly being converted into potted vertical gardens (Fig. 3.3a). By September 2019 the authorities hoped to expand 60 such existing green pillars to more than 250 in number (Indian Express 2018). Private households and commercial buildings are also following suit and actively urban greening their spaces (Fig. 3.3b). In order to fulfil the rising demand for urban greening in the capital city, small and large

(a)

(b)

**Fig. 3.3** Examples of urban greening in Delhi through non-food UA systems: (**a**) vertical gardens constructed on Metro rail pillars by the state government, and (**b**) green wall in a privately owned residential household. (Source: author)

businesses have proliferated recently, offering architectural inputs, various types of urban greening options (e.g. vertical gardens, rooftop gardens, wall planter, living walls), and looking after their maintenance. Thus, UA is a trending green business strategy and is likely to expand more in the coming years.

Overall it can be said that Delhi city is embracing and promoting food and non-food based UA opportunities at (1) individual, community, and municipal government level, (2) for farmers and UA enthusiasts of all economic and educational backgrounds, and (3) for small, medium, and large scale businesses. In the coming years, research should focus on quantifying the city-specific financial, socio-economic, environmental, and ecological cost-benefits of implemented UA strategies.

## 3.6 Conclusion

Cities, the hub of economic, infrastructural, and science and technological advancements also suffer from the ill-effects of urbanization such as high pollution, loss of green cover and biodiversity, urban heat island, and heat stress. Adequate employment is another pertinent challenge faced when the resident and migrant working population exceeds available job opportunities. Additionally, cities are highly vulnerable to climate related risks. As it stands, the economic and climate burden is likely to be faced more by the developing world cities than those in the developed world. In such an outlook, UA practices present an undervalued and untapped resource to mitigate the ill-effects of urbanization. A plethora of activities under the umbrella of UA can help engage (1) marginal workers and unemployed population, (2) youth and elderly, and (3) small, medium, or large businesses as well as hobby seekers for family sustenance, state GDP growth, environmental sustainability, and increased happiness index. Benefits of UA inclusion into the city certainly outweigh the risks. Conversion of idle lands, unfit for constructing large complexes or unused spaces in existing buildings, can be smartly used for UA practices such as small traditional farms, tree plantations, hydroponics farming, and vertical and rooftop gardening. Apart from maximizing economic returns from existing LULC when integrating them with UA, such services offer thermal comfort, increased building efficiency, aesthetic value, enhanced urban biodiversity, and fresh organic food availability from nearby sources.

However, no single UA strategy fits best. Cities in their own right and judgement should selectively adopt UA forms and practices befitting to their resource availability and specific needs. In this regard, a case study of Delhi city in India was presented. It is noted that Delhi is losing out on cultivable areas at a very fast rate to sustain rapid urbanization. By 2030, it is estimated that only 13.3% cultivable areas would remain as compared to 44.7% in 1977. However, even with continually shrinking traditional agricultural lands, decreasing animal husbandry practices, and virtually no rural population, it is encouraging to see adoption of several new UA systems in the city. The government is promoting urban greening by mass-installations of vertical gardens and green walls in dense built-up areas, supporting businesses indulging in technological UA practices like hydroponics, and encouraging small farmers switch to profitable farming practices that range from vegetable and mushroom cultivation to horticulture. Impact studies focusing on integration of UA systems in the city are still at a nascent stage and further research is needed to maximize benefits of UA to the society. Technical and technological knowledge transfer to diverse target groups (traditional farmers, businesses, and interested individuals), time to time ground based reality check (environmental constraints, economic and labour management difficulties, etc.) from farmers, and a good government support system are key components in successful integration of UA systems in any city.

## References


Akbari H, Kurn DM, Bretz SE et al (1997) Peak power and cooling energy savings of shade trees. Energ Buildings 25(2):139–148

Alig RJ, Kline JD, Lichtenstein M (2004) Urbanization on the US landscape: looking ahead in the 21st century. Landscape Urban Plan 69(2):219–234

Astee LY, Kishnani NT (2010) Building integrated agriculture: utilising rooftops for sustainable food crop cultivation in Singapore. J Green Build 5(2):105–113

Baker LE (2004) Tending cultural landscapes and food citizenship in Toronto's community gardens. Geogr Rev 94(3):305–325

Bass B, Baskaran B (2003) Evaluating rooftop and vertical gardens as an adaptation strategy for urban areas. Institute for Research and Construction, NRCC-46737, Project (A020)

Bechtel B, Alexander PJ, Böhner J et al (2015) Mapping local climate zones for a worldwide database of the form and function of cities. ISPRS Int J Geoinf 4(1):199–219

Beniston J, Lal R (2012) Improving soil quality for urban agriculture in the north central U.S. In: Lal R, Augustin B (eds) Carbon sequestration in urban ecosystems. Springer, Netherlands, pp 279–313

Bowen IS (1926) The ratio of heat losses by conduction and by evaporation from any water surface. Phys Rev 27(6):779

Census of India (2011) Provisional population totals. http://www.censusindia.gov.in/2011-prov-results/prov_data_products_delhi.html. Accessed 28 Jun 2018

Cook J, Oviatt K, Main DS et al (2015) Re-conceptualizing urban agriculture: an exploration of farming along the banks of the Yamuna River in Delhi, India. Agric Hum Values 32(2):265–279

d'Amour CB, Reitsma F, Baiocchi G et al (2017) Future urban land expansion and implications for global croplands. Proc Natl Acad Sci 114(34):8939–8944

Debbage N, Shepherd JM (2015) The urban heat island effect and city contiguity. Comput Environ Urban 54:181–194

Despommier D (2013) Farming up the city: the rise of urban vertical farms. Trends Biotechnol 31(7):388–389

Diehl JA, Oviatt K, Chandra AJ et al (2019) Household food consumption patterns and food security among low-income migrant urban farmers in Delhi, Jakarta, and Quito. Sustainability 11(5):1378

DNA (2018) Flip side of Delhi's upright gardens. Newspaper article. https://www.dnaindia.com/delhi/report-flip-side-of-delhi-s-upright-gardens-2663240. Accessed 9 May 2019

Economic Survey of Delhi (2019) Economic survey of Delhi: 2018–19. Planning Department, Government of NCT of Delhi, Delhi. http://delhiplanning.nic.in/content/economic-survey-delhi-2018-19. Accessed 6 May 2019

Eigenbrod C, Gruda N (2015) Urban vegetable for food security in cities - a review. Agron Sustain Dev 35(2):483–498

Ferreira AJD, Guilherme RIMM, Ferreira CSS et al (2018) Urban agriculture, a tool towards more resilient urban communities? Curr Opin Environ Sci Health 5:93–97

Gill SE, Handley JF, Ennos AR et al (2007) Adapting cities for climate change: the role of the green infrastructure. Built Environ 33(1):115–133

Greenpeace (2018) World air quality report 2018: region and city PM2.5 ranking. https://www.greenpeace.org/international/press-release/21193/latest-air-pollution-data-ranks-worlds-cities-worst-to-best/. Accessed 15 May 2019

Grimmond CSB, Oke TR (1999) Aerodynamic properties of urban areas derived from analysis of surface form. J Appl Meteorol 38(9):1262–1292

Guo G, Wu Z, Xiao R et al (2015) Impacts of urban biophysical composition on land surface temperature in urban heat island clusters. Landscape Urban Plan 135:1–10

Hajat S, Kosatky T (2010) Heat-related mortality: a review and exploration of heterogeneity. J Epidemiol Community Health 64(9):753–760

India Meteorological Department (2019) Monthly mean maximum and minimum temperature and total rainfall based upon 1901–2000 data: Delhi. http://www.imd.gov.in/pages/services_climate.php?adta=PDFandadtb=andadtc=../section/climate/climateimp. Accessed 15 May 2019

Indian Express (2018) MoEF to launch dust mitigation measures. Newspaper article. https://indianexpress.com/article/cities/delhi/moef-to-launch-dust-mitigation-measures-5257376/. Accessed 15 May 2019

Institute for Human Development (2019) Vision Delhi 2030 - Draft Report. http://delhi.gov.in/wps/wcm/connect/d92261804687c81fa5ccbf264798d198/Vision+Delhi+2030.pdf. Accessed 18 May 2019

IPCC (2014) Climate change 2014: synthesis report. In: Pachauri RK, Meyer LA (eds) Contribution of Working Groups I, II and III to the Fifth Assessment Report of the Intergovernmental Panel on Climate Change. IPCC, Geneva, Switzerland, pp 1–151

Jain M (2019) Urbanization and its manifestations on localized weather. [Ph.D. Thesis, Jawaharlal Nehru University]

Jain M, Dimri AP, Niyogi D (2016a) Urban sprawl patterns and processes in Delhi from 1977 to 2014 based on remote sensing and spatial metrics approaches. Earth Interact 20(14):1–29

Jain M, Dawa D, Mehta R et al (2016b) Monitoring land use change and its drivers in Delhi, India using multi-temporal satellite data. Model Earth Syst Environ 2(1):1–14

Kaufman J, Bailkey M (2000) Farming inside cities: entrepreneurial urban agriculture in the United States. Working Paper. Lincoln Institute of Land Policy, Cambridge, MA, pp 1–124

Kishtawal CM, Niyogi D, Tewari M et al (2010) Urbanization signature in the observed heavy rainfall climatology over India. Int J Climatol 30(13):1908–1916

Kovats RS, Hajat S (2008) Heat stress and public health: a critical review. Annu Rev Public Health 29:41–55

Lin BB, Fuller RA (2013) Sharing or sparing? How should we grow the world's cities? J Appl Ecol 50(5):1161–1168

Lin BB, Philpott SM, Jha S (2015) The future of urban agriculture and biodiversity-ecosystem services: challenges and next steps. Basic Appl Ecol 16(3):189–201

Lovell ST (2010) Multifunctional urban agriculture for sustainable land use planning in the United States. Sustainability 2(8):2499–2522

Lundgren K, Kuklane K, Gao C et al (2013) Effects of heat stress on working populations when facing climate change. Ind Health 51(1):3–15

Meehl GA, Tebaldi C (2004) More intense, more frequent, and longer lasting heat waves in the 21st century. Science 305(5686):994–997

Mougeot LJA (2006) Growing better cities: urban agriculture for sustainable development. International Development Research Centre, Ottawa, Canada, pp 1–97

Mukherji N, Morales A (2010) Zoning for urban agriculture. Zoning Pract 3(10):1–7

Oates J, McMahon A, Karsgaard P et al (2007) Early Mesopotamian urbanism: a new view from the north. Antiquity 81(313):585–600

Pothukuchi K, Kaufman JL (1999) Placing the food system on the urban agenda: the role of municipal institutions in food systems planning. Agric Hum Values 16(2):213–224

Pothukuchi K, Kaufman JL (2000) The food system: a stranger to the planning field. J Am Plann Assoc 66(2):113–124

Rosenzweig C, Solecki WD, Romero-Lankao P et al (eds) (2018) Climate change and cities: second assessment report of the urban climate change research network. Cambridge University Press, Cambridge, UK

Satterthwaite D (2008) Cities' contribution to global warming: notes on the allocation of greenhouse gas emissions. Environ Urban 20(2):539–549

Schär C, Jendritzky G (2004) Climate change: hot news from summer 2003. Nature 432(7017):559

Shashua-Bar L, Hoffman ME (2000) Vegetation as a climatic component in the design of an urban street: an empirical model for predicting the cooling effect of urban green areas with trees. Energ Buildings 31(3):221–235

The Better India (2017) Gurugram adopts soil-less farming and shows how easy it is to eat chemical-free! Newspaper article. https://www.thebetterindia.com/106387/gurugram-hydroponic-organic-soilless-urbanfarming-chemicalfree/. Accessed 9 May 2019

The Economic Times (2019) A soil-free success: aquaponic agriculture on the rise. Newspaper article. https://economictimes.indiatimes.com/news/economy/agriculture/a-soil-free-success-aquaponic-agriculture-on-the-rise/articleshow/68531017.cms. Accessed 15 May 2019

Thomaier S, Specht K, Henckel D et al (2015) Farming in and on urban buildings: present practice and specific novelties of zero-acreage farming (ZFarming). Renew Agric Food Syst 30(1): 43–54

Tian G, Wu J, Yang Z (2010) Spatial pattern of urban functions in the Beijing metropolitan region. Habitat Int 34(2):249–255

Touliatos D, Dodd IC, McAinsh M (2016) Vertical farming increases lettuce yield per unit area compared to conventional horizontal hydroponics. Food Energy Secur 5(3):184–191

United Nations (2014) World urbanization prospects: the 2014 revision, Highlights (ST/ESA/SERA/352). Department of Economic and Social Affairs, Population Division. http://esa.un.org/unpd/wup/Highlights/WUP2014-Highlights.pdf. Accessed 7 Apr 2015

United Nations (2016) The World's Cities in 2016 – Data Booklet (ST/ESA/SER.A/392). Department of Economic and Social Affairs, Population Division. http://www.un.org/en/development/desa/population/publications/pdf/urbanization/the_worlds_cities_in_2016_data_booklet.pdf. Accessed 9 Jan 2018

United Nations (2018) World urbanization prospects (2018 revision), Press Release. https://esa.un.org/unpd/wup/Publications/Files/WUP2018-PressRelease.pdf. Accessed 29 Apr 2019

Varzi C (2019) What is a city? Topoi. https://doi.org/10.1007/s11245-019-09647-4

World Bank (2019) Urban development overview. https://www.worldbank.org/en/topic/urbandevelopment/overview. Accessed 6 May 2019

Zhang XQ (2016) The trends, promises and challenges of urbanisation in the world. Habitat Int 54: 241–252

**Madhavi Jain** completed her PhD in Environmental Sciences from Jawaharlal Nehru University, India, in 2020. She was awarded the prestigious DST-IUSSTF BHAVAN Fellowship (2018) to pursue research work at Purdue University, USA, for 6 months. She completed her BSc in Physics (Hons) and MSc in Environmental Studies from Delhi University, India. Her research focuses on urbanization-driven land use/land cover changes in big cities and how it creates a plethora of modern urban environmental issues. Dr. Madhavi has published several papers and authored book chapters in reputed peer-reviewed international journals.

# 4 Commercial Potential for Rooftop Farming in a Major City in China

Yun-Lin Su and David W. Ow

**Abstract**

As urbanization expands metropolitan areas, the displacement of farmland pushes the production of vegetables to more distant peri-urban farms. Transporting fresh vegetables into urban markets exerts greater expenditure of carbon-based energy, not only because of longer distances but also disproportionately longer time due to traffic congestion. Growing a proportion of vegetables within city settings could reduce some of the environmental costs involved in feeding urban populations. Though much has been published on the technical feasibility of production, there is a dearth of data on whether urban farming can be economically feasible, which can vary depending on local conditions. We tested rooftop farming of leafy vegetables in Guangzhou, China and deduced that this type of farming can be commercially profitable. This chapter summarizes recent data from growing leafy vegetables on a rooftop in Guangzhou, and discusses aspects of vegetable types suitable for cultivation, vegetable quality, production capacity, and costs. Our findings suggest that rooftop farming can serve as an economically viable supplement to rural vegetable production. Considering yield, market prices, and consumer purchase preferences, a 150 $m^2$ screen house with bi-layer production could generate in the best scenario an annual yield of 6310 kg and an income of 162% of the 2018 Guangzhou's average worker's income. However, implementation would require major changes in existing government building codes and regulatory policies that affect market demand for commercial housing.

Y.-L. Su · D. W. Ow (✉)
Plant Gene Engineering Center, Chinese Academy of Sciences Key Laboratory of South China Agricultural Plant Molecular Analysis and Genetic Improvement, South China Botanical Garden, Chinese Academy of Sciences, Guangzhou, China
e-mail: dow@berkeley.edu



J. A. Diehl, H. Kaur (eds.), *New Forms of Urban Agriculture: An Urban Ecology Perspective*, https://doi.org/10.1007/978-981-16-3738-4_4

**Keywords**

Urban farming · Vertical agriculture · Leafy vegetable production

## 4.1 Introduction

Population growth will inevitably reduce per capita forest and arable land (FAO 2020a, b, c). Along with concomitant urbanization, the future holds more concrete pavement and less green space. Exasperating this trend is the increase in pollution due to human activities that reduces the quality of farmland. Food security challenges in the decades ahead must consider not only production demands, but also food safety issues resulting from degraded air, water, and soil quality.

Growing vegetables in urban settings is not new, and there are compelling reasons to propose that urban vegetable farming should be expanded to a larger scale. First, urban farming can help alleviate the loss of arable land due to urban and industrial expansion, as well as to deterioration of soil quality. Second, while many food crops can be well preserved when transported from distant locations, some vegetables, such as many leafy greens, are easily perishable and are best grown nearby. Peri-urban farms relocated farther away due to urban expansion incur higher transport costs and time to reach urban markets. Growing these vegetables within cities would insure freshness while lowering the carbon footprint from packaging and fuel. Third, peri-urban farms are not well suited for growing food, as the soil is more likely to harbor pollutants from urban activities. Chronic dietary intake of leafy greens that take up soil contaminates translates to higher health costs in the years ahead, especially impacting elderly care (Wijayawardena et al. 2016; He et al. 2013). The use of clean soil and water in controlled settings can alleviate this problem, and while it can be optional for peri-urban farms, it is most likely required for urban farming.

While many types of urban farming can be practiced, ranging from personal use of backyards to commercial production within multi-story buildings (popularly coined "vertical farming"), roof farming is one type that does not directly compete for existing land use. Aside from some space reserved for building maintenance and fire escape, a typical aerial view of any city will show that most roof space is vacant. The aggregation of space can be vast, as one estimate has placed the current available roof space in China at about 1 million hectares (Chen 2011). And by 2040, when urbanization reaches up to 75% of the population (Gao and Wei 2013), the amount of available roof space is predicted to double to ~2 million hectares. Roof farming also uses predominantly natural sunlight that can eliminate the high cost of electrical lighting.

Despite theoretic arguments in favor of roof farming, very little activity of this type is seen. Most rooftop farms are for non-profit purposes, such as for social and educational use or for enhancing living quality (Buehler and Junge 2016; Thomaier et al. 2015), and are usually of small scale with low production that cannot sufficiently supplement conventional agriculture. A high level of food production would require professional commercial operations (Buehler and Junge 2016). However, high investment costs, narrow profit margin, and uncertain marketing channels all serve as barriers for the commercialization of rooftop farming (Benis and Ferrão 2018; Sanyé-Mengual et al. 2015; Specht et al. 2014). While commercial real estate developers have taken notice of this possibility, the one question that has been asked is the experimental data, especially local data, rather than theoretical estimates, that roof farming can compete economically with conventional farming or provide a higher return on investment than other rooftop uses such as for gardens or solar panels.

To fill this research gap, we began a study in 2012 on the economic feasibility of rooftop vegetable production. Below we summarize recent findings on producing a number of leafy vegetables in a roof screen house in Guangzhou, China.

## 4.2 Study Area

The city of Guangzhou, with an area of 7430 km$^2$, has ~14 million inhabitants. It is located within a highly industrial metropolitan area of 56,000 km$^2$ with a population of around 70 million people. Now it is a part of the "Greater Bay Area" that encompasses Special Administrative Regions of Hong Kong and Macao, and nine Guangdong Province municipalities of Guangzhou, Shenzhen, Zhuhai, Foshan, Huizhou, Dongguan, Zhongshan, Jiangmen, and Zhaoqing. Within this major industrial area, arable land has been dramatically decreasing as well as becoming too polluted for ideal crop production (Li et al. 2015; Nan Fang Daily 2013; Statistics Bureau of Guangzhou Municipality 2018). In 2015, it was reported that a third of the vegetable samples sold in Guangzhou markets contained excessive pesticide residues (Jiang 2015), and a study by Chen et al. (2017) showed that up to 92% market samples in Guangzhou contained heavy metals above the MRL (maximum residue limits). While transportation infrastructure has advanced to help with timely transport of fresh vegetables from distant locations, there nevertheless is a need to consider alternative ways to produce vegetables locally and free from pollutants. The large rooftop space (~7000 hectares) in Guangzhou (Fang 2015) offers an ample opportunity for large-scale rooftop vegetable production, and we therefore started a study to test whether this farming practice can be economically profitable.

Guangzhou has a subtropical monsoon climate. Data from 2012 to 2017 showed average monthly temperature range from 12 to 30 °C (54–86 °F) with humidity from 62% to 89%. Vegetables can be grown year-around without needing a glass or polycarbonate greenhouse.

## 4.3 Rooftop Hydroponics in a Screen House

Given the warm climate of Guangzhou, a simple screen house would be adequate to keep out insect pests and protect crops from heavy rain, which peaks in the months of May to August with the daily average rainfalls of 8.0–13.9 mm in Guangzhou (2012–2017 data). Eliminating glass or polycarbonate not only reduces construction

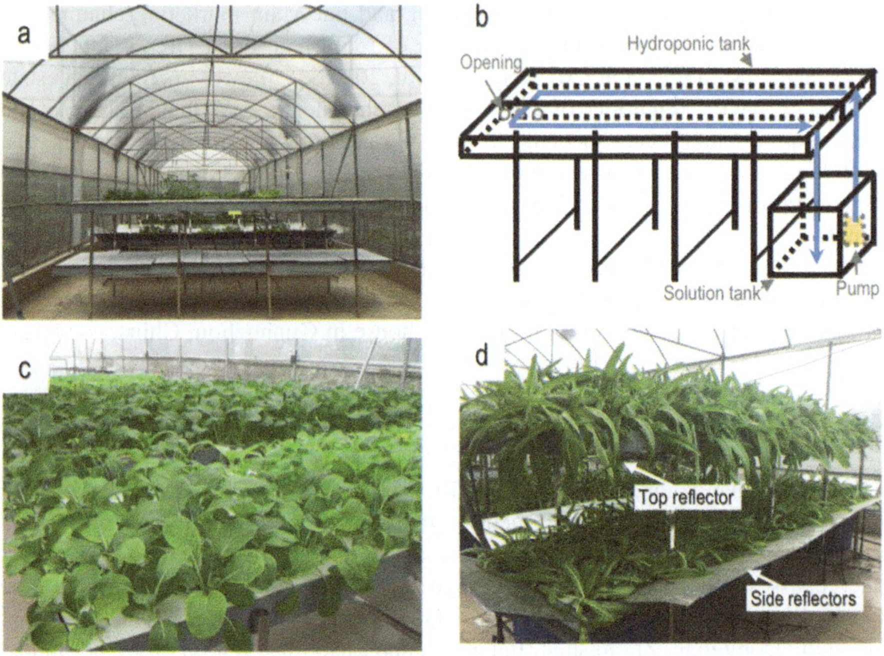


**Fig. 4.1** Rooftop screen house and hydroponic setup. (**a**) Interior of rooftop screen house. (**b**) Hydroponic tank and solution tank of a hydroponic unit. Hydroponic tank is divided by a partition, and an aquarium pump circulates solution from one side of the hydroponic tank to the other side through openings at the far end of the partition, and then back to the solution tank (blue arrows). (**c**) Single-layer hydroponic system. (**d**) Reflector-assisted double-layer hydroponic system

costs, but also lessens the possibility of dangerous materials shattering and falling from the roof. After considering other space needs for fire escape and roof maintenance, we built a 150 $m^2$ screen house (25 × 6 m; 3–5 m in height) on a 252 $m^2$ roof of a two-story building (Fig. 4.1a). Although the building is located within the South China Botanical Garden, it is similarly situated as other buildings that are near heavy traffic, as it is located 450 m southeast of 8-lane Tianyuan Road and 400 m northwest of 8-lane South China Expressway.

The screen house is comprised of a galvanized iron frame wrapped with insect screens, and an outer wrapping of plastic film on top and along the bottom half of the length of the screen house. The outer plastic film cover can be rolled up, exposing the insect screen underneath. Above the top of the screen house are sun-shade nets that can be adjusted by an electric motor for providing some control of temperature and light intensity. The temperature within the screen house is usually the same as ambient temperature, except during warm sunny days when it is generally 2 °C higher when the sun-shade is used, but can be up to 15 °C higher without the sun-shade (Liu et al. 2016). The extra warmth is fine during winter, but the higher temperature in summer months requires more heat tolerant vegetable cultivars.

Soil is not an ideal substrate for rooftop production as wet soil adds considerable weight. Water soaked concrete can also lead to leakage, and even if the soil is contained within pots, a large amount of water would constantly drain out onto the roof. Soil particles would also dirty the roof floor and choke drainage systems. Moreover, soil is a complex medium and may at times harbor soil-borne plant pathogens, requiring costly treatment or replacement. In contrast, soil-less production uses a minimum quantity of water thereby placing a substantially lighter load on roofs. Aeroponics is one option for soil-less production, where plants are periodically sprayed by a mist of nutrient solution. However, setup costs can be expensive, and spray systems can be easily clogged and difficult to clean. Unless an electrical backup system is in place, plants could become dehydrated during power failures, especially in a hot climatic region like South China. A hydroponic system would therefore be a more practical option. If spilled, hydroponic solutions can drain away. If contaminated with microbes, a hydroponic tank can be cleaned and new solutions replaced. During power outages, circulation of the solutions may cease, which reduces oxygen availability, but the submerged plant roots would not dry out.

The 25 × 6 m screen house can accommodate 16 hydroponic tanks, each 400 × 100 cm and 10 cm deep, but we chose to test a setup of 14 tanks, reserving some space for other uses. The tanks were made from polyvinyl chlorine that is more durable than extruded polystyrene foam, so damage was less likely in the process of moving and cleaning. As the tank was essentially a shallow rectangular box, it was relatively easy to clean, in contrast to many commercially available hydroponic systems made of various sizes of tubes or pipes that make cleaning the interiors difficult. The hydroponic tank was covered by polyvinyl chlorine plates with a number of holes for stabilizing individual plants. Thus, planting density could be flexibly adjusted according to the needs of vegetable cultivation by changing cover plates with different number of holes. Sponge-wrapped seedlings with a root length of ~5 cm could be inserted into cover plate holes of hydroponic tanks filled with nutrient solution. Aeration was provided by an aquarium pump recirculating the solution at a rate of ~23 L/min for 5 min every 2 h (Fig. 4.1b).

## 4.4 Feasibility of Commercial Rooftop Farming

### 4.4.1 Cost Effective Rooftop Farming

We tested the cultivation method as described above on diverse types of vegetables including cherry tomatoes, herbs, and leafy vegetables (Fig. 4.1c and Table 4.1). Except for cherry tomatoes and Italian lettuces that were planted at densities of total 33 plants per tank and 126 plants per tank, respectively, all others were planted at a density of total 224 plants per tank. As 14 tanks were placed in our 150 $m^2$ screen house, the plant densities could be translated in 3.1 plants per $m^2$ for cherry tomatoes, 11.8 plants per $m^2$ for Italian lettuce, and 20.9 plants per $m^2$ for others. However, growing cherry tomatoes was not cost effective due to the high amount of labor required for timely harvests. In contrast to cherry tomatoes, herbs and leafy

**Table 4.1** Edible crop and cultivation information

| Edible crop | | | | Production cycle (day) | Suggested growing season | Suggested production cycles/year |
|---|---|---|---|---|---|---|
| **Type** | | **Common name** | **Latin name** | | | |
| Herb |  | Caraway | *Coriandrum sativum* L. | 30 | Oct–Feb | 3 |
| |  | Mint | *Mentha haplocalyx* | – | – | – |
| |  | Scallion | *Allium ascalonicum* L. | – | – | – |
| |  | Sweet basil | *Ocimum basilicum* L. | 30 | Jan–Dec | 12 |
| Leafy vegetable |  | Ceylon spinach | *Basella alba* L. | – | – | – |
| |  | Chinese flowering cabbage | *Brassica campestris* L. ssp. *Chinensis* var. *utilis* Tsen et Lee | 30 | Oct–Apr or May–Dec, which depends on cultivars | 7 or 10, which depends on cultivars |
| |  | Chinese kale | ***Brassica oleracea var. alboglabra cv.Zhonghua*** | 45 | Jan–Dec | 8 |
| |  | Crown daisy | *Chrysanthemum coronarium* L. | 45 | Nov–Apr. | 3 |
| |  | Curly endive | *Cichorium endivia* L. var. *crispum* | 30 | Jan–Dec | 12 |
| |  | Early maturing Chinese cabbage | *Brassica pekinensis* Rupr. | 20 | Jan–Dec | 18 |
| |  | Italian lettuce | *Lactuca sativa* var. *ramosa* Hort | 45 | Nov–Mar | 3 |
| |  | Leaf celery | *Apium graveolens* var. *secalinum* | 60 | Sep–Apr | 4 |
| |  | Leaf lettuce | *Lactuca sativa* var. *longifoliaf* Lam | 35 | Oct–May | 7 |
| |  | Leaf mustard | *Brassica juncea* var. *foliosa* | 40 | Jan–Dec | 9 |
| |  | New Zealand spinach | *Tetragonia expansa* Murr. | 30 | – | – |
| |  | Pother mustard | *Brassica juncea* var. *multiceps* | 45 | Oct–Jan | 2 |
| |  | Shanghai qing | *Brassica campestris* L.ssp. *chinensis* var. *communis* | 30 | Jan–Dec | 12 |
| |  | Water spinach | *Ipomoea aquatica* | 30 | May–Oct | 6 |
| Fruit |  | Cherry tomato | *Solanum lycopersicum* var. *cerasiforme* | – | – | – |

Note: – means experiment was not carried through

vegetables generated very little waste since nearly the entire plant can be marketed. The herbs, scallion, mint, caraway, and sweet basil, were small plants that did not generate high yield (kg per $m^2$). We did not test growing them at higher density, but instead concentrated on testing several leafy greens. From 2012 to 2014, we recorded the production cycles and suitable growing seasons of six leafy vegetables popular in Guangzhou: Chinese flowering cabbage, crown daisy, Italian lettuce, leaf lettuce, leaf mustard, and potherb mustard (Liu et al. 2016). Each of the vegetables had a different growing season; for Chinese flowering cabbage, different cultivars have different optimal growing seasons. Except for leaf mustard, all other vegetables failed to grow well in the summer months when midday temperatures can reach 50 °C on the tank surfaces. Although we considered installing some form of temperature control, such as evaporative cooling fans, in the end, we found that using heat tolerant cultivars was a more cost-effective option.

Deduced annual maximum yield in the 150 $m^2$ screen house of the six vegetables: Chinese flowering cabbage, crown daisy, Italian lettuce, leaf lettuce, leaf mustard, and potherb mustard were 1040 kg (6.9 kg/$m^2$), 330 kg (2.2 kg/$m^2$), 975 kg (6.5 kg/$m^2$), 1310 kg (8.7 kg/$m^2$), 1530 kg (10.2 kg/$m^2$), and 255 kg (1.7 kg/$m^2$), respectively (Liu et al. 2016). The highest deduced yield for year-round production in the 150 $m^2$ screen house could reach 1815 kg (12.1 kg/$m^2$) by growing 3 cycles of Italian lettuce followed by 5 cycles of leaf mustard. Production cost of best case scenario for each vegetable comprised of the costs for facility, equipment, consumables, and labor. Cost for facility and equipment included the straight-line depreciation expense with no salvage value for the screen house, 14 sets of hydroponic setups, and miscellaneous tools. Consumables comprised of fertilizer, water, electricity, seeds, sponges, and packaging. Labor cost was calculated based on the Guangzhou temporary worker hourly wage multiplied by the number of hours for each vegetable, which included sowing seeds, moving seedlings into sponge and nursery tray, making nutrient solution, moving plants from nursery tray to planting tank, harvest and packaging, vegetables and waste delivering, cleaning and repair, and daily check. Since we envisioned a subscription-based sales model in which rooftop vegetables would be distributed to building residents, expenses related to advertisement and transport were not included—since advertisement would simply be notification in homeowner association newsletters, and transport would be via an elevator ride for delivery to individual households.

For an estimate of the best case scenario for maximum profit, we multiplied annual deduced maximum yield by the market price and then subtracted the cost of production (Liu et al. 2016). For market price, we used equivalent items labeled as pollution-free/green vegetables, since our vegetables were likewise produced relatively free of pollution (see Sect. 4.4.2). This best case scenario did not include the cost for rent, since it was difficult to estimate rent for space that is ordinarily not used.

Our calculations for estimated annual maximum profit from the 150 $m^2$ screen house for Chinese flowering cabbage, crown daisy, Italian lettuce, leaf lettuce, leaf mustard, and potherb mustard were ¥13,500 (~$1910), ¥6900 (~$977), ¥17,550 (~ $2490), ¥14,400 (~$2040), ¥44,400 (~$6290) and ¥6150 (~$871), respectively.

However, even with the highest earnings from growing leaf mustard, a ¥44,400 (~ $6290) annual income, or ¥3700 (~$524) per month would only be 54% of the average 2015 Guangzhou monthly compensation (salary plus benefit) of ~¥6910 (~ $978) (Zhang 2015). This shows that even though the best scenario calculations for rooftop farming can conclude that it is profitable, it nevertheless is not a very attractive commercial proposal unless a single person manages multiple screen houses.

### 4.4.2 Quality of Rooftop Vegetables: Biosafety and Nutrition

Quality is an important factor in marketability, and while most consumers can judge freshness, color, shape, and texture, they cannot see other quality factors that may be more important. We collected market vegetables and our hydroponic samples and compared their quality on biosafety, which includes accumulations of pesticides, heavy metals, and nitrate, and on nutrition, which includes contents of minerals, vitamin C, and crude fiber (total insoluble fiber) (Liu et al. 2016). In total, 31 samples of common Italian lettuce, 26 common Chinese flowering cabbage, 14 pollution-free/green Italian lettuce, 18 pollution-free/green Chinese flowering cabbage, 4 organic Italian lettuce, 5 organic Chinese flowering cabbage, 6 of our rooftop hydroponic Italian lettuce, and 5 of our rooftop hydroponic Chinese flowering cabbage samples were tested. Organophosphate and carbamate pesticide residues were analyzed by a portable testing kit for an acetyl cholinesterase inhibition assay (China state standard GB/T5009.199–2003). For analyses of heavy metals—lead (Pb), arsenic (As), cadmium (Cd), chromium (Cr) and mercury (Hg)—and minerals—potassium (K), calcium (Ca), magnesium (Mg), iron (Fe), and zinc (Zn)—heavy metals and minerals were first extracted from the dried homogenized powder of vegetables in a microwave digestion system with nitric acid. Then the solutions were subjected to inductively coupled plasma mass spectrometry (7700×, Agilent) for heavy metal content determination (Sanchez Lopez et al. 2003), and to flame atomic absorption spectrometry (ContrAA 700, Analytikjena) for mineral content measurement. Nitrate content was determined with the Griess-cadmium reduction and spectrophotometric method (Prasad and Chetty 2008). Crude fiber analysis was performed by gravimetric determination method of amylase-treated neutral detergent fiber described by Mertens (2002). Vitamin C content was estimated using a 2,6-dichlorophenolindophenol titration method reported by Harris and Olliver (1942).

Organophosphate and carbamate insecticides have been in common use for vegetable farming in the Guangzhou area (Yang et al. 2014). Even when using a simple and less sensitive biochemical assay, we were able to find 2 of 98 market vegetables that showed contamination, which should not have been sold, including one sample labeled as pollution-free/green (Liu et al. 2016). In contrast, these pesticide residues were not detected among our rooftop grown samples, as was expected since these pesticides had not been used.

Unlike growing plants in the open field or in rural greenhouses, rooftops are rarely populated by other plants that may serve as pest reservoirs. During our 7 years of experience, we found that our screen house with the 50 mesh insect screen could effectively protect vegetables from severe insect damage, especially large or medium sized insects. Some tiny arthropods such as whiteflies, spider mites, aphids, and thrips occasionally snuck in to feed on our plants, but they were effectively controlled through a combination of sticky card traps, removing infected plants, stopping vegetable production for a few days after harvest, cleaning up the screen house, and, when deemed necessary, spraying low toxicity pesticides, avermectin or imidacloprid (Wexler et al. 2014). This suggests that urban rooftop farming can produce vegetables with less pesticide contamination.

The five most serious heavy metals were below the MRL (0.1, 0.05, 0.2, 0.5, and 0.01 mg/kg fresh weight for Pb, As, Cd, Cr, and Hg) among vegetables grown in our hydroponic system (Liu et al. 2016), and the detected levels were likely derived from local tap water used for nutrient solution. In contrast, some market samples showed higher heavy metal content, with 2–5% of market samples exceeding the MRL for arsenic or lead (104–235% of the MRLs). Most likely, these metals were taken up from water and/or soil used for their propagation. Surprisingly, the metal-contaminated samples included those labeled as pollution-free/green and organically grown. With respect to nitrate content, all of our rooftop vegetables were below the MRL (3000 mg/kg fresh weight), but nearly half of Chinese flowering cabbage samples from the markets, including those sold as pollution-free/green or organic, exceeded the MRL for nitrate (101–130% of the MRL).This is probably due to excessive fertilizer application in soil-based farming systems. The data point to one indication: despite adhering to strict farming practices for certification of pollution-free/green or organic labeling, the presence of these chemical species can be attributed to environment-derived inputs. Hydroponic farming inputs are controlled through defined nutrient solutions but soil grown vegetables uptake unknown quantities of trace elements and other chemical compounds from the soil. The uptake of these chemicals cannot be predicted simply by measuring soil concentrations as their bioavailability depends on numerous factors, including but not limited to, soil pH, rainfall amount, and soil microbiome (Alloway 2013).

Another factor that consumers do not have ready access to is the nutritional content. In comparing flowering cabbage and Italian lettuce produced from our rooftop system against their market counterparts, crude fiber content did not significantly differ (Liu et al. 2016). However, compared to the average mineral contents of common, pollution-free/green and organic types of counterparts, our Italian lettuce contained higher Ca (+141%), K (+36%), Mg (+53%), and Fe (+47%), but lower Zn (−27%), while our Chinese flowering cabbage had higher Ca (+38%) and K (+10%), Mg (+1%) but lower Fe (−68%) and Zn (−54%). It is however not clear whether such differences in mineral content could be a cause for concern. For vitamin C, our rooftop vegetables showed 13–37% higher vitamin C content than their corresponding market samples. This may be related to freshness of the product, as even for fresh looking market vegetables, time elapses during long transport from harvest to market. Vitamin C is not stable and can be easily degraded after harvest. It is known that the B vitamins, phenolic compounds, β-carotene, and others are also subjected to postharvest degradation, even when kept at low temperature (Rickman et al. 2007a, b). Moreover, the increase of time and distance from "source to table" can also cause moisture loss and potential microbe-caused spoilage (Hammond et al. 2015). Therefore, a key advantage in urban rooftop vegetables may lie in these hidden qualities related to biosafety and nutrition that consumers cannot tell with a visual inspection.

### 4.4.3 Bi-layer Production

Despite finding that the hydroponic vegetable farming as described above could be somewhat profitable, the potential earnings from roof farming were still not particularly attractive compared to the average worker's income. To make it a more profitable business, we felt that it would be necessary to increase yield or lower costs. Therefore, in 2015, we tested doubling the growing area on the same footprint, by using a two-layer hydroponic system as shown in Fig. 4.1d (Su et al. 2020). Since the average light intensity of the bottom tank was only 29% of the top tank, we needed to increase bottom tank lighting. Supplemental electrical lighting was considered, but because it would require higher capital investment and operating cost, we opted instead to test using light-weight reflectors to direct ambient light to the bottom tanks. Previous studies showed that the use of inexpensive silver-tinted polyester films as reflectors reduced electricity cost up to 50% on plant tissue and algal cultures or yielded increases in the growth rate and biomass of blue-green algal cultivation (Ajayan and Selvaraju 2011; Sathi et al. 2010). Additionally, the possibility of applying reflectors in multi-layer vegetable cultivation to direct sunlight to shaded areas of a rooftop facility has been proposed previously (Jang and Chang 2013). We decided to test a simple reflector system with aluminized polyethylene terephthalate sheets to direct ambient light to the bottom layer (Fig. 4.1d). Top reflectors were glued to the bottom of the top tanks and side reflectors at a ~30° angle were erected with the support of music stands. This setup was just for testing purposes; naturally, for professional use, side reflectors would be engineered into the setup tanks by more secure means. Using reflectors, the average light intensity of bottom layer was raised to 45% of that of the top tank.

We tested 12 vegetables using top and reflector-assisted bottom layer tanks: Chinese flowering cabbage (a different cultivar from the first attempt), Chinese kale, crown daisy, curly endive, early maturing Chinese cabbage, Italian lettuce, leaf celery, leaf lettuce, leaf mustard, potherb mustard, Shanghai Qing (a type of cabbage), and water spinach. Among these vegetables, the four highest yielding in 150 $m^2$ screen house were Chinese flowering cabbage (349 kg/cycle (2.3 kg/$m^2$/cycle) in the top layer, 158 kg/cycle (1.1 kg/$m^2$/cycle) in the bottom layer), Italian lettuce (463 kg/cycle (3.1 kg/$m^2$/cycle) in the top layer, 143 kg/cycle (1.0 kg/$m^2$/cycle) in the bottom layer), leaf celery (293 kg/cycle (2.0 kg/$m^2$/cycle) in the top layer, 223 kg/cycle (1.5 kg/$m^2$/cycle) in the bottom layer), and Shanghai Qing (383 kg/cycle (2.6 kg/$m^2$/cycle) in the top layer, 158 kg/cycle (1.1 kg/$m^2$/cycle) in the bottom layer) (Su et al. 2020). Using the reflector-assisted double-layer system, growing the corresponding vegetables with the suggested production cycles per year (Table 4.1), the deduced maximum annual yields in 150 $m^2$ screen house for Chinese flowering cabbage, Italian lettuce, leaf celery, and Shanghai Qing could reach 5070 (33.8 kg/$m^2$), 1820 (12.1 kg/$m^2$), 2060 (13.7 kg/$m^2$), and 6490 kg (43.3 kg/$m^2$), respectively.

Though adding bottom tanks with reflectors only translated to the increase of 76% (leaf celery) of the yields of the top tanks, the deduced maximum annual yield of the highest yielding vegetable (43.3 kg/$m^2$) from this study was 3.6 times of the deduced maximum annual yield of our previous study (12.1 kg/$m^2$) (Sect. 4.4.1). A main reason for the much higher yield was from using Shanghai Qing but the previous study did not test Shanghai Qing, and maximum annual yield was from 3 cycles of Italian lettuce and 5 cycles of leaf mustard (Su et al. 2020). Indeed, the accumulated experience and improved environment conditions (the scheduled maintenance in 2015 included pruning of the trees around the screen house by the botanical garden, which increased ventilation and sunlight intensity, but reduced the possibility of pest infestation) may have also contributed as factors; for example, the kg/$m^2$/year deduced top tank yields of Italian lettuce in this study were higher than the previous study by 43%.

### 4.4.4 Quality of Top and Bottom Layer Grown Vegetables: Visual and Taste

The vegetables grown in the reflector-assisted bottom layer were generally smaller and less matured, which may be less attractive than those grown in the top layer; therefore, the annual yield per $m^2$ increase may not translate to increased profitability. To compare the sensory quality of top and bottom layer grown vegetables, we conducted a pilot acceptance survey ($n = 30$) similar to that described by Ojwang et al. (2016) for the four highest yielding vegetables: Chinese flowering cabbage, Italian lettuce, leaf celery, and Shanghai Qing (Su et al. 2020). The sensory perceptions of shape, color, smell, taste, and texture were evaluated on a Likert-scale from a low of 1 to a high of 7, as well as a questionnaire on purchase intentions with rankings from 1 (not willing to buy) to 3 (willing to buy). Statistical differences in ratings were not found between top and bottom grown vegetables except for the following: Chinese flowering cabbage rated 0.9 and 1 points higher for shape and color, respectively, but 0.4 points lower preference for smell; Italian lettuce rated 1, 0.9, 0.5 and 0.6 points lower for shape, taste, smell and texture, respectively; and leaf celery rated 1.2 points higher rating for color, but 0.9 and 0.7 points lower for smell and taste, respectively, when bottom grown vegetables were compared to top grown ones. This suggests that Chinese flowering cabbage would be more acceptable when grown in bottom tanks than the ones grown in top tanks, and the purchase intention surveys agreed: the percentage of people indicating they would not be willing to buy Chinese flowering cabbage grown in the top versus the bottom tanks dropped from 27% to 7% when based on the combined cues of shape and color, and dropped from 20% to 10% when based on the combined cues of smell, taste, and texture. Whereas, for the other three vegetables, the percentage of people indicating they would not be willing to buy the vegetables grown in the top versus the bottom tanks dropped by 0–3% (0% for Shanghai Qing, 3% for Italian lettuce and leaf celery) when based on the combined cues of shape and color, but increased by 10–36% (10% for Italian lettuce and Shanghai Qing, 36% for leaf celery) when based on the combined cues of smell, taste, and texture. Those results of consumer acceptance implied that the bottom tanks are suitable for planting Chinese flowering cabbage. Since the cultivar of Chinese flowering cabbage used in this study does not grow well from January to February, one cycle of Italian lettuce can be grown. Furthermore, the purchase rejection to Italian lettuce in the bottom tanks did not increase as much as leaf celery and Shanghai Qing compared to that in the top tanks, even though it was not as desirable as its top grown counterparts. A production scheme in the 150 $m^2$ screen house could therefore be an annual top-layer yield of 4590 kg (30.6 kg/$m^2$) of Shanghai Qing, and an annual bottom-layer yield of 1580 kg (10.5 kg/$m^2$) of Chinese flowering cabbage, plus 143 kg (1.0 kg/$m^2$) of Italian lettuce, or an output of 6310 kg (42.1 kg/$m^2$) of vegetables in our 150 $m^2$ screen house.

From 2017 to 2018, we calculated the potential profits generated from vegetable production with the reflector-assisted double-layer system in our 150 $m^2$ screen house. In this case, we estimated the production costs that included hypothetical rent. As there are no market data on rental use of rooftops for farming, we used two hypothetical scenarios. In the first scenario, the roof space of 252 $m^2$ that holds the 150 $m^2$ screen house would be rented at a rate of ¥50/$m^2$/year (~$7.1/$m^2$/year), which is tenfold higher than the typical ¥5/$m^2$/year (~$0.7/$m^2$/year) for vegetable production in local suburbs (Liu et al. 2016). In the second scenario, the 150 $m^2$ screen house would be rented at a residential rate for apartment rent in Guangzhou, which was ~¥624/$m^2$/year (~$88.4/$m^2$/year) (Cityhouse 2018). Market prices were also based on prices of pollution-free/green vegetables. After having multiplied annual deduced maximum yield by the market price and then subtracting the cost of production, we estimated potential annual profits for Chinese flowering cabbage (5070 kg), Italian lettuce (1820 kg), leaf celery (2060 kg), and Shanghai Qing (6490 kg) to be ¥128,000 (~$18,100), ¥56,400 (~$7990), ¥69,400 (~$9830), and ¥171,000 (~$24,200), respectively, if roof space was rented at 10× farmland rate for the roof space of 252 $m^2$. In another scenario, if the 150 $m^2$ screen house were treated as a resident apartment commanding residential rent of ¥624/$m^2$/year (~$88.4/$m^2$/year) (Cityhouse 2018), the potential annual profits for Chinese flowering cabbage, Italian lettuce, leaf celery, and Shanghai Qing would be ¥61,700 (~$8740), ¥26,400 (~$3740), ¥16,300 (~$2310), and ¥90,900 (~$12,900), respectively (Su et al. 2020). Growing Shanghai Qing year-round in the reflector-assisted double-layer system could generate the highest profit, but lower consumer preference for Shanghai Qing grown in the bottom layer could cause difficulties in sales. Based on the combination of annual yield per 150 $m^2$ screen house, market prices, and potential consumer purchase preferences, we deduced that the most profitable scenario of bi-layer production would be to plant 12 cycles of Shanghai Qing in the top tanks, and

10 cycles of Chinese flowering cabbage plus 1 cycle of Italian lettuce in the bottom tanks. This scenario would still generate a high deduced maximum annual yield of 6310 kg and an annual income of ~¥167,000 (~$23,600) (¥13,900/month, ~$1970/month) if the roof rented at 10× farmland rate, or ~¥87,000/year (~$12,300) (¥7250/month, ~$1030/month), if the 150 $m^2$ screen house was rented at the city apartment rate. These income estimates would be equivalent to 162% or 84% of 2018 Guangzhou average worker's monthly compensation of ¥8600 (~$1220) per month (Zhou 2018).

## 4.5 Conclusion and Perspective

From our testing of rooftop farming from 2012 to 2018, we deduced that the double-layer production system in our 150 $m^2$ screen house can reach maximum annual yield of 6490 kg from 12 year-round cycles of Shanghai Qing. But considering consumer purchase intentions, a lower deduced yield of 6310 kg would be more realistic from a combination of 12 cycles of Shanghai Qing from the top layer, and 10 cycles of Chinese flowering cabbage plus 1 cycle of Italian lettuce from the bottom layer (Su et al. 2020). This production capacity could ensure the self-reliance of a city with regard to green vegetables. According to the China Nutrition Society's dietary guidelines, each adult consumes at least 0.3 kg of vegetables per day. The deduced yield of 6310 kg per year can meet the vegetable needs of 58 adults. If every 252 $m^2$ of roof space (where the 150 $m^2$ screen house is built) can supply 58 adults (4.3 $m^2$ per adult) with 0.3 kg/day vegetables, then the annual vegetable demands of 14 million people in Guangzhou would only require 60.2 million $m^2$, equivalent to ~86% of Guangzhou's ~70 million $m^2$ roof space (Fang 2015).

The net income from the double-layer production system in the 150 $m^2$ screen house would be up to 162% or 84% of the average income in Guangzhou in 2018 if rent is collected, respectively, at 10× farmland rate, or at Guangzhou residential rate (Su et al. 2020). It should be noted that net income was estimated based on market prices at the time of the study, which could change depending on supply and demand. For example, if the supply of Shanghai Qing, Chinese flowering cabbage, and Italian lettuce was to increase due to success of rooftop farming, that could lower demand and hence their market value. Consequently, over time, it may be necessary to readjust which types of vegetables to grow for market profitability. Moreover, consumers would naturally want more varieties than just these three vegetables.

Information on food safety and nutrition are difficult to access for the average consumer. Labels such as pollution-free/green or organic can provide consumers with some assurance on how the vegetables are grown and the expected safety standards. Many large cities are facing pollution problems and the uptake of contaminants in vegetables can be ameliorated through stricter control on the inputs, such as quality of water, amount of fertilizer, and pesticides. In that regard, the use of hydroponic production in controlled settings would be an appropriate alternative, whether that be in screen houses in peri-urban zones or on urban rooftops. Whereas the former option requires longer "transport miles," the latter option could be as simple as delivering daily harvests by routing reusable bags of vegetables through building elevators. Freshness of the vegetables can be better ensured, translating to improved nutrition and taste.

Despite the consumer's desire for safer, fresher, and more nutritious food, and the private sector's need for profitability, rooftop farming still faces many obstacles. Urban greening has become a focus in many cities. For example, Guangzhou (The Standing Committee of Guangzhou People's Congress 2012) stipulates that the green space rate of new construction projects should reach at least 30% of total area, but roof farming has not contributed to achieve this 30% rate. A major obstacle lies in the lack of government regulations for this new type of business. Revised building codes would be needed to insure proper architectural and engineering designs. And even if rooftop farms were built, a developer would need to coordinate with businesses to purchase or rent the facilities for farming purposes. Finally, the ever increasing demand in China for urban housing due to continued urbanization means that a developer could sell all of the building units it builds, without needing to market new attractive features such as rooftop farming. Conversely, the high demand and high prices of urban homes mean that consumers would care less about fresh, safe, and nutritious vegetables than the home's purchase price. Until government takes initiation to make rooftop farming a reality, it will likely remain an interesting academic exercise, and with a missed opportunity for building new neighborhoods with roof farming as part of the current urbanization trend.

**Acknowledgements** We are grateful for funding from the Chinese Ministry of Science and Technology (2016YFD0101904) and Key Projects of Chinese Academy of Sciences (KSZD-EW-Z-013; QYZDY-SSW-SMC010).

## References

Ajayan KV, Selvaraju M (2011) Reflector based chlorophyll production by *Spirulina platensis* through energy save mode. Bioresour Technol 102:7591–7594. https://doi.org/10.1016/j.biortech.2011.05.013

Alloway BJ (2013) Bioavailability of elements in soil. In: Selinus O (ed) Essentials of medical geology. Springer, Dordrecht

Benis K, Ferrão P (2018) Commercial farming within the urban built environment – taking stock of an evolving field in northern countries. Glob Food Secur 17:30–37. https://doi.org/10.1016/j.gfs.2018.03.005

Buehler D, Junge R (2016) Global trends and current status of commercial urban rooftop farming. Sustainability 8:1108. https://doi.org/10.3390/su8111108

Chen Z (2011) Vertical greening, an important way to build an ecological and low-carbon city. http://www.xbmiaomu.com/miaomuhangyexinwen15898/. Accessed 17 Feb 2018

Chen Z, Huang L, Zhou C et al (2017) Characteristics and evaluation of heavy metal pollution in vegetables in Guangzhou. Environ Sci 38:389–398

Cityhouse (2018) Trend in rental cost in Guangzhou. http://gz.cityhouse.cn/information/hamarket/lease.html. Accessed 23 Jul 2018

Fang M (2015) The cost and ecological benefit analysis of green roof in Guangzhou. Dissertation, South China University of Technology

FAO (2020a) The world bank. Arable land (hectares per person). https://data.worldbank.org/indicator/AG.LND.ARBL.HA.PC?view=chart. Accessed 20 Feb 2020

FAO (2020b) The world bank. Forest area (sq. km). https://data.worldbank.org/indicator/AG.LND.FRST.K2?view=chart. Accessed 20 Feb 2020

FAO (2020c) The world bank. Population, total. https://data.worldbank.org/indicator/SP.POP.TOTL?view=chart. Accessed 20 Feb 2020

Gao C, Wei H (2013) Prediction study on the urbanization trends of China. Modern Econ Sci 35: 85–90. https://doi.org/10.3969/j.issn.1002-2848.2013.04.010

Hammond ST, Brown JH, Burger JR et al (2015) Food spoilage, storage, and transport: implications for a sustainable future. Bioscience 8:758–768

Harris LJ, Olliver M (1942) Vitamin methods: the reliability of the method for estimating vitamin C by titration against 2,6-dichlorophenolindophenol. 1. Control tests with plant tissues. Biochem J 36:155

He B, Yun Z, Shi J et al (2013) Research progress of heavy metal pollution in China: sources, analytical methods, status, and toxicity. Chin Sci Bull 58:6–12

Jang ST, Chang SJ (2013) Exploration of a light shelf system for multi-layered vegetable cultivation. Korean J Ecol Environ Archit 13:61–66. https://doi.org/10.12813/kieae.2013.13.2.061

Jiang T (2015) Hongkong media: 1/3 vegetable samples in Guangzhou markets while less samples in Beijing and Shanghai markets contained excessive pesticide residues. http://www.cankaoxiaoxi.com/china/20150116/631496.shtml. Accessed 31 Aug 2019

Li JW, Li C, Huang CL et al (2015) Investigation of heavy metal pollution in farmland soil in Panyu district of Guangzhou. China Environ Protect Ind 8:65–69

Liu T, Yang M, Han Z et al (2016) Rooftop production of leafy vegetables can be profitable and less contaminated than farm-grown vegetables. Agron Sustain Dev 36:41. https://doi.org/10.1007/s13593-016-0378-6

Mertens DR (2002) Gravimetric determination of amylase-treated neutral detergent fiber in feeds with refluxing in beakers or crucibles: collaborative study. J AOAC Int 85:1217–1240

Nan Fang Daily (2013) Chinese ministry of agriculture: over 20% of farmlands in the Pearl River Delta exceeded the maximum residue limit for heavy metals. http://health.huanqiu.com/green/2013-07/4118366.html. Accessed 31 Aug 2019

Ojwang D, Nyankanga R, Imungi J et al (2016) Cultivar preference and sensory evaluation of vegetable pigeon pea (*Cajanus cajan*) in eastern Kenya. Food Secur 8:757–767. https://doi.org/10.1007/s12571-016-0592-8

Prasad S, Chetty AA (2008) Nitrate-N determination in leafy vegetables: study of the effects of cooking and freezing. Food Chem 106:772–780. https://doi.org/10.1016/j.foodchem.2007.06.005

Rickman JC, Barret DM, Bruhn CM (2007a) Nutritional comparison of fresh, frozen and canned fruits and vegetables. Part 1. Vitamins C and B and phenolic compounds. J Sci Food Agr 87: 930–944. https://doi.org/10.1002/jsfa.2825

Rickman JC, Bruhn CM, Barret DM (2007b) Nutritional comparison of fresh, frozen, and canned fruits and vegetables II. Vitamin A and carotenoids, vitamin E, minerals and fiber. J Sci Food Agric 87:1185–1196. https://doi.org/10.1002/jsfa.2824

Sanchez Lopez FJ, Gil Garcia MD, Sanchez Morito NP et al (2003) Determination of heavy metals in crayfish by ICP-MS with a microwave-assisted digestion treatment. Ecotoxicol Environ Saf 54:223–228. https://doi.org/10.1016/S0147-6513(02)00050-7

Sanyé-Mengual E, Anguelovski I, Oliver-Solà J et al (2015) Resolving differing stakeholder perceptions of urban rooftop farming in Mediterranean cities: promoting food production as a driver for innovative forms of urban agriculture. Agric Hum Values 33:101–120. https://doi.org/10.1007/s10460-015-9594-y

Sathi RKR, Koduru UD, Pathan AAK (2010) Utility of a reflector for energy saving in plant tissue and algal culture laboratories. Curr Sci India 99:569–570

Specht K, Siebert R, Hartmann I et al (2014) Urban agriculture of the future: an overview of sustainability aspects of food production in and on buildings. Agric Hum Values 31:33–51. https://doi.org/10.1007/s10460-013-9448-4

Statistics Bureau of Guangzhou Municipality (2018) Area of cultivated land at year-end in main years. http://210.72.4.58/portal/queryInfo/statisticsYearbook/index. Accessed 10 Sept 2019

Su Y-L, Wang Y-F, Ow DW (2020) Increasing effectiveness of urban rooftop farming through reflector-assisted double-layer hydroponic production. Urban For Urban Green 54:126766. https://doi.org/10.1016/j.ufug.2020.126766

The Standing Committee of Guangzhou People's Congress (2012) Guangzhou greening regulations. http://www.rd.gd.cn/lfzw/zxfg/jdfg/201204/t20120424_124823.html. Accessed 17 Feb 2018

Thomaier S, Specht K, Henckel D et al (2015) Farming in and on urban buildings: present practice and specific novelties of zero-acreage farming (ZFarming). Renew Agric Food Syst 30:43–54. https://doi.org/10.1017/S1742170514000143

Wexler P, Abdollahi M, Peyster AD et al (eds) (2014) Encyclopedia of toxicology, 3rd edn. Elsevier, Amsterdam

Wijayawardena MAA, Megharaj M, Naidu R (2016) Chapter three - Exposure, toxicity, health impacts, and bioavailability of heavy metal. In: Sparks DL (ed) Advances in agronomy. Academic, London, pp 175–234

Yang M, Liu T, Han ZG, Wei JJ (2014) Pesticide survey and residue test of Guangzhou vegetables. J Green Sci Technol 4:193–195

Zhang L (2015) Release of Guangdong salary survey annual report from Southern talents. http://www.cssn.cn/dybg/gqdy_sh/201509/t20150918_2372483_1.shtml. Accessed 26 Mar 2019

Zhou C (2018) Release of Guangdong salary survey report: Guangzhou average monthly salary increased by 1,393 yuan/month. http://news.ycwb.com/2018-10/11/content_30108934.htm. Accessed 26 Mar 2019

**Yun-Lin Su** received her BSc (2009) in Plant Protection from South China Agricultural University and her PhD (2015) in Agricultural Entomology and Pest Control from Zhejiang University through a joint PhD training program with Wageningen University. During her PhD years in the Netherlands, she was exposed to the concept of green roofs that inspired her interest in roof farming. After returning to China, she joined the research on rooftop agriculture in Dr. David W. Ow's lab. Dr. Su hopes to continue on a research career that can help address environmental problems related to food safety.

**David W. Ow** after his PhD from Harvard in 1983, became a visiting scholar to Shanghai, where he witnessed the coupon rationing of food. Upon his return to the USA in 1984, his experience in China shifted his focus to agriculture, and he embarked on a career in plant genetic engineering. His engineering of a glowing plant from transfer of a firefly gene has become a textbook example and his work on developing site-specific recombinase technology has been used in the development of a commercial maize cultivar. More recently, he has focused on engineering rice with reduced uptake of toxic cadmium.

# Land Use Models, Drivers, Institutional Arrangements, and Major Discourses in Promotion of Urban Agriculture in India

5

Rajesh S. Kumar, Shilpi Kundu, Bindu Gauri Kottaram, and S. Gopakumar

**Abstract**

Nearly 40% of India's population is projected to be living in cities and urban settings by 2030. The increasing urban population not only strains the urban systems but also drains the resilience of sub-urban areas supporting such urban areas with adverse impacts on vital socio-economic development indicators. Urban Agriculture (UA) offers considerable potential to mitigate food, nutrition, health employment, and income security concerns in urban areas. India's policy on urban transformation mainstreams a resilient city approach encompassing realization of zero hunger (Sustainable Development Goals-2) and sustainable cities and communities (Sustainable Development Goals-11). It is observed that UA has been gaining wider acceptability in Indian cityscapes, in the recent times. We argue that UA needs to be actively promoted in urban settings by providing an enabling dynamic policy environment particularly from the perspective of input side factors. As the land value in cities and urban areas is prohibitively high, there is a need for exploring land use models and such critical inputs that can support UA on a sustainable scale. This chapter applies discourse analysis to identify and understand various land use models, drivers, institutional arrangements, and major discourses promoting UA in selected Indian cities and

R. S. Kumar (✉)
Indian Forest Service (IFS), New Delhi, India

S. Kundu
Department of Agricultural Extension & Information Systems, Sher-e-Bangla Agricultural University, Dhaka, Bangladesh

B. G. Kottaram
Agri Business School for Women, Krishi Vigyan Kendra, Indian Council for Agricultural Research, Coimbatore, India

S. Gopakumar
Directorate of Education, Kerala Agricultural University, Thrissur, Kerala, India

 
J. A. Diehl, H. Kaur (eds.), *New Forms of Urban Agriculture: An Urban Ecology Perspective*, https://doi.org/10.1007/978-981-16-3738-4_5

urban areas. In this backdrop, we studied the UA in India with the following questions: (1) What models of UA can potentially address constraints on land availability? (2) What major supply side factors facilitate such models? (3) What institutional arrangements are driving UA?

Following standard discourse analysis methods, data were collected through secondary literature review coupled with semi-structured interviews with various actors and stakeholders in the domain of UA in the country. This research profiles a set of validated land use models and/or strategies supporting UA with potential for up-scaling and replication and explores how innovative institutional arrangement is a key to push UA on sustainable scale, and what may constitute the major pillars for wide scale enhancement of UA in cities and urban areas. We conclude the chapter by suggesting a way forward for promotion of UA in the country and in the region, where such comparable contexts exist.

**Keywords**

Urban agriculture · Land use models · Institutions · Drivers · Discourses · India

## 5.1 Urbanization and Urban Agriculture

The post-millennial times have been witnessing unprecedented urbanization across the globe and more prominently in Global South (Roberts 2020; Prasad 2019; Cook et al. 2015). It is reported that six out of ten in the global milieu will be living in urban settings and many of the ten million plus population megacities will be located in the developing geographies of the globe by the year 2030 (UN-team 2019). In India, nearly 40% of its 1.21 billion plus population is projected to be living in cities and similar urban settings by 2030 (Prasad 2019). However, the food and nutritional insecurity in Indian cities have been already much concerning due to prevalence of anaemia and undernourishment among vulnerable sections of societies in considerable proportions (Sahasranaman 2016). As per reports, one in three, under-five stunted children live in urban areas (Jena 2019). The pervasiveness of such socio-economic development gaps in urban areas can potentially undermine the efforts to realize Sustainable Development Goal (SDG) 2.2, on eliminating all forms of malnutrition in urban settings, and SDG 11, aiming to realize inclusive, safe, resilient, and sustainable cities (Tacoli 2019).

The unprecedented rate and scale of urbanization across the landscapes coupled with ever increasing rural urban migration, underline the need for increased food supplies (Cook et al. 2015). Besides, the unsustainable escalation of consumption demand has also been draining out the resilience of the rural hinterlands sustaining such cities and urban areas (Dubbeling 2017). Over the time, Urban Agriculture (UA) has been widely advocated as a complement to the basket of solutions to mitigate food security concerns and other development challenges in urban landscapes (Sahasranaman 2016). UA denotes the practice of growing crops and rearing grazing livestock in urban, sub-urban, and peri-urban areas (UNDP

1996). UA is largely promoted owing to its potential to foster local value chains, promote circular economy (Ellen Macarthur Foundation 2019), facilitate supportive social relations, enhance environmental quality in urban areas, etc. (Heather 2012). In the recent times, UA has been gaining much traction in Indian urban landscapes with cities such as Mumbai, Delhi, Kolkata, Chennai, Bangalore, Pune, Thiruvananthapuram, Kochi, Hyderabad, and Chennai taking the lead (Sahasranaman 2016).

UA is largely promoted and facilitated in the cities and urban areas by a host of actors and stakeholders, viz. governmental departments, city corporations, municipal bodies, private agencies, entrepreneurs, start-ups, communities, individuals, etc. (Anonymous 2019a, b, c; Sahasranaman 2016). UA has been increasingly complemented to India's strategy on urban transformation, which endeavours to mainstream a resilient city approach in sync with Sustainable Development Goals (Anonymous 2018a). Even though UA has been gaining wider acceptability and coverage in India; the prevailing trends, however, demand enhanced facilitation at institutional, technological, and other input side factors besides addressing the demand side concerns in order to make UA viable, incentivized, and attractive to urbanites (Sahasranaman 2016).

The promotion of UA as a functionally integrated feature of urban landscapes is, however, fraught with concerns around availability of disposable land for UA, secured land tenure, availability of institutional arrangements, and governance mechanism besides other input side factors (Martellozzo et al. 2014; Cook et al. 2015). As the land value in cities and urban areas is prohibitively high, there is a need for exploring land use models and such critical inputs that can support UA on a sustainable scale. To that end, we attempted to identify and understand various land use models, drivers, institutional arrangements, and major discourses promoting UA in selected Indian cities and urban areas. Our analysis was guided by the following questions: (1) What models of UA can potentially address constraints on land availability? (2) What major supply side factors facilitate such models? (3) What institutional arrangements are driving UA? In order to address the research questions, we considered the UA activities in the cities and urban areas mentioned above to identify the major features of land use models, governance arrangements, and discourses promoting UA in those cities.

We believe that such research will be highly relevant in shaping evidence-based policy prescriptions to promote UA in the country, besides filling the gaps in recognizing validated land use models supporting UA in Indian cities and urban areas. Besides, we also expect that this paper will add fresh insights into the literature on UA in the context of India. In the backdrop of these concerns, we discuss the following questions in this chapter: (1) What models of UA can potentially address constraints on disposable land availability for UA? (2) What major supply side factors facilitate such models? (3) What institutional arrangements are pushing UA in the Indian cities and urban areas? This research profiles a set of validated land use models and/or strategies supporting UA with potential for up-scaling and replication, and explores how innovative institutional arrangement is a key to push UA on a sustainable scale and what may constitute the major pillars for wide scale enhancement of UA in cities and urban areas.

## 5.2 Scenario Analysis Framework

At the landscape level, UA can generate desirable economic outcomes by harnessing un-utilized or under-utilized land into productive landscapes (Kaufman and Bailkey 2000). But UA is highly influenced by input side urban resources such as land, labour, irrigation water, etc. as well as by the prevailing urban conditions such as policy environment, competing land uses, market conditions etc. (Cook et al. 2015). This narrative about UA underscores the need for recognizing it as an integral and well-embedded productive system in the urban setting (De Bon et al. 2010). However, this assumes greater complexity as many UA models are currently practiced across different landscapes. These range from subsistence models to commercial market fed UA systems with significantly differing relations with basic input side factors. In order to scale up and integrate UA as part of an urban sustainability paradigm, there needs to be researched evidence available which provides in-depth, empirical features of UA models and their interaction with the urban conditions and input side factors.

Driven by the above-mentioned concerns, we carried out a scenario analysis in order to capture the prevailing trends in UA practices in India with the specific interest on land use models supporting UA, its drivers and institutional arrangements bearing on them, and the major discourses pushing such models. In order to capture the land use models available for UA, we considered the UA practices in vogue in megacities, cities, and other urban landscapes in India. Our sample frame included the cities of New Delhi, Chennai, Mumbai, Thiruvananthapuram, Pune, Bangalore, Coimbatore, Chandigarh, Jaipur, etc. These cities were selected based on the availability of literature on UA in the public domain. However, the study was focused further down to a select sample of cities based on the abundance of literature in the open domain.

The study was designed after Discourse Analysis Method (DAM). In our study, we followed the spirit of the definition of DAM as propounded by **Norman Fairclough as** "*a social practice which constructs social identities, social relations and the knowledge and meaning systems of the social world . . . [which] both reflects and produces the ideas and assumptions relating to the ways in which personal identities, social relations, and knowledge systems are constituted through social practice*" (Nielsen and Nørreklit 2009, p. 204).

### 5.2.1 Discourse Analysis

In the above background, the data exploration involved scanning both scientific and grey literature available in the domain followed by structured interviews with domain experts and UA practitioners in key selected cities in the country. The published literature available on UA in India, over the platforms such as Google Scholar, Web of Science, and other online resources was collated and then filtered using key words such as *UA in India, urban land use and UA, policies on UA in India, and land use policy*. The search output was further screened for availability of at least one explanation on UA and land use in India. The data search continued until the already selected topics and discourse elements were found repeated. The data analysis then followed a multi-stage approach comprising a pre-analysis stage, a structured analysis stage, followed by a text analysis using the Jäger framework (Wodak 2007). In the pre-analysis stage, relevant materials from the domain literature were culled out and subsequently screened for comprehensiveness. This stage was followed by a structured analysis in which we examined the documents in-depth searching for elements of various argumentations in the context of the research questions. Through this process, we segregated the themes/constructs that frequented on UA in India, as emerging discourse elements for further aggregation and analysis. The process enabled us to develop basic assumptions on the features of UA practice in general and its construct in India from a land use perspective.

### 5.2.2 Semi-structured Interviews

The study objectives of identifying drivers and institutional arrangements, governance aspects of UA in the study cities, led us to the framing of a set of open ended interview questions. The questions were subjected to iteration and evolution with every interview in order to validate new information received, as well as to reflect on the discourse themes emerging. The adoption flexible pattern for the semi-structured interviews enabled us to gather data on the UA models, drivers, and their governance arrangements whilst ensuring consistency and flexibility to discuss and cross validate new elements of discourse emerged (Schensul and LeCompte 1999). The interviews were carried out Skype and/or telephone and the data were recorded through exhaustive note taking by digital/manual means.

The interviews were held with domain experts and practitioners selected from key cities widely practising UA in the country. Their selection was based on (1) prominent role played by them in the promotion of UA in terms of provisioning of various input factors and (2) active practising of UA in key cities. The interviewees were requested at the outset to share their experience on UA in general and subsequently guided the questions what drove their interest in UA, the extent of land under cultivation, access to land and land tenure, cropping models adopted, incentives and institutional support they received, value chains associated, access to financial resource/support, technological support received from different sources, overall experience, and suggestion for expanding UA in their area as well as in other such comparable areas. Through the interviews we tried to gather the contours of major discourses on UA in the backdrop of their motivation for engagement with UA, how they overcome the barriers in securing land area for UA, access to adequate amount of quality irrigation water for crops, cultivation model, the marketing arrangements for UA products, returns from UA, etc.

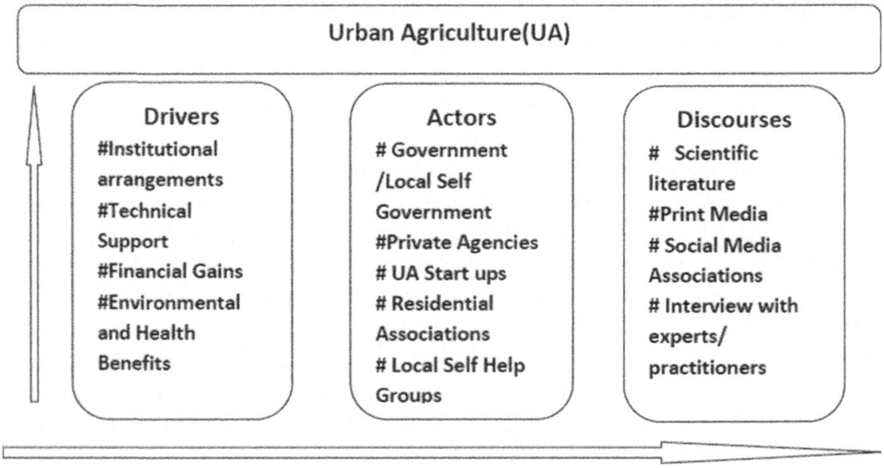


**Fig. 5.1** Analytical framework adopted to explore the construct on urban agriculture. (Source: authors)

### 5.2.3 Data Analysis

In the study, the data analysis proceeded concurrently with data collection in order to modify the search for discourse elements and to focus on the emerging areas in the subsequent searching. Coding of discourse major features and sub-features was carried out manually. Subsequently, we proceeded to identify the major elements of the discourse on UA which were further aggregated for figuring out major constructs on UA. The analytical framework we applied in the study is given in Fig. 5.1.

## 5.3 Results

We observed through this study that the engagement of urbanites with UA is highly influenced by institutional support, favourable land use arrangements, operation of local value chains, etc. In this section we describe various models of UA with a particular focus on land use. The section throws light on land use, land use policy, tenure, renting and leasing models and reflects on the major discourse of UA in India as well.

### 5.3.1 Land Use Models in UA

As observed elsewhere in the world, the relations between land and UA in cities and urban areas can be seen ranging from a classical backyard subsistence cultivation model to profit-oriented greenhouse-based rooftop production centres (Mougeot

**Table 5.1** Urban densities and per capita land availability in Indian cities

| Sl No | City | Area (in sq km) | Population | per capita land (in sq km) |
|---|---|---|---|---|
| 1 | Delhi NCR | 1484 | 11,034,555 | 0.0001 |
| 2 | Bangalore | 709 | 8,443,675 | 0.0001 |
| 3 | Visakhapatnam | 689 | 2,982,904 | 0.0002 |
| 4 | Hyderabad | 650 | 6,731,790 | 0.0001 |
| 5 | Mumbai | 603 | 12,442,373 | 0.0000 |
| 6 | Indore | 530 | 1,664,086 | 0.0003 |
| 7 | Jaipur | 485 | 3,046,163 | 0.0002 |
| 8 | Ahmadabad | 464 | 5,577,940 | 0.0001 |
| 9 | Chennai | 426 | 4,646,732 | 0.0001 |
| 10 | Bhubaneswar | 422 | 837,737 | 0.0005 |
| 11 | Pune | 331 | 3,124,458 | 0.0001 |
| 12 | Kolkata | 205 | 4,496,694 | 0.0002 |

2000). However, in order to realize a healthy urban population, it is widely suggested that nearly a third of urban land area needs to be brought under UA in order to locally provision the prescribed 300 g of vegetables per capita (Sachdeva et al. 2013). But the availability of disposable land for UA in small cities is reported to be nearly 10% of the city area only, and that available in large cities is miniscule due to high cost of land. As such the per capita urban space and its availability for UA operate as limitations for prospects of UA in several cities and urban areas. The per capital land availability across the major Indian cities reveals the very low level of availability of land resources for allocation for utilization for UA and such allied land uses (Table 5.1).

The comparison of the per capita availability of land available in the cities reflects the constraints to allocate dedicated land parcels for UA. Nevertheless, the urban areas or locations with comparatively lower economic value for land will certainly be a promising consideration for active implementation of UA policies. Despite the constraints, the studies flag that UA in India can be much facilitated by improving access to available land as well as actively promoting it in private backyards, rooftops, idle lands, balconies, communal spaces, urban parks, terrace farming, etc. (Maćkiewicz et al. 2018). Considering the potential that can be harnessed through these options, we briefly discuss various land use models operating in the context of UA in India in the following sub-sections.

### 5.3.2 Land Use Policy, Tenure, and UA

The draft land use policy of the Department of Land Resources, Government of India provides optimum utilization of land resources while addressing the concerns such as sustainability, adverse land use conflicts, requirements of provisioning high quality ecosystem services, food security concerns, protecting natural history, heritage requirements, etc., in the country (Annonymous 2013a, b). While the policy underlines the need of land for urbanization, the requirement for high quality ecosystem services, catering to the needs of the farming community for food security, is also made integral to the land use policy. This policy provision complements the need for provisioning land resource for UA besides integrating it in the overall context of urbanization in India.

It is also very relevant to observe that several state and local governments have made arrangements for provisioning land for UA in cities. The UA models promoted by the Pune Municipal Corporation, Maharashtra; Greater Chennai Corporation, Tamil Nadu; and the Delhi Development Authority, Delhi have the institutional arrangements for facilitating access to urban land for UA. And such arrangements are observed to not only promote UA but also secure the tenure of the land engagement for UA in the cities. Recognizing the need for local production of food and vegetables in the urban areas, particularly in the context of the on-going COVID-19 pandemic, the Government of Kerala announced an innovative institutional arrangement for mass promotion of urban agriculture. This policy move provides for institutional arrangements to harness locally available vacant land for UA engagement through kitchen gardens, terrace cultivation, community projects, besides accelerating mechanized interventions. Under, this initiative, the local self-governments are given the mandate to initiate community-based steps for cultivation of vacant land provided the landowners are not been able to cultivate the land themselves. Such measures have incredible potential to secure land for UA whilst ensuring tenure and ownership of land use (Babu 2020).

### 5.3.3 Land Renting and Lease Models

Land renting has been one of the arrangements to gain access to land in peri-urban areas for UA and animal husbandry by the farming communities. Current literature evidence many such land renting models prevalent in several peri-urban areas in the country. Some of such models are briefly discussed below.

#### 5.3.3.1 Peri-urban Model

In this model, the land in the outskirts of the city or beyond is leased to urban agriculturists for agriculture or animal husbandry. Such models are prevalent in many cities in the country, including the densely populated capital city of Delhi. Cultivation in the floodplains of rivers in cities is also a fairly common feature in the landscape of the cities in India. A study by Lintelo et al. (2002) mentions leasing land for agriculture has been prevalent for a long time in the peri-urban areas and river floodplains of Delhi. It is reported that nearly 700 small urban farming families sustain on floodplain cultivation on the riverbanks of river Yamuna in Delhi (Cook et al. 2015). These are mostly commercial cultivators and they cultivate on small land parcels in the riverbanks (Fig. 5.2). These land parcels are owned, rented, or occupied unauthorized by the urban agriculturists. As per the prevailing institutional arrangements these urban agriculturists organized under farmers' cooperative societies and get land from development agencies such as Delhi Development

**Fig. 5.2** Image of urban agriculture in the floodplains of Yamuna River in Delhi. The image shows a tomato crop grown in rows. (Source: author)

Authority to raise agricultural crops in the floodplains of the river Yamuna. UA in such floodplains focuses on short duration or seasonal vegetable crops such as spinach, cauliflower, radish, tomatoes, onions, different varieties of guards, etc. (Guy 2017).

However, the floodplain cultivation has been shrinking in Delhi due to the demand for land for various development activities. Nevertheless, it is also reported that farmers in the outskirts of Delhi rent land parcels of nearly 2–3 ha or more and at fixed rents in cash. Lintelo et al. (2002) observed that sub-marginal farmers with less than half a hectare of land preferred leasing land to urban agriculturists under formal or informal arrangements for affordability considerations at a micro-enterprising level. It is remarkable to note that such local level land tenure models are facilitating access to land for agriculturalists in the cities for cultivation with benefits of access, inclusion, scale, and economy besides providing occupational avenues to both resident and migrant agricultural workers.

### 5.3.3.2 Indian Railway's Mumbai Model

Indian Railway (IR) is the fourth largest rail network in the world with a cumulative track length of nearly 70,000 km dotted with more than 7000 railway stations. The IR owns considerable land at the stations and along the right of way of the rail tracks. The un-utilized land available in and around the railway establishments in cities and urban areas are potential candidate areas for practising UA. A recent initiative of the Central Railway, Mumbai, immensely demonstrates the potential of such models. According to the land leasing model advanced by the IR in Mumbai under the Grow More Food Campaign, the IR land with potential for crop production on the sides of tracks is leased to its employees at a nominal rate for raising vegetable crops. The model was widely received among its employees for providing access to land for cultivation at very nominal cost, and it improved the financial status of the employees due to additional income generated through UA. Under this model, the field employees of the IR cultivated short duration crops okra, beans, spinach, radish, etc. and managed to sell vegetables worth INR 2000 (approximately $27 USD) per month per cultivator. The model also earned is robustness due to the win-win approach configured to secure the interests of the IR in protecting its highly valuable estate from encroachment, as well as in attracting the employees for securing food security and improved family income (Anonymous 2016; Unnithan 2001). Thus, it is evident that by escalating this model to the whole network, it can facilitate UA in a phenomenal way whilst improving the local economy.

#### 5.3.3.3 City Farming Model, Pune City

The Pune Municipal Corporation launched this model in 2008 to encourage the denizens to cultivate vegetables in the city land to develop an attitude towards quality consumption and joy of UA. Under this model, access to land for cultivation of vegetables is provided by the local self-government by identifying and designating certain land parcels for vegetable cultivation by the citizenry. The local body acquired a three-acre plot in Salisbury park for UA and provided seeds, fertilizer, and irrigation water free of cost to interested persons. Besides, this model also provided marketing arrangements to sell the vegetables as well. However, the project is reported to have suffered for want of continued institutional support (Jadhav 2012).

Yet another model advanced by the local self-government worth adding here is the UA initiative of Mumbai Port Trust (MPT). MPT, under this model, developed rooftop farms on the buildings of the port trust. Although it was conceived initially as an initiative for kitchen waste management, it evolved into a UA model supporting cultivation of various fruits and vegetables like tomatoes, okra, radish, beet root, guava, pomegranate, etc. The MPT model later received accolades at national level for active promotion of UA. These two models clearly evidence that land can be identified even in high intense land use zones and can be made available for UA. However, what makes the model sustainable is basically the institutional arrangement that anchors the model and the governance leadership that model enjoys.

#### 5.3.3.4 Community Farming Model

The current literature evidence that across the country several community faming models have evolved over time in the cities and urban areas. These locations are spread over different cities in the country. Here, we reflect three such models operating in three locations, viz. Mumbai, Cuttack, and Alappuzha. In Mumbai, garbage dumps were cleaned up and converted into community farms with the assistance of NGOs for organic production of vegetables, fruits, and food crops. The initiative not only identified land area for UA, but also added synergy to the environmental management of the city, transforming lives of slum-dwellers. On a similar note, in Cuttack, Orissa, slum residents have been encouraged to engage in UA in community farms to meet their own nutritional requirements, as well as to sell the surplus to the markets (Navya 2017).

The third case we want to discuss is that of community vegetable cultivation in Kanjikuzhi, Alappuzha District in the State of Kerala. In this initiative, through a decision of the local self-government in 1995, all the households in the jurisdiction were mandated to raise vegetables in every available space including terraces and home gardens. As a result of the community engagement, a self-sustaining organic model for vegetable cultivation evolved over time in the locality. This model not only supports the local community to meet their demands for vegetables, but also provides arrangements to sell surplus vegetables to nearby cities and urban centres. The community sustains the UA momentum through efficient horticultural extension services and supply of production materials and inputs, all complemented with an efficient marketing arrangement (Priya 2019). The case of Kanjikuzhi is an evidence that communities can identify land resources for UA as well as shape UA into a community enterprise to realize enhanced food security and family income.

The UA landscape is immensely dotted with various models of community cultivation of vegetables, which includes marketing aggregation models as well (Fig. 5.3). For instance, under Pachakudukka satellite cultivation model promoted by an NGO in Kerala, students bring vegetables and fruits cultivated in their homesteads to their school for sale. The proceeds of the sale are deposited in local public banks in the name of the students. Currently, this model has evolved into a partnership of students from nearly 1000 homes and 2000 schools in Kerala (Raman 2019). These case studies evidently emphasize that governmental and non-governmental agencies can have tremendous potential to develop pragmatic land use policies/engagements to promote inclusive UA in cities and peri-urban areas.

#### 5.3.3.5 Rooftop and Terrace Model

Rooftop agriculture is the cultivation of agricultural, floricultural, medicinal, and aromatic crops, and sometimes animals on the rooftops for domestic consumption or providing to the local value chains (Fig. 5.4). Such models are known for their efficiency to combine environmental and energy conservation concerns such as water conservation, water recycling, preservation of local biodiversity, ethno-socio-botanical cultural relationships, climate friendliness, etc. (Dubbeling and Massonneau 2012). The current discourses indicate that by and large, UA as rooftop cultivation is one of the predominant models for intra-city UA which can potentially obviate the requirement of dedicated urban land area for UA (Kanchala 2016). The terrace garden model represents a case worth endorsement on a wider scale for immediate availability of rent-free, as well as for the convenience of cultivation and crop protection. It is reported that a house with an area of 1330 square feet can

**Fig. 5.3** Image of mixed crop nursery for community faming in Coimbatore City. The images have vegetable seedlings, medicinal plants, tree crops. (Source: author)

provide enough area to produce 200 kg vegetables annually. As such, this model provides an inclusive option to meet the recommendation of the Indian Medical Association to consume nearly 300–350 kg of vegetables per three to four-member sized families (Kanchala 2016). The Organic Terrance Garden (OTG) model widely practised in Bengaluru City, Karnataka exemplifies the versatility of the model for wide adoption and practising. The OTG model essentially anchors in the philosophy that *organic is the way to go and what you grow is what you eat*. It engages urbanites to cultivate fruits and vegetable, chemical free, even in small home gardens and thus promotes healthy eating practices and health promoting physical activities among urbanites (Nair 2012).

#### 5.3.3.6 UA in Home Gardens Model

Home gardens represent one of the oldest hallmarks of human domestication of crop and animal husbandry in the cultural evolution of mankind. Essentially, home gardens represent a system of subsistence farming to meet the food requirements of homesteads. These high productivity systems are also widely acclaimed for conserving local cultivars, application of indigenous knowledge, and locally compatible land resource management practices. These gardens have been the first call to address the requirements of nutritional and medicinal security of communities. Home gardens as practised in the State of Kerala represent an average size of

**Fig. 5.4** Image of rooftop cultivation in Ernakulam City, Kerala. The images show okra, cow pea, curry leaf crop grown in grow bags. (Source: author)

0.22 ha (John 2014) (Fig. 5.5), and often integrate animal husbandry and permaculture principles (Agarwal 2018). The interviews with UA practitioners revealed that such models can yield a net profit of nearly INR 5000 (approximately $68 USD) per month from an area of 1000 square feet by adopting a multi-species production model. All the above are evidence that the home garden model is a potential candidate for wide expansion of UA in cities and urban areas in scale and land use efficiency.

#### 5.3.3.7 Localized Intensive Land Use Model

This model aims to secure greater coverage and inclusion in UA while capturing local advantages to localize production. A case worth discussion in the context is the Vegetable Development Programme (VDP) of the Government of Kerala in 2012. The VDP had the twin policy objectives of promoting subsistence and commercial farming. However, the core programme strategy remained to make available unused space for vegetable cultivation/UA in homesteads, educational institutions, government and non-government institutional premises. The policy implementation strategy included a set of interventions ranging from provision of subsidies and free distribution of vegetable seeds to technical support if the farmers formed clusters of a minimum of 15 people. The strategy also reserved an allocation of 10% of the budget for UA (Anonymous 2018b).

**Fig. 5.5** Image of homestead cultivation in Thiruvananthapuram City, Kerala. The image depicts Amaranthus sp., coconut trees, eggplant, okra, cow pea, curry leaf grown in inter-cropping. (Source: Personal contact of first author with Dr Vijayadra Bhas)

### 5.3.4 Major Drivers of UA in India

In India, UA has been found widely practised in major cities such as Bangalore, Mumbai, Chennai, Kolkata, and Delhi and is fast emerging as a feature of urban landscape in many other cities. In the previous section we discussed various land use models and their features across several cities and urban areas in India. As it could be observed, the conceptualization of UA models, their implementation and governance play a critical role in the successful evaluation and development of UA in the cities. The developmental trajectory of the UA is found to be critically influenced by the interaction of various actors and stakeholders in the governmental, non-governmental agencies, social entrepreneurs, local self-government, and communities. In this section we discuss various drivers behind the prevailing construct of UA in the country and how the interplay between the actors and stakeholders influenced the development of UA by adducing cases from different cities and urban localities in the country.

#### 5.3.4.1 Promotion by Government Agencies

The prevailing discourse on UA gives strong evidence of the impact of pragmatic policies at national, sub-national, and local levels for the promotion of urban agriculture in various cities in India. At the national level, *Rashtriya KirshiVikasvYojana/Vegetable Initiatives for Urban Clusters* promotes vegetable farming initiatives. At regional level, several initiatives launched have been launched by state? Governments and local governments promote UA. A few of them were discussed as case studies in the section. In general such policy instruments are observed to address both demand side and supply side concerns along requirements for mass campaign to anchor the schemes deep in the community on sustainable scales. However, it is important to note that such comprehensive measures to promote urban vegetable cultivation not only promote UA, but also secure value addition in the supply chains besides, enhancing income streams for urban cultivators (John 2019).

### Case Example of Kerala

The State Government of Kerala has been actively promoting homestead and terrace cultivation through various schemes and programmes. Generally, under these initiatives, UA is promoted through provisioning inputs such as grow bags, seedlings, targeted support to institutional vegetable gardens, rain shelter cultivation groups, federated organizations, and farmer producer organizations. Besides, these initiatives promote micro-irrigation, interventions for productivity enhancement, marketing, and promotion of high-tech framing (Anonymous 2019a, b, c). In this context, we have considered the case of the State Horticulture Mission of Kerala for discussion. It is observed that the mission had embarked on increasing urban vegetable cultivation with the collaboration of households to expand rooftop vegetable cultivation to nearly 15,000 households in the city of Thiruvananthapuram and other municipalities in the District. The beneficiaries of the mission are identified in collaboration with the Federation of Residents' Associations Thiruvananthapuram (FRAT) at household level on the basis of willingness to partner with the programme. The mission also envisages supporting the participating households with planting inputs in collaboration with the FRAT (Levenston 2011).

Another institutional model for governance of UA in the cities is that of Harita Groups, set up under the Charitable Societies Act, 1955 for the promotion of urban vegetable cultivation through development of a federation of urban clusters in Kerala. The scheme, launched by the State Department of Agriculture, aims to confederate clusters of three to five resident welfare associations in the urban areas to constitute Harita Groups. The scheme provides a financial assistance of INR 50,000 (approximately $680 USD) towards administrative and operational expenditures. These implementation arrangements are given sustainability through provisioning technical and marketing support services under the state agricultural organization at the district level (Radhamony 2018). Alongside the State Horticulture Mission, the State Department of Environment and Climate Change is also engaged in the promotion of UA with the overarching idea of availing terrace gardens for biodegradable waste control. This strategy essentially involves providing 25 grow bags, vegetable seedlings, and an automatic timer equipped micro-irrigation systems at subsidized cost to the interested urban agriculturists for combining composting and UA on convergence mode (Suchitra 2015).

In yet another local body initiative on UA governance, the Corporation in the Thiruvananthapuram City facilitate UA in the city with very pragmatic institutional and financial arrangements. In order to give sustainability to the initiative, the local body has allocated nearly 30% of the financial resources for the year 2019–2020 for project related to agricultural working groups.

**Case Example of Hyderabad**

Although UA is a new feature in the city landscape, it is reported to be fast catching up wide popularity among city dwellers. More than 4000 homesteads in the peri-urban areas of the city are reported to have become self-reliant for meeting their family needs for vegetables. The city's engagement with vegetable cultivation is based on an incentive model that provides subsidy directly to the denizens interested in growing vegetable crops. Under this model, the Horticulture Department offers the denizens a subsidy of INR 360 (approximately $4.90 USD) a contribution when the beneficiary invests INR 1200 (approximately $16.30 USD) to grow a cluster of vegetables in the homestead. Essentially, the subsidy kit provided under the scheme involves four silatin round beds, red earth, farmyard manure, 14 bags, and other supporting materials to raise vegetable crops in the home yard. The model is reported to have picked up well in multiple pockets in the city (Awasthi 2013).

**Case Example of Chennai and Coimbatore**

The Government of Tamil Nadu floated the *Do IT YOURSELF* kit programme in 2014 (TNAU Agritech Portal—Horticulture 2014) for inclusive promotion of roof-top vegetable gardens in the cities of Chennai and Coimbatore. The cultivation kit provided under the scheme, comprised of grow bags, coco peat, polythene spreading sheets, water soluble fertilizers, bio-fertilizers, bio-pesticide, bio-fungicides, a hand sprayer, rose cane, a digging fork, a trowel, seedling trays, a vegetable seed kit, and manual booklet. The programme implementation converges the engagement of Greater Chennai Corporation, Agriculture and Horticulture Departments of the State Government, TN Corporation for Development of Women, and Corporate bodies. Such organizational convergence is reported to promote UA in the schools in the city. This model houses a unique provisioning arrangement under which the technical inputs for programme implementation are given by the state agencies' materials for cultivation, training is provided by the women development corporation, and financial support is provided by corporate bodies (Ramachandran 2019). Major initiatives are briefly described in Table 5.2.

#### 5.3.4.2 Social Networks for UA

Social networks supported by social media platforms are observed to play a facilitative role in the promotion of UA and organic cultivation as a social movement in many Indian cities and urban areas. For instance, social networking groups such as *Kitchen Garden* and *Krishibhoomi*, *Bangalore based Organic Terrace Farming* (*OTG*) have been found significantly attracting membership and are actively involved in sharing knowledge and experience on UA among the members (Anonymous 2020a). Several of such online associations are also found to coalesce under the banner of *Kitchen Garden* Forum. The forum aims to spread awareness and facilitate support available for UA from the government such as subsidies and technological services. A similar case is reported from the Mararikulam village in

**Table 5.2** A profile of major initiatives on UA across the cities

| State | Initiative | Focus area/s | Target group/s | Institutional arrangement | Resource support provided |
|---|---|---|---|---|---|
| Kerala | Niravu organic village Model (Yomichan 2020) | Restoring the tradition of agriculture by focusing on growing native crops (pulses, pepper, desi paddy, coconut, banana, mango, eggplant, okra), creation of native seed bank | | Registered under the Companies Act | |
| Kerala | Sustainable vegetable production initiative of City Corporation, Thiruvananthapuram (Anonymous 2020b) | Grow healthy fresh vegetables locally | Residents | Local self-government | Services and subsidized kits to grow up to 30 different varieties of vegetables |
| Kerala | Initiative of promoting rooftop cultivation of vegetables in urban and peri-urban areas (Devi 2017) | Focus on urban clusters: use grow bags to cultivate cauliflower, tomato, bottle gourd, bitter gourd, etc. on terraces and on sticks and poles | 25 member Farmer Interest Groups (FIG) in peri-urban areas registered under the Companies Act and under Vegetable and Fruit Promotion Council of Kerala | Kerala state department of horticulture | State Department provides training of terrace cultivation |
| Kerala | Organic terrace farming movement (Agarwal and Sinha 2017) | Vegetables | Terrace farm owners | Networks of farmers that are locally registered | Training workshops are organized for cultivators for capacity building, government subsidies, technology transfers, and schemes |
| Kerala | My Home, My Vegetable (Levenston 2020) | Aims to bring 12,500 ha of land under cultivation towards becoming self-sufficient in food production | Subscribers of New Indian Express Newspaper | Complementing with 'Subhiksha Keralam' project for ensuring food security | Seed packets are distributed along with the newspaper |

(continued)

**Table 5.2** (continued)

| State | Initiative | Focus area/s | Target group/s | Institutional arrangement | Resource support provided |
|---|---|---|---|---|---|
| Telangana | Hyderabad Urban Farming Initiative (Awasthi 2013) | Vegetable cluster cultivation in the homesteads | City dwellers | Horticulture Department, Govt. | Subsidy of INR 360 at the investment of INR 1200 by the cultivator. Subsidy kit includes cultivation inputs |
| Karnataka | Garden City Farmers Trust, Bangalore (Agarwal and Sinha 2017) | | Motivated citizens | Garden City Farmers Trust, Bangalore | Cultivation inputs provided to needy farmers through crowd funding |
| Maharashtra, Pune | City farming project (Ali and Srivastava 2017) | Vegetables and crops | Corporation area dwellers | Pune Corporation | Land allotment provisions for cultivators, Public Private Partnerships |
| Maharashtra, Mumbai | Urban Farming Forum Initiative (Jhangiani 2016) | Terrace farming of vegetables and green crops | Communities, housing societies, and corporations | The Energy and Resources Institute (TERI)/State Urban Development Department/Navi Mumbai Municipal Corporation (NMMC) | Facilitate volunteering in urban community farms across the city |
| Maharashtra, Mumbai | Mumbai Post Trust Initiative (Hildebrandt 2016) | Organic cultivation of vegetables | Kitchen staff members | Mumbai Post Trust, Mumbai | Cultivation space and organic mineral inputs |
| Chennai, Tamil Nadu | Valam collective Initiative (Levenston 2017) | Organic farming | Rural farmers | Urban-rural joint ventures | Joint ventures invest and work in the land owned by farmers |
| Chennai, Tamil Nadu | Resilient Chennai Initiative (Ramachandran 2020) | Vegetable gardens in schools | Selected schools, owners of houses and apartments | Public private initiative of M/s APVN and Greater Chennai Corporation, State Agriculture Department | Land for cultivation, technical expertise and inputs, caretakers for gardens, private financial support |

Kerala, where local vegetable cultivation was patronized by a Facebook community located in another urban centre (Priya 2019).

#### 5.3.4.3 Enabling Marketing Arrangements for UA Products

Availability of efficiently operating local value chains is crucial for the efficient performance of various UA models. These arrangements range from sale of excess UA products at the homesteads as farm gate sale to organized marketing arrangements. Many of the practitioners such as rooftop cultivators, back yard cultivations, hydroponics units, agricultural scientists, agricultural extension experts, policy makers related to agricultural policy formulation, and urban planners who participated in the interview highlighted the roles played by local contacts and social media platforms in feeding their produce to the value chains. However, the operation of well-organized institutional arrangements for marketing or local/regional value chains is found to fetch a premium price for UA products. For instance, participating students under the grow vegetable campaign earned INR 275,000 (approximately $3663 USD) over a period of 9 months for their products as the NGO partner facilitated the operation of supply chain (Padre 2019).

We now turn to present the governance of community-based marketing arrangements for UA and how it gives positive reinforcement to the UA models operating the locality. A case worth discussion is the local committees set up by the local self-government in the peri-urban areas of Kanjikuzhi, Alappuzha District, and Kerala. Under this governance arrangement, surplus products from each household are channelled to the local markets and value chains by a Panchayat Development Society set up by the Local Self-Government, and it sells the products in the outlet network along the national highway passing through the district (Priya 2019).

Similarly, presence of State backed marketing arrangements for a UA product is also found to facilitate the wider adoption of various UA models across the cities and urban areas. A case worth discussion in this context is of the Green Friends Brigade (Harita Mitras) organized under the Green Kerala Haritha Keralam Scheme of the Government of Kerala. Under this arrangement, vegetable products raised by the urban clusters are marketed through a network of local volunteers called Harita Mitras who are engaged under the scheme. It is also remarkable to observe that the schemes are given sustainability through a yearly budget from the State Government. It is evident from the UA landscape that the marketing arrangements involved are varied and range from efforts at UA practitioner level to organized supply chains (Radhamony 2018).

#### 5.3.4.4 Start-Ups in UA

Start-ups have been increasingly making their presence on multiple frontiers in India. The UA business ecosystem has also benefited by the presence of several start-ups spread over different cities of India. Due to the new models of service delivery, UA has been receiving considerable currency among urban denizens. The interested entrants to UA domain are often reported to be facilitated by a growing

**Table 5.3** A list of start-ups and organizations engaged in the promotion of UA[a]

| City | Agency | Services provided |
|---|---|---|
| Thiruvananthapuram | City Corporation (Sreekumar 2019) | Services and subsidized kits to grow up to 30 vegetable varieties |
| Thiruvananthapuram | Karshika Karma Sena joint venture of Kudappanakunnu Gram Panchayat and Krishi Bhavan | Technical advisory services |
| Kerala Urban | Kudumbasree Mission (Women's self-help groups) (Venugopalan 2014) | Technical services in a limited way |
| Jaipur | The Living Greens (Agarwal 2019) | Set up and maintain organic urban rooftop farms |
| Mumbai | iKheti (Dey 2017) | Set up and maintain organic urban rooftop farms |
| Pune | iKheti (Shah 2012) | Workshops, consultancy, and gardening resources; set up and maintain organic urban rooftop farms |
| Delhi | Khetify, Edible Routes (Parekh and Khare 2017) | UA services in general |
| Hyderabad | Homecrop, UrbanKissan | Modular vegetable gardens on rooftops, modular vegetable gardens on their rooftops |
| Bangalore | Greentechlife and Squarefoot Farmers (Gundmi and Makam 2019) | |
| Chandigarh | Pindfresh (Anonymous 2019c) | Hydroponic do-it-yourself (DIY) kits, replacement nutrients, organic seeds, and inert mediums |
| Coimbatore | Indian Superheroes | Rent patches of organic farms to grow and harvest crops with guidance from knowledgeable organic farmers |
| Chennai | Urban Farmers | Customized kits for setting up rooftop farms and home gardens |
| Chennai | Urban Chennai Farm | Promotes local production of pesticide free vegetables for communities |
| Pune | Abhinav Farmers | Input and technical services |

[a]Note: The list is indicative only and not exhaustive

network of start-ups offering multiple services ranging from consultancy to on-the-farm management services. A review of the current literature reveals the presence of a multitude of start-ups and companies offering different categories of services. A short summary of the various current start-ups and services provided is listed in Table 5.3.

## 5.4 Discussion

The current analysis of the major discourses on UA in India, presents a profile of the various land use models, major drivers, and the governance arrangements available for promotion of UA in the country. The key findings of this research are: (1) UA in India can be scaled up in cities and urban area with suitable land use models supporting UA; (2) an enabling policy ecosystem is available for harnessing the potential of UA for provisioning high quality ecosystem services to urban areas; (3) various land renting/leasing models are uniquely supporting UA; (4) emerging techniques such as rooftop and terrace UA models are potential candidates for obviating the requirements of free land for UA in the cities; (5) institutional arrangements are crucial to accelerate the expansion of UA; (6) social networks and UA start-ups are found to add synergy by promoting intensive use of available land for UA; and (7) enabling marketing arrangements provide great buoyancy to UA.

Insofar as the access and availability concerns on land for UA is considered, it may be recognized that different UA models presently operational in the cities and urban areas evidence a huge level of untapped potential to support UA. An efficient harnessing of such potential already available in the landscape can immensely complement the efforts to realize sustainable resilient urban landscapes as envisaged by the prevailing urban development paradigm. Although accessing land for UA by agriculturists is far from simple and less complicated, it may be cognized that UA has been able to address the aspirations of the urbanites to exercise control on production of food for their consumption, as well as their interest in various co-benefits generated by UA in an increasing manner. This argument is much evidenced by the rate, scale, and focus given on UA in the cities like Bangalore, Thiruvananthapuram, Cochin, and Pune.

Even though wide scale expansion of UA in cities is fraught with multiple issues on the supply and demand sides, various models currently in operation in different cities and urban areas of the country provide encouraging evidence for scaling up UA in cities with a strong localization approaches harnessing locally available resources and opportunities for value chain development. Besides, UA can be further facilitated by tweaking the land tenure laws and rules to provide for aggregation of available vacant land whilst securing land title and tenure in the cities and urban areas. Such developments in the policy ecosystem may encourage active investments in UA and the associated value chains by urban farming communities, farmer groups, and start-ups to realize economies of scale. The evidence and experience generated by several cities in Kerala constitute a great deal of optimism in this direction.

Arguably much facilitative arrangement is expected by the urban agriculturists in the interface between availability and accessibility of land for UA. The already evolved models of land use for UA present a great deal of cautious optimism. Such models may be considered for scaling up in the urban landscape with greater inclusion complemented by efficient precision agriculture technology, extension, innovation, institutional support, and value chain development access to credit and crop risk management on account of the risks posed by such as cyclones, heatwaves, and ruptures in supply chains (Cabannes 2012).

The emerging UA trends as we described in this paper give hope that UA can be considered for integration with the city systems in terms of scale, localization, and inclusion in order to add resilience and sustainability to urban systems. The diversity of land utilization models, drivers, and the discourses on UA supports the aspiration to address various concerns on food security, income enhancement, promotion of innovation and entrepreneurship, urban employment, circular economies, and sustainable development in the urban landscape. However, multiple gaps need to be bridged at the levels of institutional arrangements, technological and financial access, and efficient value chain development to elevate the construct of UA from a patchwork of different land use models driven by distinct engagement approaches and objectives. In this background, we suggest the following framework for wide scale enhancement of UA in the cities and urban areas (Fig. 5.6).

However, in the prevailing scenario such an arrangement is found missing in several urban areas resulting in the practice of UA in a sporadic manner. Besides, institutional arrangements regarding access to public and private land, land tenure, land title security, land pooling, and financing, marketing, promoting, value addition, and branding are found fraught with multiple gaps. In the light of these discussions, we propose four pillars for promotion of UA in the urban landscapes, summarized in Fig. 5.6. The analysis of various land use models linked to UA evidence that UA is an organized and institutionally supported feature of urban landscape contrary to the general perception of UA happening in a random manner. Although UA in Indian cities and urban areas is a relatively recent phenomenon, the

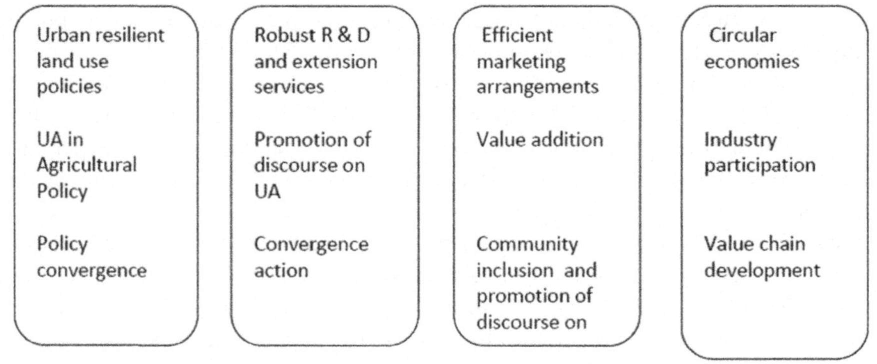


**Fig. 5.6** Major pillars for wide scale enhancement of UA in the cities and urban areas

presence of several actors and stakeholders in the domain indicates the fast widening acceptability of UA in the cities and urban areas despite several constraints.

The discourse analysis also reveals that sustainability of UA in the urban landscape is not a straitjacket proposition, there are several barriers to expansion and intensification of UA in the cities as gathered through this study. Firstly, the governance quality of the institutional arrangements for promotion is a prerequisite that needs to be ensured. It was observed in the study that vertical and horizontal integration of the line departments in the provisioning of resources, extension of technology, and resilient value chains have been contributing to the expansion of UA experience in the cities and urban areas of Kerala. Similar evidence was registered in the community UA models considered in the study. However, it is to be mentioned that the Pune UA Model and Hyderabad UA Models suffered in their sustainability due to gaps in the governance arrangements. Secondly, the concerns around access to technological options available to undertake UA in scale and profit are areas that need to be addressed. These concerns are majorly around access to efficient package of practices of crop cultivation and crop husbandry, vertical greening technologies, precision agriculture, integrated pest management, research, and extension. Thirdly, the long-term sustainability of UA models operating at economies of scale demands availability of efficient value chain arrangements to channel the production to urban consumption centres. Besides, access to and availability of efficient financial sources are found constraining the scale of adoption of UA in the cities. The prevalence of such a scenario nonetheless implies that the dilemmas, trade-offs across diverse approaches to UA, and land use will remain until such gaps are addressed under a dynamic comprehensive urban land use policy. Nevertheless, the current discourse on UA underscores the potential of localization and innovation in UA to realize the optimum results of UA in cities and urban areas.

The above discussion indicates that presence of pragmatic and tenable access to land, secure land tenure, and efficient value chain is crucial to realize UA in cities and urban areas in a large way. Similarly, availability of robust institutional arrangements for governance of UA is also essential not only to facilitate UA in the urban landscape but also to ensure its sustainability. The current analysis evidence that the interplay of several actors (both government and private) and stakeholders is significantly facilitating expansion of UA in several cities and urban areas in India.

## 5.5 Conclusion and Suggestions

Urbanization has been progressing at unprecedented rates across the globe and peri-urban areas supporting cities are being over-stretched for resource provisioning to sustain the cities. The emerging paradigm on urban resilience anchors on sustainable urban landscapes to meet the complex development issues that are being faced by the cities and urban areas. These development challenges range from basic urban food security and nutrition to mitigating climate change. UA has been widely advocated a part of the basket of solutions to realize sustainable urban landscapes whilst addressing several socio-economic development concerns in cites. However, access to costly urban land for UA is often highlighted as bottleneck in wide scale practising of UA in many cities and urban areas.

The current research by following a discourse analysis has explored various initiatives and developments on promotion of UA in India focusing on land use models, drivers, and institutional arrangements across different cities and urban areas in India. The findings of the study indicate that several UA land use models prevalent across the study cities. These land use models are driven by pragmatic land use policies, institutional arrangements governing UA, efficient value chains, readily available faming technologies and equipment, technology extension, and networking. However, the analysis surfaced that the drivers on UA are operating at different levels in different cities and urban areas, with consequences on the extent of UA and its sustainability. The prevailing scenario is also fraught with issues around gaps in governance and institutional arrangements in different cities, wide scale technology extension campaigns, and maintenance of efficient value chains.

We conclude that UA can potentially complement the efforts of the city governments in realizing resilient sustainable urban landscapes. However, to realize its potential in a greater way, we suggest the following (1) Enhanced policy support by the local self-government to facilitate aggregation of un-utilized or under-utilized land in the urban areas for UA on tenure basis; (2) Policy reforms to promote various models of UA in institutional areas, residential colonies, city parts, and gardens with the active collaboration of stakeholders; (3) Identification and promotion of incentives for promotion of UA in cities and urban areas, as well as for development value chains based on products of UA in such landscapes; (4) Enhanced promotion of discourses on UA in the urban societies through multimedia highlighting the complementary role of UA in realizing sustainable resilient urban landscapes; (5) Positioning efficient UA technology extension services to develop UA models suitable to agroecology and space constraints as well as development of local markets for product aggregation, storage, and sales. This can be further strengthened with new and emerging technologies such as AI and Internet of Things (IoTs); (6) Promotion of start-ups for aggregation, marketing and distribution, and development of value added value chains, and provision of efficient capital for emerging entrepreneurs in UA; and (7) Promotion of research, education, and capsule and certification courses on UA in the agricultural Universities to advance UA.

The findings from this multi-city research on UA in India evidence that UA is increasingly practiced in the fast-expanding city and urban landscapes of India. The urban development policy of the country emphasizes on landscape resilience in the pursuit of urbanization including the provisioning of various ecosystem services required for healthy and progressive urban living. Although securing land for UA is fraught with competing demand for land for urbanization and urban systems, the current scenario indicates the availability of a basket of options to access land resources and land area for UA in the cities. It is also important to highlight that much of such options are institutionally ensured by the government agencies in collaboration with local self-governments and private players. Besides, the business ecosystems of UA in cities and urban areas are further brightened by the presence of a large number of start-ups, NGOs, civil society organizers, and online communities that facilitate UA land use on efficient lines through provisioning various production factors.

**Acknowledgements** We take this opportunity to thank all those policy makers, practitioners, scientists, Civil Society Organizations, who have helped us in the study by sharing their experience and views on UA.

**Declaration** The views/analysis/observations presented in the current paper are personal and are based on the domain experience of the authors and may not be linked/reflected to the official positions announced by the organizations under which the authors are presently employed with or employed in the past.

## References

Agarwal S (2018) An urban garden which produces 17kg of fruits and vegetables per day. DownToEarth, Hyderabad. https://www.downtoearth.org.in/blog/food/an-urban-garden-which-produces-17kg-of-fruits-and-vegetables-per-day-58892

Agarwal S (2019) Portable farming system. The Living Greens, Jaipur. https://thelivinggreens.com/

Agarwal HP, Sinha R (2017) Urban farming - a sustainable model for Indian cities. Int J Emerg Technol 8(1):236–242

Ali F, Srivastava C (2017) Futuristic urbanism-an overview of vertical farming and urban agriculture for future cities in India. Int J Adv Res Sci Eng Technol 4(4)

Anonymous (2013a) National Land Utilisation Policy framework for land use planning & management. Ministry of Rural Development, Department of Land Resources, Government of India, New Delhi. https://smartnet.niua.org/sites/default/files/resources/draft_national_land_utilisation_policy_july_2013.pdf

Anonymous (2013b) National Land Utilisation Policy framework for land use planning & management. Department of Land Resources, Ministry of Rural Development; Government of India, New Delhi

Anonymous (2016) Facts and figures 2016--2017. Indian Railway, Ministry of Railway, Government of India. http://www.indianrailways.gov.in/railwayboard/uploads/directorate/stat_econ/IRSP_2016-17/Facts_Figure/Fact_Figures%20English%202016-17.pdf

Anonymous (2018a) Strategies for New India @75. NITI Aayog, Government of India, New Delhi. https://niti.gov.in/sites/default/files/2019-01/Strategy_for_New_India_0.pdf

Anonymous (2018b) Thirteenth five year plan 2017– -22, Second years programme 2018--19, Green book. Finance Department, Kerala

Anonymous (2019a) Agriculture Policy: Vision 2020. Indian Agricultural Research Institute, New Delhi. http://www.planningcommission.nic.in/reports/genrep/bkpap2020/24_bg2020.pd

Anonymous (2019b) Annual plan 2019--20, sectoral programme - details of schemes, agriculture and allied sectors. Kerala Agriculture, Kerala. http://keralaagriculture.gov.in/wp-content/uploads/2019/06/English-Write-up.pdf

Anonymous (2019c) The commercial of a commercial system. Pindfresh, Chandigarh. https://www.pindfresh.com/

Anonymous (2020a) Krishibhoomi. Facebook. https://www.facebook.com/groups/krishibhoomi/

Anonymous (2020b) Corp gives push to organic farming, provides free kits to 20,000 houses. Indian Express

Awasthi P (2013) Urban agriculture in India and its challenges. Int J Environ Sci Dev Monit 4(2): 2231–1289

Babu G (2020) Covid-19: Kerala to boost agriculture amid fears of prolonged impact. Business Standard, Chennai. https://www.business-standard.com/article/economy-policy/covid-19-kerala-to-boost-agriculture-amid-fears-of-prolonged-impact-120042201474_1.html

Cabannes Y (2012) Financing urban agriculture. Environ Urban 24(2):665–683. https://doi.org/10.1177/0956247812456126

Cook J, Oviatt K, Kaur H, Main D, Bret J (2015) Re-conceptualizing urban agriculture: an exploration of farming along the banks of the Yamuna River in Delhi, India. J Agric Hum Values 32(2):265–279

De Bon H, Parrot L, Moustier P (2010) Sustainable urban agriculture in developing countries. A review. Agron Sustain DevSustainable Development 30(1):21–32

Devi LUR (2017) Shanlax. Int International Journal 6(1):45–49

Dey S (2017) This Mumbai ecopreneur has been turning city dwellers into urban farmers for over half a decade. The better India, Mumbai. https://www.thebetterindia.com/99172/ikheti-priyanka-amar-shah-urban-farming/

Dubbeling M (2017) Growing urbanization goes hand in hand with growing consumer demand for food. Meeting of the Minds. https://meetingoftheminds.org/sustainable-cities-require-urban-agriculture-20941#:~:text=Urban%20and%20peri%2Durban%20agriculture%20will%20not%20be%20able%20to,roles%20in%20enhancing%20this%20potential.. Accessed 2 May 2020

Dubbeling M, Massonneau E (2012) Rooftop agriculture - a climate change perspective. RUAF Foundation, Kastanjelaan, Leusden, the Netherlands

Ellen Macarthur Foundation (2019) Cities and circular economy for food. www.ellenmacarthurfoundation.org: https://www.ellenmacarthurfoundation.org/assets/downloads/Cities-and-the-circular-economy-for-food-1.pdf. Accessed 1 May 2020

Gundmi A, Makam V (2019) Squarefoot farmers. Yellow Place, Bangalore. https://yellow.place/en/squarefoot-farmers-bangalore-india

Guy TA (2017) Urban farming in India—is it serious. Delhi, India

Heather KL (2012) The environmental benefits of urban agriculture on unused, impermeable and semi-permeable spaces in major cities with a focus on Philadelphia, PA. https://repository.upenn.edu/cgi/viewcontent.cgi?article=1044&context=mes_capstones. Accessed 1 May 2020

Hildebrandt S (2016) Urban gardening in Pune and Mumbai, India. Growing in cities—interdisciplinary perspectives on urban gardening. Institue Social Planning, Organizational Change and Urban Development, Basel, p 22

Jadhav R (2012) Urban agriculture idea brings farms to high-rises. The Times of India, Pune. https://timesofindia.indiatimes.com/city/pune/Urban-agriculture-idea-brings-farms-to-high-rises/articleshow/13145685.cms?utm_source=contentofinterest&utm_medium=text&utm_campaign=cppst

Jena M (2019) The urbanization of malnutrition. Inter Press Service. http://www.ipsnews.net/2017/09/the-urbanization-of-malnutrition/

Jhangiani D (2016) A growing movement: bringing farming to the city. The Hindu

John J (2014) Homestead farming in Kerala: a multi-faceted land-use system. J Found Agrarian Stud 4(1)

John O (2019) Green revolution revisited—vision for a better future. Climate South Asia Network, Kerala

Kanchala S (2016) Facilitate urban farming in Chennai to supply for old age homes. Milaap org, Chennai. https://milaap.org/fundraisers/urbanchennai

Kaufman J, Bailkey M (2000) Farming inside cities: entrepreneurial urban agriculture in the United States. Lincoln Institute of Land Policy, Cambridge, MA

Levenston M (2011) Rooftop farming set to cover 15,000 city homes in Thiruvananthapuram, India. City Farmer News, Kerala. https://cityfarmer.info/rooftop-farming-set-to-cover-15000-city-homes-in-thiruvananthapuram-india/

Levenston M (2017) India: Chennai mini urban-rural joint venture. City Farmer News-New Stories From 'Urban Agriculture Notes'. https://cityfarmer.info/india-chennai-mini-urban-rural-joint-venture/. Accessed 4 June 2020

Levenston M (2020) India: TNIE launches 'My Home, My Vegetable' initiative. City Farmer News-New Stories From 'Urban Agriculture Notes'. https://cityfarmer.info/india-tnie-launches-my-home-my-vegetable-initiative/. Accessed 7 Jun 2020
Lintelo DT, Marshal F, Bhupal D (2002) Urban food: the role of urban and peri urban agriculture in India: a case study from Delhi. Food Nutrition and Agriculture, Delhi, pp 1–11
Maćkiewicz B, Asuero PR, Pawlak K (2018) Reclaming urban space: a study of community gardens in Poznan. Quaestiones Geographicae 37(4):131–150. https://doi.org/10.2478/quageo-2018-0042
Martellozzo F, Landry J-S, Plouffe D, Seufert V, Rowhani P, Ramankutty N (2014) Urban agriculture: a global analysis of the space constraint to meet urban vegetable demand. Environ Res LettEnviron. Res. Lett. 9:064025
Mougeot LJA (2000) Thematic Paper 1. In: Urban agriculture: definition, presence and potentials and risks
Nair A (2012) Organic terrace gardening. Citizen Matters, Bengaluru. https://bengaluru.citizenmatters.in/3969
Navya KP (2017) What stops Indian cities from growing their own food? Citizen Matters. http://citizenmatters.in/indian-cities-farming-and-farmers-crisis-4532
Nielsen AE, Nørreklit H (2009) A discourse analysis of the disciplinary power of management coaching. Soc Bus Rev 4(3):202–214. https://doi.org/10.1108/17465680910994209
Padre S (2019) Kids get to grow and sell veggies in Kerala schools. Civil Society, Kerala. https://www.civilsocietyonline.com/education/kids-get-to-grow-and-sell-veggies-in-kerala-schools/
Parekh S, Khare K (2017) Khetify solutions. Khetify, New Delhi. https://khetify.journoportfolio.com/
Prasad S (2019) Why the world should be watching India's fast-growing cities. World Economic Forum. https://www.weforum.org/agenda/2019/01/india-urbanization-why-the-world-should-watch/. Accessed 2 May 2020
Priya LS (2019) What India can learn from Kerala's only vegetable-sufficient panchayat, Kanjikuzhi. The better India, Kerala. https://www.thebetterindia.com/175780/kerala-organic-vegetable-farming-village-kanjikuzhi-india/
Radhamony PS (2018) Proceedings of the Director of Agriculture, Directorate of Agricultural Development and Farmers' Welfare. VikashBhavan, Thiruvananthapuram
Ramachandran KM (2019) Chennai urban horticulture vegetable gardening. Avpn Asia, Chennai. https://avpn.asia/apfx/chennai-urban-horticulture-vegetable-gardening/
Ramachandran K (2020) Chennai urban horticulture vegetable gardening. AVPN. https://avpn.asia/apfx/chennai-urban-horticulture-vegetable-gardening/. Accessed 30 May 2020
Raman GK (2019) Green shoots of hope with Pachakudukka project. The Hindu, Kerala. https://www.thehindu.com/news/national/kerala/green-shoots-of-hope-with-pachakudukka-project/article28726227.ece
Roberts LC (2020) Global south communities and urbanization. Cornell Research. https://research.cornell.edu/news-features/global-south-communities-and-urbanization. Accessed 1 May 2020
Sachdeva S, Sachdev TR, Sachd R (2013) Increasing fruit and vegetable consumption: challenges and opportunities. Indian J Commun Med Indian Journal of Community Medicine 38(4):192–197. https://doi.org/10.4103/0970-0218.120146
Sahasranaman M (2016) Future of urban agriculture in India. Institute for Resource Analysis and Policy, Hyderabad. http://www.irapindia.org/
Schensul J, LeCompte M (1999) Designing and conducting ethnographic research. AltaMira Press, Walnut Creek, CA
Shah PA (2012) Grow your own edible garden, at home. Money Control, Pune. https://www.moneycontrol.com/news/business/personal-finance/ikheti-grow-your-own-edible-garden-at-home-962563.html
Sreekumar SK (2019) Corporation of Thiruvananthapuram. http://www.corporationoftrivandrum.in/
Suchitra M (2015) Rooftop revolution. More and more people in food-deficit Kerala are taking up terrace farming to grow organic vegetables and fruits. India. https://www.downtoearth.org.in/coverage/rooftop-revolution-50161

Tacoli C (2019) Food and agriculture-sustainable markets-urban. International Institute for Environment and Development. https://www.iied.org/urban-food-insecurity-malnutrition-are-about-more-just-food
UNDP (1996) Urban agriculture, Food, jobs and sustainable cities Publication Series for Habitat II UNDP. United Nations Development Programme, New York
Unnithan S (2001) Central Railways leases its land to employees for growing vegetables in Mumbai. Mumbai. https://www.indiatoday.in/magazine/offtrack/story/20010806-central-railways-leases-its-land-to-employees-for-growing-vegetables-in-mumbai-773951-2001-08-06
UN-team (2019) Sustainable development goals SDG 11: sustainable cities and communities. United Nations in India, India. https://in.one.un.org/page/sustainable-development-goals/sdg-11/
Venugopalan K (2014) Influence of Kudumbasree on women empowerment – a study. J Bus Manage (IOSR-JBM) 16(10):35–44. www.iosrjournals.org
Wodak R (2007) Critical discourse analysis. In: Clive Seale DS (ed) Qualitative research practice. Sage Publications Ltd., New Delhi, p 185
Yomichan S (2020) Niravu Organic Village - a study on Niravu Organic Village model, Vengeri, Kozhikode District, Kerala. Acta Sci Agric 19

**Rajesh S. Kumar** is a career civil servant and a member of the Indian Forest Service (IFS). He is currently serving as a Director in the National Institution for Transforming India (NITI Aayog), the premier think tank of Government of India. He brings expertise in urban forest governance, climate change mitigation, resilient urban landscapes, and community-based approaches to nutrition security. He has a Master's degree in Climate Change and International Development from the University of East Anglia (UEA), UK. One of his research interests is in promoting evidence-based policy interventions for resilient food production systems in urban settings.

**Shilpi Kundu** is an Assistant Professor in the Department of Agricultural Extension and Information Systems, Sher-e-Bangla Agricultural University (SAU), Dhaka, Bangladesh. She is currently studying institutional effectiveness in climate resilient agriculture in coastal Bangladesh as part of her doctoral research in Griffith University, Australia. She has Master's degrees in agricultural extension from SAU, Dhaka, and in landscape ecology from Swedish University of Agricultural Sciences, Alnarp, Sweden. Her research interests are focusing on promotion of climate resilient agriculture systems through robust institutional arrangements, application of emerging science communication tools to promote ecosystem-based sustainable food production systems, biodiversity conservation in agrosystems, etc.

**Bindu Gauri Kottaram** is a faculty and an agriculture entrepreneur in Agribusiness School for Women, Krishi Vigyan Kendra, Indian Council for Agricultural Research, Coimbatore, India. She is actively promoting value chain development in urban agriculture products with the involvement of women self-help groups through capacity building, product innovation, and marketing. She works extensively for the promotion of millet-based sustainable food production systems to secure food security among the indigenous communities inhabiting the biodiversity-rich Western Ghats in Kerala. She has a Master's degree in Environmental Science from Pondicherry University, India, and is interested in sustainable food production models.

**S. Gopakumar** is the Director, Directorate of Education, Kerala Agricultural University, Thrissur, Kerala, India, and teaches ecosystem management and sustainable cities for the undergraduate and postgraduate students in the Faculty of Forestry. He has a doctoral degree from the Forest Research Institute University, India, and has published papers on home garden diversity in Kerala, mapping of ecologically fragile lands, phytosociological assessment of terrestrial ecosystems, forest policy, value addition in agroforestry products, value chain development, and agroforestry models. His cross-cutting research interests cover areas such as species diversity studies, vegetation ecology, biodiversity conservation, community forestry, urban forestry, and crop physiology.

# Protecting Peri-urban Agriculture: A Perspective from the Pacific Islands

# 6

Sarah W James

**Abstract**

With the rapid pace of climate change and rising levels of non-communicable diseases, peri-urban agriculture is increasingly seen as critical for urban food security and resilience globally. Farmland on the urban fringe, however, is under threat across the world with pressure for housing and commercial developments as cities grow. In its focus on an often-neglected urban environment, that of a Pacific Island capital, this paper highlights the importance of understanding the unique manifestations of these apparently universal issues in the development of effective protection mechanisms. Port Vila, the capital of the Republic of Vanuatu, is one of the fastest growing cities in the Pacific Islands with urban expansion threatening prime farmland. It is also vulnerable to food and nutrition insecurity due to its high risk of natural disasters, isolated island geography, and heavy reliance on imported food. These factors are all compounded by the threat of climate change with predicted impacts to Vanuatu including rising sea levels, increased temperatures, and changing rainfall.

Complex post-colonial land arrangements, however, mean that mechanisms for protection of peri-urban agriculture utilised in the Global North, such as land use zoning, cannot be simply transferred to this tropical context. The Vanuatu example illustrates the importance of adopting a context specific approach in addressing seemingly universal challenges. In this context, this paper explore two mechanisms have been proposed to protect Port Vila's peri-urban farmland: a tax and an independent decision-making panel. Acknowledging the unique manifestations of this issue in Vanuatu, analysis of the potential effectiveness of these mechanisms also provides insights with relevance to cities beyond the Pacific. In drawing attention to the significance of political and economic

S. W. James (✉)
Australian National University, Canberra, ACT, Australia
e-mail: sarah.james@anu.edu.au



J. A. Diehl, H. Kaur (eds.), *New Forms of Urban Agriculture: An Urban Ecology Perspective*, https://doi.org/10.1007/978-981-16-3738-4_6

pressures, the paper concludes that such factors must be addressed for success of any protection mechanism. Strong political will, transparency in decision-making, and economic incentives to maintain peri-urban farms are essential to ensuring farmland is not only available (i.e., not developed) but that it is also accessible to and utilized by those who want to farm.

**Keywords**

Vanuatu · Pacific Islands · Urban · Land use · Peri-urban agriculture

## 6.1 Introduction: Peri-urban Agriculture in the Pacific—Vanuatu

Vanuatu is an archipelago of over 80 islands in the South West Pacific, with a population of over 270,000 people spread over 12,281 km$^2$ (Fig. 6.1). Sitting within Melanesia, it gained independence from Britain and France in 1980. The formal economy is based largely on tourism and agriculture, with agriculture accounting for approximately 20% of GDP (Vanuatu Government 2019). Primary exports are non-food agricultural crops, or "cash crops" including kava, cocoa, coffee, coconut (primarily copra), and spices (vanilla and pepper).

As with many Pacific Island Countries (PICs), Vanuatu, has historically been viewed as food secure with a predominantly rural population enjoying plentiful rains, rich volcanic soils, and abundant space for farming (Quantin 1982). But, as with many other PICs, this "pacific idyll" no longer reflects reality for many indigenous Ni-Vanuatu (Indigenous peoples of Vanuatu) (Connell 2007). While around 75% of the Vanuatu population still live in areas classified as rural (VNSO 2017), this designation belies the significant demographic shifts that have occurred in recent decades. Rapid urbanization in the capital city Port Vila has been fed by rural-urban migration and high population growth. With growth rates of over 10% concentrated on the urban fringe, urbanization has spread into areas officially classified as "rural" (Trundle and McEvoy 2015). Urbanization has brought with it increasing levels of poverty as well as lack of resources such as housing and land for the burgeoning urban population (Storey 2006). These social changes intersect with the threat of climate change and increasing natural disasters, and a remote island geography, to make Port Vila, the capital city of Vanuatu, highly vulnerable to food insecurity. Increasing and securing peri-urban food production are therefore important for the city's long-term food and nutrition security. Urban spread, however, threatens prime agricultural land on the peri-urban fringe.

Such tension between farms and housing or other development can be mapped onto the peri-urban regions around the world. Increasing urbanization and urban population growth has resulted in housing and commercial development encroaching onto what is often prime farmland close to urban markets. With increasing concern about urban sustainability and climate change as well as rising rates of non-communicable diseases there are rising calls to protect this farmland for

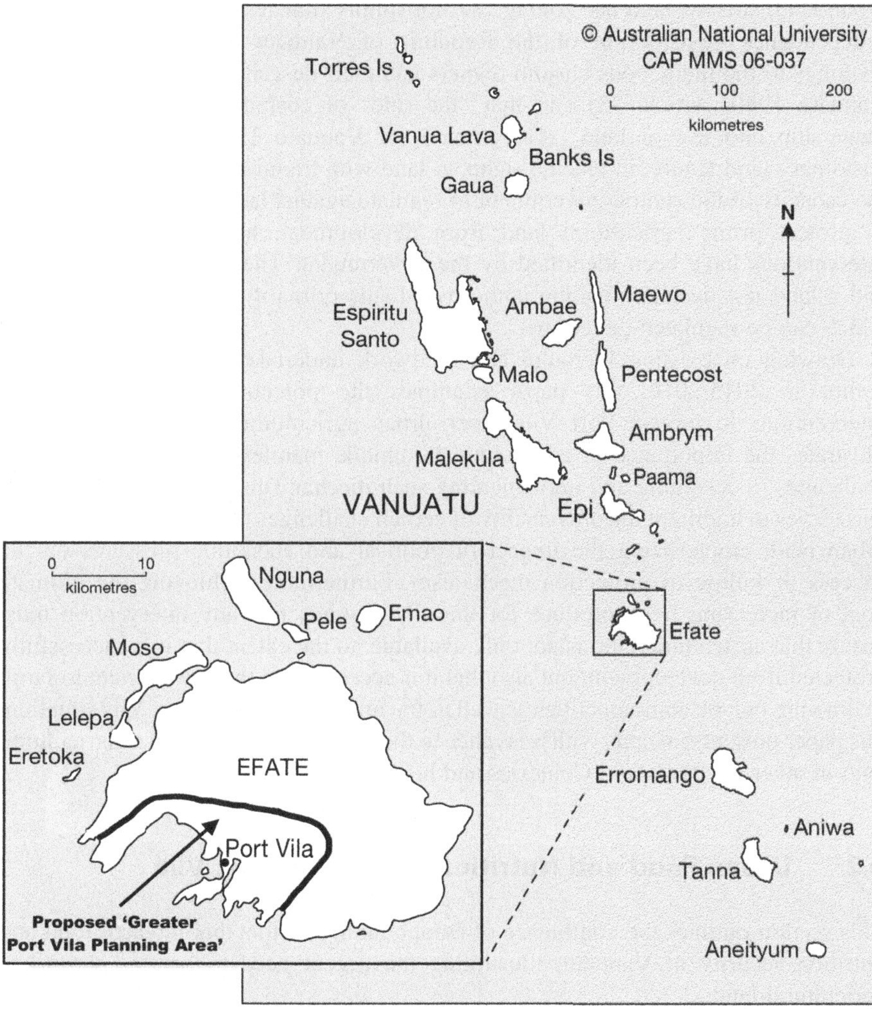


**Fig. 6.1** Map of the Vanuatu Archipelago and the Island of Efate. (Source: Republic of Vanuatu 2007)

food and nutrition security into the future. Where they exist, including cities such as Sydney, Toronto, Berlin, measures implemented to protect peri-urban farmland are typically land use planning mechanisms such as zoning (James 2016a).

Despite the pervasiveness of this issue globally, the situation in Port Vila illustrates the importance of understanding and addressing the unique manifestation of these issues in a particular context for the development of effective mechanisms to address them. The population of Port Vila face multiple vulnerabilities to food insecurity due to their isolated geography, reliance on imported foods, and frequent natural disasters. However, planning mechanisms commonly used to protect peri-urban land, such as land use zoning, are not simply transferable to the Pacific. The Independence Constitution of the Republic of Vanuatu declares that all land "belongs to the indigenous custom owners and their descendants" (Government of Vanuatu 1980, Article 73) and that "the rules of custom formed the basis of ownership and use of land" (Government of Vanuatu 1980, Article 74). This customary land tenure, in placing rights to land with traditional land owners, limits the capacity of the central government to regulate against land use change. Seeking to protect prime agricultural land from development, however, a number of interventions have been identified by the government. These mechanisms—a tax and a land use decision-making authority—focus primarily on lands under lease, which can be regulated under law.

Drawing on existing literature and fieldwork undertaken in Port Vila by the author in 2015/2016, this paper examines the potential of these proposed interventions to protect Port Vila's peri-urban agricultural land. In doing so it illustrates the importance of addressing the unique manifestations of these global challenges in designing and implementing such mechanisms. Conversely, the analysis serves to highlight the universality of certain challenges to the protection of peri-urban land, emphasizing the impact of political and economic pressures on the success or failure of protection mechanism. Furthermore, achieving the ultimate goal of increasing fresh produce for the city requires that any intervention must ensure that agricultural land is not only available, to the extent that it is successfully protected from development, but also that it is accessible to those who want to farm. In drawing out the commonalities as well as the uniqueness of the Port Vila situation, this paper presents insights with relevance to the protection of peri-urban agricultural land in other Pacific Island Countries and beyond.

## 6.2 Urban Food and Nutrition Security in Port Vila

This section outlines the confluence of compounding factors that threaten food and nutrition security in Vanuatu, illustrating the urgent need to protect peri-urban agricultural land.

### 6.2.1 High Rates of Urbanization

In contrast to a dominant perception of a primarily rural population enjoying "subsistence affluence" (Government of Vanuatu 2010), Vanuatu has one of the highest rates of urbanization in the Pacific, and more than twice the global average (UN Habitat 2015). At 4%, the official rate of Port Vila's population growth is double the national growth rate of 2% (Trundle and McEvoy 2015); and accounts for close to 19% of the national population according to the 2016 mini-census (VNSO 2017). Despite highlighting the strong urban growth in Port Vila, the official figures underplay the reality of the urbanizing landscape. As they account for only the official municipal urban areas, which are relatively small, offical figures suggest that around three quarters of the national population still live in rural areas (Beca 2015). This fails to account for the extensive growth beyond urban borders. The municipal area and close surrounding suburbs, that are often used to signify urban Port Vila, account for 24.3 square kilometers (Beca 2015). The urbanizing areas that surround Port Vila, which have been included in the proposed "Greater Port Vila Planning Area" (refer to Fig. 6.1), account for an estimated 234 square kilometers (Beca 2015). Growth was particularly concentrated in peri-urban areas with an average annual growth rate of 10%, with certain areas experiencing rates of 30% and 60% growth (Beca 2015; Trundle and McEvoy 2015). Based on the latest 2016 mini-census, the Greater Port Vila Planning Area accounted for close to 75,000 people[1] or approximately 27% of the national population (VNSO 2017); a population density of approximately 320 people/km$^2$ compared to the 19 people/km$^2$ nationally (VNSO 2007). In other words, more than one quarter of Vanuatu's national population is now concentrated in one small geographical area. This is not yet counting the other urban settlements in the islands of Espiritu Santo or Tanna (Fig. 6.1), which would bring the urban population even higher. The rapid increase in urbanization in recent decades has creating a high demand for fresh local food at the same time that housing construction spreads over the farmland that produces it.

### 6.2.2 Import Reliance/Nutrition Transition

These factors, including growing populations working in the cash economy without access to farmland, contribute to a rapid shift from traditional to imported processed food in urban areas. While the so-called nutrition-transition is evident across Vanuatu, it is in the urban population that this change is most prominent (Martynm et al. 2015). Dietary changes are typified by increased consumption of nutritionally poor imports that are high in fat and/or sodium (chicken wings, tinned tuna and meats, instant noodles, cabin biscuits) or relatively low nutritional value (white rice) (Martynm et al. 2015). The extent of import reliance is illustrated by statistical analysis from the Secretariat of the Pacific Community (SPC) which indicates white rice is the single biggest item of household expenditure in Vanuatu, at 7.6% of household spending, followed by motor vehicles at 5.9% (SPC 2013).

The nutrition-transition contributes to the rise in non-communicable diseases in PICs such as Vanuatu. Statistics from the Vanuatu National Statistics office (VNSO) illustrate the triple burden of malnutrition in Vanuatu with 28.5% of children under 5 suffer from stunting, 10.7% of children under 5 underweight, and 49.5% of women and 36% of men overweight or obese (VNSO 2013). Overall, 70% of mortality in Port Vila hospitals is attributed to non-communicable diseases (NCDs) (VNSO

[1] The calculation for the "Greater Port Vila Area" is based on the populations of Port Vila Municipal Council (50,995) and Mele (4711) Ifira (1186), Pango (2326), Erakor (8918) and Eratap (6640) area councils.

2013), with an estimated seven amputations a month due to Type II diabetes (Daniel 2019).

Rates of NCDs and obesity are particularly high in urban areas (Dancause 2010). A report by UNICEF in 2009 found that Port Vila was the most deprived in terms of food and health of all Vanuatu (Government of Vanuatu 2010). Indeed, Hughes and Lawrence (2005) go so far as to argue that malnutrition is a product of urbanization across PICs. In the process of urbanization, people typically move away from a relatively subsistence lifestyle with access to a food garden into the cash economy. It has been estimated that over 80% of the Ni-Vanuatu population of Port Vila has migrated from elsewhere (Aruntangai 1995 in Storey 2003). Away from their customary lands, urban Ni-Vanuatu often have little capacity to produce their own food. With high numbers of residents in the inner urban areas without access to gardens, many rely on imported foods as they are cheaper than market produce (James 2017). A study from Jones and Charlton (2015) indicated that those on low incomes would have to spend 40% of their food budget to get sufficient fruit and vegetables for health requirements. With the highest rates of poverty in the nation, with over 30% of the Port Vila population estimated to be under the poverty line (Government of Vanuatu 2010), many people would not be able to afford sufficient healthy food.

### 6.2.3 Limited Access to Local Food

The increase in consumption of imported foods in recent decades has occurred at the same time as a decrease in the availability of local foods. Indeed Mackenzie-Reur and Kulakit Galgal (2018) assert that the overall production of staple root crops (such as taro, yam, and cassava) has not significantly increased since Independence in 1980, despite high population growth. In 1983, 0.9 kg of root crop per capita was produced per day compared to 2007 when only 0.5 kg per capita was produced (Mackenzie-Reur and Kulakit Galgal 2018). Analysis by the Government of Vanuatu based on the last agricultural census (2007) indicated that local food production, inclusive of all food crops, was only sufficient to feed approximately a small proportion of the population without reliance on imports (Vanuatu Government 2019).

Different factors could have contributed to the limited production of root crops. One factor could be the focus on the production of non-food export or "cash" crops (including crops such as coconut, kava, cocoa, coffee, and spices) (Mackenzie-Reur and Kulakit Galgal 2018). With these crops representing key national exports, cash cropping has been promoted to generate foreign exchange earnings and reduce the trade deficit. Conversely, the increase in cheap food imports is also likely to be a driver in the limited production of local food, reducing its price competitiveness (Hughes and Lawrence 2005).

### 6.2.4 Climate Change

Vanuatu is one of the most vulnerable countries in the world to climate change, one of the Pacific Island "canaries in the coal mine" for the effects of climatic change (Cass 2018). By the end of the century, Vanuatu is predicted to experience sea level rise and salt water inundation, rising average temperatures, as well as days of extreme heat with greater variation in rainfall including more intense droughts and rain (PACCSAP 2013). These changes will likely have a substantial impact on food production capacity in Vanuatu, requiring adaptation of current practices to maintain food supply.

Furthermore, the current reliance on food imports, particularly imported rice, also increases Vanuatu's vulnerability to food insecurity within the global context of climate change. Predictions from the International Panel on Climate Change (IPCC 2018) are that crop failures will increase globally. If there was a rice crop failure in one of the main rice exporting countries it would cause price inflation in exports to Vanuatu. Rice would no longer be a cheap calorie source and there would be a gap in national food supplies. Climate change will also affect ability to produce food in Vanuatu, with increased weather variability as well as potential saltwater inundation of low-lying areas (IPCC 2018). These factors highlight the necessity of ensuring that all the prime farmland that exists in Vanuatu is available for agricultural production, to secure sufficient food production to feed the local population in an uncertain future.

### 6.2.5 Natural Disasters

In addition to the threat of climate change, Port Vila has been determined to be *the most* at-risk city in the world to natural disasters (Bündnis Entwicklung Hilft 2019). Vanuatu experiences cyclones, storm surges, landslides, flooding, and droughts, which are likely to become more intense as a result of climate change (PACCSAP 2013). Situated in the Pacific "Ring of Fire," it is also highly exposed to geophysical threats such as volcanic eruptions, earthquakes, and tsunamis, as well as human, animal and plant diseases, and human-caused disasters. In the first quarter of 2020 alone, Vanuatu experienced travel and movement restrictions in reaction to the COVID-19 pandemic, severe Tropical Cyclone Harold which caused widespread destruction, and increased volcanic activity of Mt. Yasur on Tanna Island that decimated crops and local food supplies.

A particularly significant natural disaster in recent years was Tropical Cyclone Pam, which hit Port Vila in March 2015. It was the first category five cyclone to make a direct hit to the Capital. Cyclone Pam had significant short-term effects on food production and potentially longer-term impacts, increasing the reliance on imported foods. On the islands of Efate and Tanna, another key food producing island, it was estimated that over 90% of food crops were destroyed (Wentworth 2019). As a result, the availability of fresh produce to the Port Vila markets plummeted (Wan Smolbag 2016). There was scarce food available, and what was available increased in price exponentially, making it prohibitively expensive for most people (Wentworth 2019). A number of factors, including the cost of inter-island shipping, meant that food from islands that had not been affected by the cyclone was not relocated in sufficient quantities to the capital. Furthermore, the limited produce that was transported was very expensive. This resulted in an increasing reliance on imports in the short term, including food aid primarily comprised of rice and tinned goods such as tuna (Wentworth 2019). Available data suggests that this may have exacerbated the nutrition-transition, in the medium term at least, with significant increases in rice importation in the years following Tropical Cyclone Pam (2015–2018) (VNSO 2019).

The effect on food security due to other natural disasters has also been significant in different ways. Historically it has been the case that many evacuees from natural disasters such as volcanic eruptions have chosen to resettle near the major settlements, to be close to perceived opportunities for work and other resources. There are many such communities in Port Vila and, after the 2017 volcano eruption on the island of Ambae, in and around the urban settlement of Luganville in Santo. This disaster driven rural to urban migration, in turn, puts further stress on urban settlements.

### 6.2.6 Policy Neglect

These intersecting factors highlight the need for greater access to fresh produce for Port Vila's rapidly expanding population. Historically, however, there has been little attention to food security, particularly urban food security, by development actors, although this is starting to shift. This lack of attention can partially be attributed to the perception that there is plenty of available fertile land across Vanuatu for food production. A detailed land capacity assessment, undertaken in 1982 following Vanuatu's Independence, indicated that approximately 40% of Vanuatu was arable productive land (Quantin 1982). The report concluded that the extent of productive land available would enable Vanuatu to produce enough to feed its growing population. This conclusion about Vanuatu's capacity for self-sufficiency appears to have underpinned a relative neglect of food security issues, with agriculture related policy and programming instead focusing on increasing production of cash crops for export (Mackenzie-Reur and Kulakit Galgal 2018).

The assumption that existence of prime agricultural land means that it is available and accessible for those who wish to farm does not reflect reality. The large-scale leasing of land to non-indigenous settlers post-independence, has made much prime agricultural land, particularly on the island of Efate, inaccessible to locals. Prior to Independence in 1980, there was extensive alienation of land to non-custom owners, particularly to non-indigenous settlers under French and English colonial rule (Farran 2009). Concern over loss of land into non-indigenous hands was a key driver of the fight for independence from colonial rule.

While the Independence Constitution vested ownership of land to customary owners, it has not ensured custom owners the full control over their land that was intended (Regenvanu 2008). Many non-indigenous land holders who had "owned" land under colonial rule were able to secure leases following Independence, for a period of up to 75 years (the life of a coconut tree) (Farran 2009). In subsequent decades, particularly in the late 1990s and early 2000s, further extensive leasing of large areas of land, described by some as "land grabbing" occurred particularly on the rich coastal lands on Efate (McDonnell 2015). Much of this was due to significant overreach by successive Ministers of Lands who approved leasing of lands without appropriate consent from customary owners (McDonnell 2015). Today, as a result, significant areas of land remain in non-indigenous hands.

### 6.2.7 Lack of Land to Farm

On the larger island of Efate, where Port Vila is located, an estimated 70% of what is classified as good agricultural land is under lease (Fig. 6.2) (Lebot and Simeoni 2014). These agricultural leases, many carried over from colonial-era cattle farms and coconut plantations, are disproportionately used for commercial production rather than food crops. Lebot and Simeoni (2014) estimated across Vanuatu the area under production for food crops as 7511 ha compared to over 75,000 ha for coconut plantations and 15,000 ha for beef pasture. However, they note that many of the large agricultural leases are not currently used for agricultural production. They suggest that instead the land is often being held for land speculation.

The effect of the land lease situation is that Efate has very limited available land for local food production—estimated at 1.4 ha per household based on 2009 census data (Lebot and Simeoni 2014). This is a fraction of the estimated 3–4 ha that a household requires to produce enough food to feed themselves (Dr Vincent Lebot, pers. comm.). Due to the rapid population growth, land available to farm per household is likely to have fallen even further in subsequent years. It is also important to note that lands designated "available" by Lebot and Simeoni (2014) were those not under lease. They are, however, still under customary ownership and as such can only be accessed through an agreement with the customary owner. Such an agreement may be difficult to reach or provide little security of tenure. As a result, even this limited land identified as not under lease or developed may still be inaccessible to Port Vila residents who may wish to farm to produce food. This is particularly a concern for the high number of Port Vila residents who do not originate from the island of Efate, and therefore do not have their own custom lands to farm on.

The lack of available agricultural land on Efate has significant implications for the supply of food to the Port Vila market, on which many residents rely for fresh food (James 2016a). A high proportion of the food sold at the Port Vila markets is from Efate (VNSO 2007; Mael 2011). Key sites for semi-commercial agricultural production are around Efate, particularly in the peri-urban areas significantly affected by urban encroachment due to growth rates over 10% (Trundle and McEvoy 2015). There is prime agricultural land on other islands of Vanuatu (Quantin 1982), and during different seasons they can provide substantial supplies the Port Vila markets.

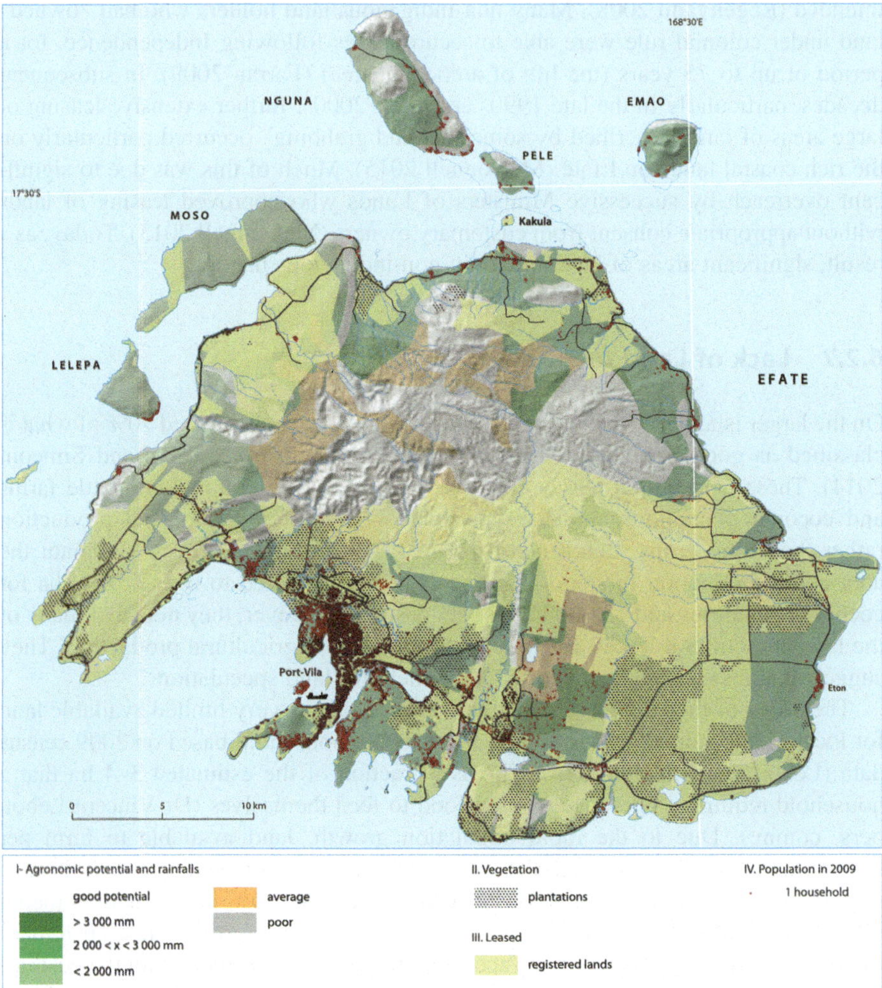


**Fig. 6.2** Land use map of Efate Island. (Source: Lebot and Simeoni 2014)

However, the cost and logistical challenges of inter-island transport have limited the volume of produce overall that comes from other islands relative to Efate. The risk of relying on supply from other islands in Vanuatu to feed Port Vila was illustrated in the aftermath of Cyclone Pam in 2015. When Efate farmers were decimated, the limited volumes of food sourced from other islands meant that the Port Vila market was expensive and undersupplied (Wan Smolbag 2016). These factors suggest that if the peri-urban land around Port Vila is not protected, then the supply to the main market will be threatened.

This overview of factors influencing food and nutrition security in Port Vila highlights the importance of ensuring the capacity to produce fresh food in close proximity to the capital, particularly into the future. Ensuring access to sufficient peri-urban agricultural land for production, however, is problematic. Barriers include high rates of urbanization leading to an increasing demand for housing, resulting in encroachment on farmlands, and a high percentage of land under lease to non-indigenous interests, resulting in the majority of prime agricultural lands unavailable for people wishing to farm.

## 6.3 Challenges to Land Use Planning in the Pacific

Around the world a variety of land use planning measures, particularly zoning, have been implemented—albeit to varying success—to protect peri-urban agriculture in places such as Australia, North America, and Europe (Han and Go 2019). Applying such interventions in the Vanuatu context presents a number of challenges, in particular the unique customary land tenure there that is similar to many PICs. As ultimate control over land is vested in the individual custom owners, mechanisms such as land use zoning cannot simply be enforced by a central government. This presents significant challenges in seeking to manage urban growth outside the limitations of designated urban boundaries, within which land is primarily public (Storey 2006). Attempting such zoning would require the consent of each of the (multiple) affected customary land owners, which would be time consuming and, ultimately, unlikely to be successful due to different views on the value of development. The higher financial returns from leasing land, either formally or informally, for housing or commercial development make it a desirable option for many landowners, as has been seen in other peri-urban areas in Efate (Trau et al. 2014). These examples indicate that even if agreement was achieved it would likely remain insecure due to the potential for a breakdown of consensus. The lack of authority of the central government over land outside municipal boundaries makes the protection of peri-urban agricultural land through planning mechanisms such as land use zoning almost untenable.

The extensive tracts of land that are held under leases on Efate and other islands, however, represent land that can be regulated by the government. As noted, on the island of Efate the land under lease constitutes a considerable area of arable land and retaining it for agricultural use would make a significant impact on the land accessible for agriculture. With a focus on land under agricultural lease, current government legislation has proposed two mechanisms for protection of peri-urban agricultural land from development: a tax and an independent decision-making panel.

### 6.3.1 Tax

New legislation for the Vanuatu Department of Agriculture and Rural Development was passed in 2018, which sought to discourage the transfer of agricultural leases to residential or other uses through taxation. In Part 5 of the Agriculture Act (2018) the Government imposed a tax of 25% of the value of an agricultural lease for any changes to land use class or subdivision for any uses other than agriculture on prime agricultural land. It was determined that 25% of the value of the lease was potentially a significant amount of money for the landholder and would act as a deterrent to changing of leases. This tax has yet to be implemented as of December 2021, and is likely to prove challenging as it is anticipated to require amendments to legislation under the Ministry of Lands.

### 6.3.2 Land Management Planning Committee

Barring change in legislation, another avenue could be pursued to protect peri-urban agricultural lands. While not ensuring a particular outcome in the way levying a tax on all lease changes would consistently penalize all lease changes to agricultural leases, the structures developed in the Land Reform (Amendment) Act of 2013 provide a means for greater protection of prime agricultural land. The Land Reform (Amendment) Act of 2013 shifted the authority over lease decisions and changes to leases from Minister of Lands to the Land Management Planning Committee (LMPC) (McDonnell 2015). According to the Act, the LMPC should consist of seven people: the Chairperson, Director of Lands, Director of the Department of the Environment; Director of the Vanuatu National Cultural Centre and three senior planners. Its main functions include considering and submitting to approval by the Minister applications for lease, subdivision or change of least type; leases on state land; and advise the Minister of Lands on land policy and lease making. Under the Land Reform Act (2013) the Minister of Lands is not allowed to approve a lease agreement, or changes to a lease, without approval by the LMPC and its independent Chairperson. This change was significant as it shifted oversight of land leasing from the Minister of Lands to an independent body (McDonnell 2016). The aim of this shift was to de-politicize decisions around land leasing by creating an independent committee to pass judgment on leasing requests (Garae 2014). These changes were made as part of sweeping land law reforms under the Hon. Ralph Regenvanu, when he became Minister of Lands in 2012 (McDonnell 2015). Prior to these reforms, there had been significant concerns about ministerial overreach by previous Ministers of Lands, including leases being granted to expatriate interests without the permission of the appropriate custom owners (Regenvanu 2008; Farran 2009).

The Land Reform (Amendment) Act (2013) positions the LMPC in a critical role for protecting agricultural land, as it has the authority to grant change of lease requests. This would include changes to lease classes and capacity to subdivide. The five land lease classes under the Land Leases Act (CAP 163) include Agricultural, Residential, Commercial, Industrial, and Special leases. Currently the approval process of a request to change a lease class or to subdivide provides an opportunity for a recommendation by the Department of Agriculture and Rural Development. A recommendation by the Department of Agriculture against approving the change in lease class or subdivision can be made on the basis that the land is prime agricultural land that should be retained in agricultural use. This clause has not been greatly utilized to date, in part due to the lack of a clear approval process. A formalization of the processes would allow for greater input into the protection of agricultural class leases. However, it is not clear the extent to which the LMPC is compelled to adhere to a recommendation by the Department of Agriculture to protect farmland. That the final decision would still lie with the LMPC suggest this option does not have the same regulatory strength as a compulsory tax to protect peri-urban farmland. However, how successful it was would ultimately depend on implementation. It is also important to note that leaseholders are only affected by the new processes if they choose to change the conditions of their leases, such as subdivide or change lease type. As McDonnell (2016) notes, it is therefore possible for the leaseholders looking to develop to wait and see if a new government overturns the legislation.

While both mechanisms, tax and independent decision-making committees, are potentially useful in protecting peri-urban agricultural land around Port Vila, there are a number of barriers to the implementation of either option. Significant coordination across government ministries and change of legislation as noted above are potentially needed. One example of the coordination and agreement required between the government departments is the designation of areas as "prime agricultural lands." Such designation is necessary as the basis for either a tax or an LMPC decision. Currently a clear and public designation of what constitutes "prime agricultural lands" in Vanuatu does not exist. It could be created, however, based on land use capability maps such as those produced by Lebot and Simeoni (2014). Their existing maps would need to be updated with leasing and population information, agreed upon by the relevant authorities as the definitive map of prime agricultural land, and published publicly. Agreement on such designations would potentially be hindered by technical issues, including the existence of different land use maps and access to required data. It would also require effective cooperation between the relevant government departments and strong political will to enforce designations against another key barrier to the successful implementation of any mechanism: pressure for development.

## 6.4 Pressure for Development

Any restriction on transferring or subdividing agricultural class leases to another land use is likely to create a backlash from leaseholders and developers. Changing from an agricultural to a residential or commercial class lease would allow for profitable development or re-sale of the lease, as would subdivision for development purposes. This includes leaseholders who have held large agricultural leases in Efate since Independence. Such leaseholders might be motivated to take direct legal action or seek political influence to challenge a restriction on their capacity to subdivide or develop their lease. The long history of conflict over rights to land, and the more recent example of political and economic influence to obtain land leases without proper authority from the custom owners, suggest that any changes to restrict develop and profit are likely to be met with strong resistance.

While Vanuatu's customary land tenure system makes the protection of agricultural land a unique challenge in many ways, the pressure for housing or commercial development on peri-urban land is common to cities around the world. The desire to capitalize on the higher value of land zoned for development by landowners and developers, as well as the assertion that it will create more affordable housing, has underpinned this push for development (Bunce 1998; Amati and Yokohari 2006; James 2016a). From the Toronto's Golden Horseshoe to Sydney's former Greenbelt, peri-urban lands zoned for agricultural use have faced, and often been lost to, mounting pressure for development (Bunce and Maurer 2005; James 2016a; Han and Go 2019). In the case of Sydney, for example, the Greenbelt was ultimately disbanded as approval for developments by smaller councils slowly ate away at the land protected by zoning (Freestone 1992). These examples illustrate that the challenge for protecting peri-urban agriculture land is not only the development of an effective mechanism, it is also the political will to uphold and enforce the protective measures against economic and political pressure for development.

In addition to putting into place a mechanism to protect agricultural land with the goal of increasing urban food and nutrition security, it is also important that any land protected is actually used for production. In their analysis, Lebot and Simeoni (2014) indicated that it is likely that much of the agricultural leases around Port Vila are not under full production. Instead, it is being left insufficiently utilized. The reason is unclear but it may be due to land speculation or production being unprofitable. The problem of land successfully protected for agriculture not being used for food production is, however, common to other peri-urban areas such as Sydney or Berlin (James 2016a). In these locations extensive areas of agricultural land protected under land use zonings are instead used for non-food purposes such as turf (lawn) production or horse agistment (Zasada et al. 2013; James 2016a). This illustrates that while land may be available for farming, i.e., not developed, this does not mean it is actually used for food production. Addressing the economic factors that influence the way in which available land is used is an important issue to take into account in any intervention. It is necessary to protect farmland, but it is also crucial to protect and promote farming as a practice, to address the end goal of increased food and nutrition security. One option in the Vanuatu context would be to encourage subdivision of large leases for agricultural purposes, rather than residential. This would provide more security for small lessors to develop farms and provide some economic return for the large leaseholders. Subdivision could be subsidized, such as through a market mechanism like in North America (Harman et al. 2015), although how this would be implemented in the Vanuatu context would require further analysis. Another aspect of improving the economic viability of farmland would be to increase the return to farmers on food production (James 2016b). This aspect is, however, beyond the scope of this paper.

The issue of how to ensure peri-urban agricultural land, if protected, remains accessible and utilized by those wishing to undertake food production is a critical one for Port Vila and cities around the world. Arguably it requires incentives to balance out the loss of potential profit offered by other forms of development. How these dynamics are dealt with will determine whether peri-urban agricultural land will be protected in Vanuatu.

## 6.5 Conclusion: Adopting a Context Specific Approach to Universal Challenges

Despite the persistence of the "pacific idyll," the rapid urbanization of the Greater Port Vila area creates a vulnerability to food and nutrition security. The rapid on-set of climate change and the high risk of natural disasters only serve to exacerbate the threats to food insecurity in Port Vila and highlight the need to ensure food production remains in the city region. Ensuring the availability of peri-urban agricultural land for farming presents a number of challenges, however. The tension between use of land for agriculture or, often more profitable, uses such as housing or commercial development in Vanuatu echoes the situation in many cities across the globe.

The particularities of land use politics in Vanuatu highlight how a universal issue uniquely manifests in a particular geographical context. Understanding and addressing these unique manifestations are essential in the adoption and implementation of any mechanism to protect peri-urban agriculture. There are, however, mechanisms available for the protection of peri-urban agricultural land that hold promise. For example, as described in this chapter, the option of working within the leasing system to impose either a tax on changing from an agricultural class lease or facilitate arbitration by an independent committee, such as the LMPC, allows for a context specific response.

This paper also highlights that the success of any mechanism in ensuring continued agricultural production on the peri-urban fringe will largely be determined by political and economic factors. Lessons from international case studies, as well as previous problems with land leasing in Vanuatu, illustrate that restricting those who wish to transfer agricultural leases to other uses is likely to generate contestation and challenge by leaseholders or prospective developers. These actors may seek to contest such restrictions through political or legal means.

Furthermore, these factors have been shown to influence not only if land remains available for agriculture but, more significantly, if it is actually accessible to and utilized by those who want to undertake food production. International experience highlights the importance of increasing the economic benefit of retaining land for farming. Such incentives can encourage use of agricultural land for farming rather than it simply being retained (unused), perhaps with the hope of future development. One option for Port Vila would be subdivision of agricultural leases for smallholder farming.

The success of any of the discussed mechanisms in protecting peri-urban agricultural land will ultimately depend on whether there is sufficient political will to protect it under what is likely to be intense pressure for development. This, in turn, is vital to ensuring long-term food and nutrition security for Port Vila.

## References


Amati M, Yokohari M (2006) Temporal changes and local variations in the functions of London's green belt. Landsc Urban Plan 75(1–2):125–142

Beca International Consultants (2015) Urban growth trends report - Port Vila and Luganville: risk mapping and planning for urban preparedness. Vanuatu Meteorological and Geo-hazards Department, Port Vila, Vanuatu

Bunce M (1998) Thirty years of farmland preservation in North America: discourses and ideologies of a movement. J Rural Stud 14(2):233–247
Bunce M, Maurer J (2005) Prospects for agriculture in the Toronto region: the farmers' perspective. Neptis studies on the Toronto metropolitan region. Neptis Foundation, Toronto
Bündnis Entwicklung Hilft (2019) World risk report 2019. https://reliefweb.int/sites/reliefweb.int/files/resources/WorldRiskReport-2019_Online_english.pdf
Cass P (2018) Pacific Islands canary coal mine climate change says minister. Kanivatonga News, New Zealand. https://kanivatonga.nz/2018/03/pacific-islands-canary-coal-mine-climate-change-says-minister/
Connell J (2007) Islands, idylls and the detours of development. Singap J Trop Geogr 28:116–135
Dancause KN (2010) Modernization and population health: a natural experimental model of health transition in Vanuatu, South Pacific. PhD Dissertation, State University of New York at Binghamton, Ann Arbor
Daniel B (2019) VCH confirms diabetes on the rise, 25th February 2019. Vanuatu Broadcasting and Television Corporation, Port Vila, Vanuatu
Farran S (2009) Land leases: research: ministerial leases in Efate, Vanuatu. J South Pacific Law 6(2) http://www.paclii.org/journals/fJSPL/vol06/6.shtml. Viewed 8 Aug 2019
Freestone R (1992) Sydney's Green Belt 1945–1960: stop laughing, this is serious. Aust Planner 30:70–77
Garae L (2014) Minister says success of land laws depends on performance of CLOs. Daily Post, Port Vila, Vanuatu. https://dailypost.vu/news/local_news/minister-says-success-of-land-law-depends-on-performance-of/article_bc84247a-d636-55ff-b4cf-06f52410d672.html
Government of Vanuatu (1980) Constitition of the Republic of Vanuatu, Port Vila, Vanuatu
Government of Vanuatu (2010) Millennium development goals 2010 report. Vanuatu Prime Ministers Office, Government of Vanuatu, Vanuatu
Han A, Go M (2019) Explaining the national variation of land use: a cross-national analysis of greenbelt policy in five countries. Land Use Policy 81:644–656
Harman BP, Pruetz R, Houston P (2015) Tradeable development rights to protect peri-urban areas: lessons from the United States and observations on Australian practice. J Environ Plan Manag 58(2):357–381. https://doi.org/10.1080/09640568.2013.859130
Hughes R, Lawrence M (2005) Globalisation, food and health in Pacific Island countries. Asia Pac J Clin Nutr 14(4):298–306
IPCC (International Panel on Climate Change) (2018) Special report on global warming of 1.5 degrees. Intergovernmental Panel on Climate Change, Geneva
James S (2016a) Farming on the fringe: peri-urban agriculture, cultural diversity and sustainability in Sydney. Springer, New York
James SW (2016b) Beyond 'local' food: how supermarkets and consumer choice affect the economic viability of small-scale family farms in Sydney, Australia. Area 48:103–110. https://doi.org/10.1111/area.12243
James S (2017) Food security in Port Vila, Vanuatu. World Vision Vanuatu, Port Vila
Jones H, Charlton K (2015) A cross-sectional analysis of the cost and affordability of achieving recommended intakes of non-starchy fruits and vegetables in the ćapital of Vanuatu. BMC Public Health 15:301. https://doi.org/10.1186/s12889-015-1644-2
Lebot V, Simeoni P (2014) Vegeculture project: complex cropping systems based on vegetatively propagated crops in the humid tropics. CIRAD, France
Mackenzie-Reur V, Kulakit Galgal K (2018) Building the evidence base on the agriculture-nutrition nexus. The Technical Centre for Agricultural and Rural Cooperation, Wageningen, Netherlands
Mael JT (2011) Vanuatu domestic market study: the potential impact of increased tourist numbers on the domestic market for selected fresh vegetabl produce. Food and Agricultural Organisation of the United Nations, Fiji
Martynm T, Yi D, Fiti L (2015) Identifying the household factors, and food items, most important to nutrition in Vanuatu. Food and Agricultural Organisation of the United Nations, Fiji and Government of Vanuatu, Port Vila

McDonnell S (2015) Vanuatu embraces landmark reforms, East Asia Forum, 14th March. pp 1–3. https://www.eastasiaforum.org/2015/03/14/vanuatu-embraces-landmark-reforms/
McDonnell S (2016) My land, my life: power, property and identify in land transformations in Vanuatu. PhD Thesis, Australian National University, Canberra
Pacific Australia Climate Change Science and Adaptation Planning Program (PACCSAP) (2013) Current and future climate of Vanuatu. https://www.pacificclimatechangescience.org/wp-content/uploads/2013/06/15_PACCSAP-Vanuatu-11pp_WEB.pdf
Quantin P (1982) Vanuatu agronomic potential and land use map. Ministry of Land and Natural Resources, Vanuatu and Office De La Recherche, Scientifique Et Technique Outre-mer, Orstrom, Paris
Regenvanu R (2008) Issues with land reform in Vanuatu. J South Pacific Law 12(1):63–67
Republic of Vanuatu (2007) Chief Roi Mata's domain: nomination by the Republic of Vanuatu for inscription on the World Heritage List. Vanuatu Cultural Centre, Port Vila
SPC (2013) Household expenditure patterns in Vanuatu. Secretariat of the Pacific Community, Fiji and Vanuatu National Statistics Office, Port Vila
Storey D (2003) The peri-urban Pacific: from exclusive to inclusive cities. Asia Pac Viewp 44(3): 259–279. https://doi.org/10.1111/j.1467-8373.2003.00214.x
Storey D (2006) Urbanisation in the Pacific, State, Society and Governance in Melanesia. Australian National University, Canberra
Trau A, Ballard C, Wilson M (2014) Bafa Zon: localising world heritage at chief Roi Mata's domain, Vanuatu. Int J Herit Stud 20(1):86–103
Trundle A, McEvoy D (2015) Greater Port Vila climate vulnerability assessment - full report (2015). United Nations Human Settlements Programme Cities and Climate Change Initiative, Fukuako, Japan
UN Habitat (2015) Urbanisation and climate change in small developing island states. United Nations Habitat, Nairobi
Vanuatu Government (2019) Gudfala Kaikai policy. Vanuatu Government, Port Vila
VNSO (2007) Agricultural census. Vanuatu National Statistics Office, Port Vila
VNSO (2013) Vanuatu living conditions survey. Vanuatu National Statistics Office, Port Vila
VNSO (2017) 2016 Post TC pam: mini-census. Vanuatu National Statistics Office, Port Vila
VNSO (2019) Overseas trade data (Unpublished). Government of Vanuatu, Port Vila, VU
Wan Smolbag Theatre (2016) Urban nutrition program for El Nino drought response: final report. Wan Smolbag Theatre, Port Vila, Vanuatu
Wentworth C (2019) Unhealthy food aid: food security programming and disaster responses to Cyclone Pam in Vanuatu. Anthropology Forum. https://doi.org/10.1080/00664677.2019.1647830
Zasada I, Berges R, Hilgendorf J, Piorr A (2013) Horsekeeping and the peri-urban development in the Berlin Metropolitan Region. J Land Use Sci 8(2):199–214

**Sarah W James PhD** has 15 years research and policy experience relating to food security and healthy and sustainable food systems in Australia and the Pacific. She has wide-ranging experience across academia, government, NGOs, and CSOs. Her PhD focused on policy and planning for peri-urban agriculture and was published by Springer, titled *Farming on the Fringe: Peri-urban agriculture, cultural diversity and sustainability in Sydney.* She is currently based in Vanuatu, focusing on disaster risk reduction and climate change impacts on food systems in the Pacific Islands.

# Part II

# Water/Waste

# Engineering Perspective of Water Use for Urban Agriculture

# 7

Stephen Fisher

**Abstract**

Engineering perspectives of water in urban agriculture must be adequately considered in planning. Examination of water use for urban agriculture from an engineering perspective can help guide urban agriculture policy and inform the urban planner's implementation of it. In particular, water stress has become a global denominator in wet and dry climates alike, particularly in urban settings. Urban centers all over the world are also experiencing accelerating rates of failure in water treatment and distribution infrastructure. In this chapter, I characterize urban water use and how water use can impact a sometimes delicate local and regional status quo. I compare how much water it takes to grow, process, and deliver fresh vegetables for large commercial farms in peri-urban areas and small-scale gardens in urban areas. Finally, I examine advantages and disadvantages of some common practices and technologies that use urban water for growing urban vegetables from the perspectives of planned districts, building systems, soils, and irrigation.

**Keywords**

Green infrastructure · Engineering · Policy · Gray water

## 7.1 Characteristics of Water in Global Cities

As the saying goes, "water is life." Nearly every city, from ancient to modern, is founded with a water supply. As cities became larger, the amount of land and water required to feed their populations naturally increased. Even ancient Romans faced

S. Fisher (✉)
Resilient Sustainable Infrastructure, Tetra Tech, Inc., Denver, CO, USA
e-mail: stephen.fisher@tetratech.com



J. A. Diehl, H. Kaur (eds.), *New Forms of Urban Agriculture: An Urban Ecology Perspective*, https://doi.org/10.1007/978-981-16-3738-4_7

water stress related to population growth and urbanization similar to what many global cities experience today. In fact, it is now regarded that the Roman's renowned aqueducts and water works, that enabled food production, population growth, and urbanization to a vulnerable degree, played a part in the empire's collapse (Dermody et al. 2014). Meanwhile, in the opposite hemisphere, history shows that ancient native American civilizations' ability, or lack thereof, to manage their water supply and food production meant extinction or thriving, thus requiring tough decisions to live within the limits of their resources.

The Industrial Revolution and the agricultural revolution(s) that followed enabled infrastructure to meet increased demand for water and food such that larger and larger farms outside urbanized areas were established. As cities outgrew their initial water supplies, distant water diversions and connections to water conveyance networks allowed cities to thrive in areas that would otherwise have been impossible. Los Angeles, for instance, is fed almost three times more water from outside its immediate watershed than the next most-connected city (Boston, Massachusetts), while on a per capita basis, Pretoria, South Africa is highest, receiving 829 L/day (McDonald et al. 2014) from sources outside its immediate watershed. The 185 most-populous cities in the world occupy about 1% of the global land mass but retrieve water from watersheds covering a stunning 41% of global land mass (McDonald et al. 2014).

### 7.1.1 Water Stress and Sustainable Development

It is clear that technological innovation has played a key role in slaking and feeding our cities. But society and policy have also played a role. Indeed, many argue that there should be no excuse for food and water scarcity experienced by the poor and vulnerable populations of the world, and there are numerous local and global responses.

For example, several of the U.N. Sustainable Development Goals (SDG) address agriculture and water use directly. The form and scale of urban agriculture could be a key to addressing these goals. Beneficial use of water for food (as it is a basic need) is almost incomparable to other beneficial uses, with urban agriculture perhaps even more so because it tends to be smaller scale, more accessible to the poor, and socially more equitable than imported food from large farms. Increasing food production using less water (SDG 6) and reversing land degradation and increasing the sustainable management of soils (SDG 15) can be addressed in urban agriculture just as it can in large-scale farms, where SDGs 6 and 15 are primarily targeted. Other collateral SDGs include ending poverty (SDG 1); ending hunger (SDG 2); achieving gender equality and empower all women and girls (SDG 5); and combating climate change (SDG 13) (FAO 2019).

Residents of many global cities experience water stress (McDonald et al. 2014). If we regard with equal importance water for urban agriculture as for all other urban water uses, agriculture is both a cause and consequence of water stress. Thirty-six percent of 185 global cities with a population greater than 750,000 experience some degree of water stress (defined for groundwater as abstraction/recharge >1, and for surface water as water use/water available >0.4), affecting about 438 million urban dwellers (McDonald 2014). So, it is worth examining all urban uses of water compared to the water demand of urban agriculture. In this chapter, I first look at sources of water for urban agriculture. Later, I look at a fine-grained spectrum of how urban agriculture uses water.

### 7.1.2 Water Resources

Water sources for cities globally are overwhelmingly—almost 80%—from surface water, such as rivers and lakes (McDonald et al. 2014). Contributing the other 20%, groundwater also plays an important role, both as a supplement to seasonal swings in surface water supplies and, in some cases, groundwater is a sole source. Because water sources occupy a significant land mass, water quality is highly vulnerable to land use in the related watersheds. For agriculture, this is normally not an issue unless the water is saline or has other inorganic contaminants. In most cities, especially the largest cities in the developing world, there are several types of water available. Raw water is untreated water, either from surface sources, an aqueduct, or a well. Most cities have water treatment facilities and a potable water distribution system, but generally deliver to only a fraction of the urban population. So, while in some urban areas, only expensive, potable water is available for any use, other urban areas have cheap or free access to raw water.

### 7.1.3 Wastewater as a Resource

Most urban areas operate centralized wastewater treatment facilities to varying degrees. Typically, the portion of treated wastewater to total wastewater is low. A majority of developing countries treat less than 15% of the wastewater they produce (Jiménez and Asano 2001; Malik et al. 2015). But there are numerous instances of urban agriculture using every type of wastewater to some degree of success. Wastewater types include:

- Black water—untreated sewage with animal and human fecal components.
- Gray water—untreated sewage without animal and human fecal components, for example, from laundry, dish washing, and bathing.
- Stormwater—runoff collected from storm drains, roofs, gutters, streets, etc.
- Treated wastewater—effluent from a treatment plant ranging from minimal treatment (screening, primary oxidation) to maximum treatment (tertiary, reverse osmosis, filtration).

Wastewater in the urban setting is generally free (although some investment may be needed to transport it to a useable location), flows are relatively steady, it contains nutrients for plant growth (Fig. 7.1), and it is somewhat evenly distributed around the

**Fig. 7.1** Human fecal sludge for agriculture: fecal sludge being delivered (**a**) and discharged (**b**) to a dewatering (drying) bed (**c**) in Accra, Ghana. The signs on either side of the truck indicate the value of fecal sludge as a resource (Hamilton et al. 2014)

city. Disadvantages include numerous pathogenic threats to human health for farmers as well as consumers and contaminants (especially from industrial discharges) that can stunt or kill plant growth and poison the soil. The World Health Organization (WHO) has compiled an extensive list of contaminants in wastewater and the effects on plant growth to guide policy and use surrounding urban wastewater (WHO 2006).

Oddly, it is in the developed world that application of wastewater in urban agriculture may be most difficult. This may be owed to generally more robust and risk-averse regulations and enforcement where the policy goal is to eliminate nearly all risk to human health. This curtails irrigation with wastewater greatly. The effect is that in urban areas in the developed world, use of potable water (and to a lesser, nascent degree, highly treated and regulated wastewater), clearly competes with total water resource allocation, making urban agriculture irrigation incrementally more difficult to justify. Mostly clandestine instances of growing large quantities of food in the urban context using wastewater without negative human health consequences indicate that nuanced and detailed regulations and guidance are needed to balance human health, food equity, and water resource politics (Hamilton et al. 2014).

In the meantime, many programs around the world aim to promote best growing and handling practices to mitigate pathogenic threats (Hamilton et al. 2014; Buechler et al. 2006). It has been proposed that a common typology of wastewater source and end uses would assist policy makers in developing regulation that balances human

**Fig. 7.2** Common informal sources of wastewater. Sources of wastewater for irrigating urban vegetable plots in Accra, Ghana. (**a**) Graywater collection for irrigation of an adjacent crop. (**b**) Sewer-mining to flood-irrigate a crop. A hole punctured in the bottom of the pipe is stoppered with a rag (placed on top of the pipe in this photo) that can be removed when irrigation is required. (**c**) Open street drain carrying stormwater, sewage, and graywater. (**d**) Manual collection of water from an urban stream immediately downstream from a raw sewage discharge point (Hamilton et al. 2014)

health with natural resources. These include direct, end-of-pipe irrigation (Fig. 7.2), dilution of wastewater with raw water, domestic wastewater, industrial effluent, and stormwater (Buechler et al. 2006). Van der Hoek et al. (2002) combine source and use into three policy areas: direct land application of untreated wastewater; direct land application of treated wastewater piped to fields; and indirect use of wastewater where it has first flowed to or through another water body such as a pond, lake, canal, tank, or river.

## 7.2 How We Use Water to Produce Food

At first glance, it would seem that agriculture uses massive amounts of water to produce food. This is partly correct since about 70% of fresh water available globally is used for agriculture (FAO 2015). However, it may be more accurate to say that

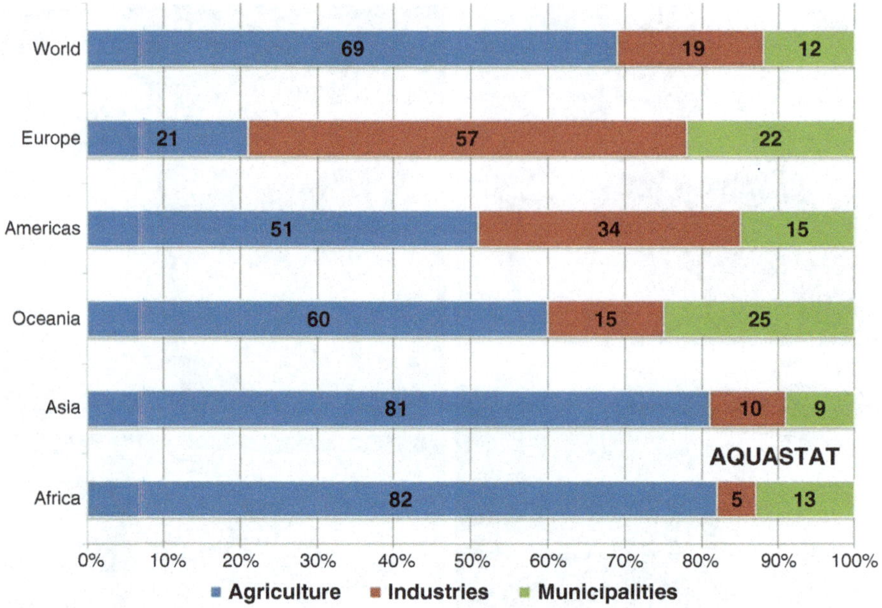


**Fig. 7.3** Water withdrawal ratios by continent (FAO 2015)

massive amounts of water are applied but not necessarily taken up by crops. Overall global irrigation efficiency, defined as the ratio of water applied to agronomic uptake capability, is about 56% (FAO 2014). This leaves room for much innovation. Figure 7.3 illustrates how inextricable water use and food production have become globally. Almost every city in world must essentially operate like Rome once did, growing some food locally, but importing the majority of food from the hinterlands and global regions. In fact, municipalities on almost every continent consume the smallest share of water—the water resources used to produce their food have footprints elsewhere.

The agricultural sector is often blamed as one of the most inefficient users of water (OECD 2016) but the reasons for this are more than agricultural practice. If we assume that just under half the world's population still lives in a rural setting—we became mostly urban dwellers in 2007 (Orsini et al. 2013)—and most are involved in subsistence agriculture or, to a lesser degree, large farm operations, then this helps frame the analysis. In the rural setting, raw water is cheap or freely available. When water is cheap, efficiency is generally low. Which has led to a problem of the commons. Generally, greater water efficiency tracks with higher economic values of water (FAO 2004). Conversely, some of the most efficient uses of water are in urban areas where water is more expensive. Therefore, appropriate valuing of water is important for wise use of water resources and could provide relief to water stress in the decades to come. Of course, this has its limits because water is a commons resource and should be equitable and accessible for vulnerable populations.

Based on the author's professional experience and literature review, data that supports evaluation of water allocation to uses within the cities are abundant; but data that teases out urban water use allocated to an "agriculture" category is greatly lacking. Typical use categories in municipal water resources planning include landscape, domestic, industrial, and commercial. A water use classification document assessing utilities in five US cities (Phoenix, Arizona; Tampa Bay, Florida; San Francisco Peninsula area, California; New York City, New York; and Austin, Texas) listed 38 categories of commercial, industrial, and institutional uses, and 7 categories of domestic uses. Only one of five water providers (Phoenix, Arizona) had a category that indicated water use specifically for urban agricultural production (WRF 2015). However, despite the lack of direct water use data, it is still possible to estimate water used for urban agriculture by back-calculation using specific yield data combined with estimates of urban food self-sufficiency. One such method is the life cycle assessment.

### 7.2.1 LCA a Valuable Tool

The life cycle assessment (LCA) method inventories all the associated inputs and outputs for a particular product or process, and then estimates impacts to resources, environmental quality, the economy, and society (EPA 2006). Here we find a helpful methodology that illuminates water use for urban agriculture. In LCA-speak, water consumption is referred to as "depletion," after taking any sources of "free" water (e.g., repurposed wastewater) into account. Processes that LCA inventories take into account for vegetable production are illustrated in Fig. 7.4.

There are numerous studies that inventory and assess impacts for the production of all manner of fresh vegetables, field crops, and animals (for example, Hayashi et al. 2006; Heller and Keoleian 2003; Gössling et al. 2011; Kulak et al. 2013; Mogensen et al. 2008). The literature illustrates that many variables influence the metric of interest, e.g., urban agriculture water use, for any particular crop. For example, Thoma (2018) surveyed water depletion for growing tomatoes under three conditions—open-field, plastic strings and stakes, and greenhouse—and found a range of specific water depletion to grow the tomatoes ranging from 0.062 to 0.013 $m^3$/kg tomato, or nearly 600% variation!

While there are different methods to aggregate all food produced in all ways for large geographic regions, that is beyond the scope of this chapter. With so much variation between climate, growing conditions, soil, irrigation, growing format, and farmer practices, to name only a handful, such aggregations are not very useful or applicable for small geographic areas like cities. In such cases, individual case studies are almost always needed to estimate yields and resource use more accurately.

To illustrate these points, I combine a range of water demand for vegetable crops presented in a self-sufficiency study in Cleveland, Ohio (Grewal and Grewal 2012) and from a number of international sources found in Stoessel et al. (2012) and US Department of Agriculture (USDA) Economic Research Service (ERS) (2010) to

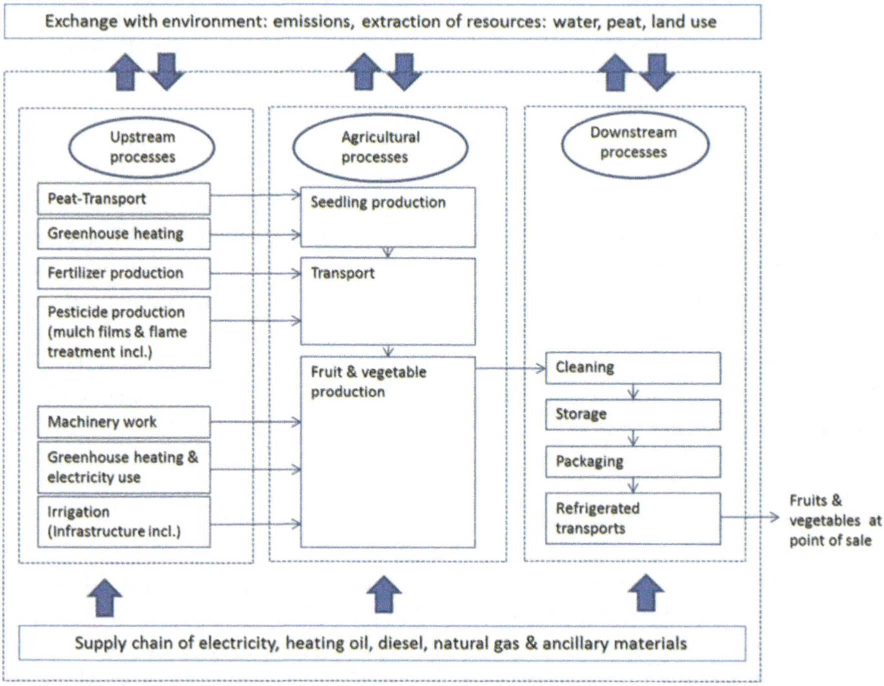


**Fig. 7.4** LCA components for vegetable production: system boundaries for cradle-to-gate fruit and vegetable production (Stoessel et al. 2012)

synthesize and estimate the amount of urban water needed to feed a population. Table 7.1 is intended as a planning tool to give order-of-magnitude precision to important policy questions, such as, "Can a city feed itself?" or "How much should a city allocate water for urban agriculture?"

### 7.2.2 Using Life Cycle Inventories to Guide Policy Goals

Promoting one crop over another so that water resources can be conserved, both intra-urban and across the urban-rural boundary, is an important policymaking outcome. A useful evaluation method for policymaking is the comparison of water stress caused by production of particular crops. Stoessel et al. conducted such a comparison for fresh vegetables consumed in Switzerland, where imported food has water footprints all over the globe (Stoessel et al. 2012). Using the Water Stress Index (WSI) of Pfister et al. (2009), Stoessel plotted an illuminating figure showing the WSI weighted by retail sales (indicating popularity) per kg of crop. As shown in Fig. 7.5, not only are some crops much more water intensive than others, there is also great variability in water use for each particular crop.

**Table 7.1** Estimated range of land and water requirements for fresh vegetable production

| Note | a | a | b | b | c | d | e | f | g |
|---|---|---|---|---|---|---|---|---|---|
| Product | Irrigation demand—low, $m^3$/kg | Irrigation demand—high, $m^3$/kg | Yield—commercial/rural, kg/$m^2$ | Yield—intensive urban, kg/$m^2$ | U.S. demand, kg/yr | Commercial/rural land Reqt per 1 M people, ha | Intensive urban land Reqt per 1 M people, ha | Low irrigation requirement per 1 M people, $m^3$/yr | High irrigation requirement per 1 M people, $m^3$/yr |
| Lettuce | 0.006 | 0.185 | 3.50 | 7.70 | 6.2 | 177 | 81 | 37,200 | 1,147,000 |
| Asparagus | 0.013 | 3.777 | 0.33 | 0.31 | 0.5 | 152 | 161 | 6500 | 1,888,500 |
| Spinach | 0.008 | 0.037 | 2.04 | 2.44 | 0.7 | 34 | 29 | 5600 | 25,900 |
| Broccoli | 0.012 | 0.073 | 1.77 | 8.95 | 2.7 | 153 | 30 | 32,400 | 197,100 |
| Cauliflower | 0.026 | 0.056 | 2.03 | 6.30 | 0.7 | 34 | 11 | 18,200 | 39,200 |
| Potato | 0.000 | 0.325 | 4.44 | 7.32 | 16.0 | 360 | 219 | – | 5,200,000 |
| Green bell pepper | 0.005 | 0.038 | 3.50 | 2.93 | 4.5 | 129 | 154 | 22,500 | 171,000 |
| Tomato | 0.002 | 0.106 | 6.31 | 12.45 | 8.4 | 133 | 67 | 16,800 | 890,400 |
| Cucumber | 0.008 | 0.161 | 1.72 | 15.62 | 3.1 | 180 | 20 | 24,800 | 499,100 |
| Leek, onion, carrot | 0.048 | 0.073 | 5.00 | 8.50 | 6.0 | 120 | 71 | 288,000 | 438,000 |

The table shows there is immense variability in yield and, therefore, water use. In Grewal's (2012) study, agricultural growing formats were differentiated, in order of increasing yield, into "conventional urban gardening," "commercial rural farming," "intensive urban gardening," and "hydroponic rooftop gardening." The highest yield was approximately 17 times the lowest yield. In conclusion, the authors state that yield is the key factor in determining food self-reliance. Further granularity into not only the growing format, but individual crops, further reveals variability in water use that can be helpful for choosing which crops to grow. This is discussed in the following section

*ha* hectare, *kg* kilogram, *m* meter, *yr* year, *M* million

[a]The irrigation demands shown represents typical non-local sources for a retailer in Switzerland. The countries represented include Belgium, Central America, Costa Rica, France, Greece, Holland, Hungary, Israel, Italy, Mexico, Morocco, Peru, Spain, and Switzerland; each with a unique growing method (Stoessel et al. 2012)

[b]The yields shown are U.S. large commercial farm production and for intensive, small-scale, urban production in Seattle, Washington, USA (Grewal and Grewal 2012)

[c]Retail availability is used as a surrogate for consumer demand, but not consumption. Consumer demand for particular vegetables varies substantially by country and people group (USDA ERS 2010)

[d]The land requirement was calculated using large commercial, rural farm yield figures

[e]The land requirement was calculated using intensive urban yield figures

[f]The irrigation requirement was calculated using the lowest water intensity figures from Column A

[g]The irrigation requirement was calculated using the highest water intensity figures from Column B

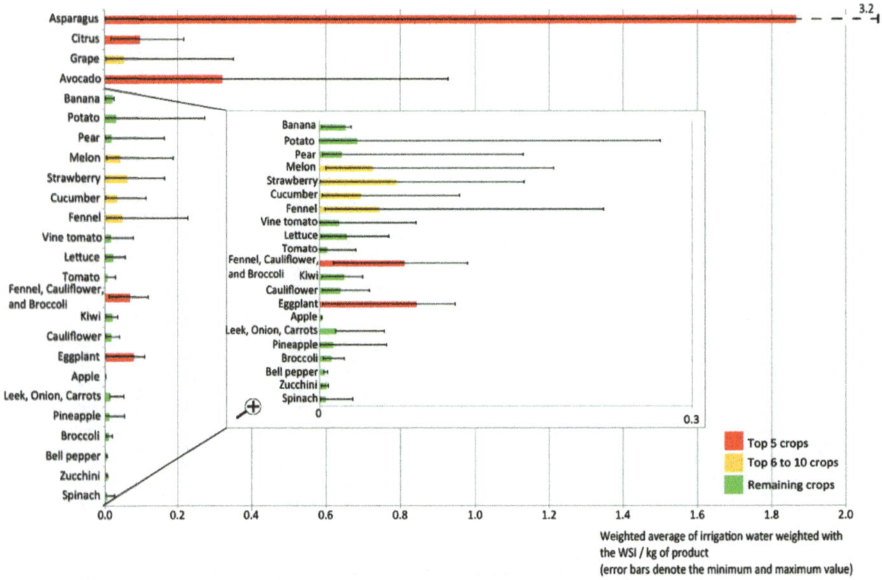


**Fig. 7.5** Water stress of selected fruits and vegetables weighted by consumption popularity sales-amount weighted water stress (irrigation water ($m^3$)·WSI) per kg (Stoessel et al. 2012)

## 7.3 Spectrum of Urban Agriculture in Practice

Urban agriculture has been characterized by closeness to markets, high competition for land, limited space, use of urban resources such as organic solid wastes and wastewater, a low degree of farmer organization, mainly perishable products, and a high degree of specialization (Van Veenhuizen 2011). Such conditions manifest as organic, ad-hoc, market-driven, and community-driven development over time and are often in response to lack of financial resources, healthy food, connection to nature, access to markets, jobs, and other metrics (Kloppenburg Jr et al. 2000). These metrics serve as guideposts for policy development.

### 7.3.1 Urban Agriculture as a Policy Outcome, or Not

In contrast to a general lack of intentional government policy related to urban agriculture, there are instances of urban agriculture that can be seen as the *result* of government policy. A most notable example is the organized efforts in Havana, Cuba, where nearly 100% of some vegetable crops has been provided by urban agriculture on only 12% of its land (Orsini et al. 2013). In other cities and regions, city form and massive urbanization have driven irrigated urban vegetable production. Urban agriculture in Accra, Ghana provides up to 90% of the city's vegetable demand. Urban-grown lettuce, for example, feeds about 250,000 people daily (Van Veenhuizen 2011). In West Africa, market oriented irrigated agriculture occupies between 20 and 650 ha in each major city, producing 60–100% of the locally consumed perishable vegetables (Drechsel and Keraita 2014).

These, and the examples in the previous sections, show evidence that urban agriculture can in fact feed entire urban populations with locally available water and land. Where urban areas are more challenged with both water and land, urban agriculture responds with innovation. The Milan Urban Food Policy Pact (MUFPP) consists of 167 cities representing an urban population of nearly half a billion people around the world (FAO 2018). One of the key MUFPP areas is "Food Production," including urban-rural linkages. This large stakeholder group aims to catalogue the benefits and impacts of urban agriculture to its stakeholders, from farmers to government to large agri-business. From multilateral efforts such as these, a policy framework has risen with which to make informed decisions.

The policy innovation in the urban agriculture area, resulting from these data and informed decisions, can affect water resources, water quality, markets, agricultural zoning, and building codes. Site-specific combinations of policy and technology hold the most promise to increase urban agriculture in places where it does not currently or minimally occurs. Such site-specific factors may include customary diet, how agriculture is regarded by socio-economic classes, availability of irrigation water, availability of markets, infrastructure and land, soil conditions, climate conditions, and zoning laws.

### 7.3.2 Innovation in Policy and Technology: An Effective Combination

Innovation in policy and technology go hand-in-hand with urban agriculture, one affecting the other. Technological innovation has driven up the incidence of urban agriculture and its yield through viable instances of aquaculture, hydroponics, aquaponics, zero-acreage farming (Thomaier et al. 2014), drip irrigation, rooftop gardening (Sanyé-Mengual et al. 2015), vertical farming (Despommier 2010), controlled-environment agriculture (Benke and Tomkins 2017), and conventional greenhouses. Policy can accommodate and incentivize use of these technologies, sometimes only available to businesses, investors, and large-scale interests. But policy can also encourage community self-sufficiency (resilience) in impoverished areas. Next, I describe two instances of agricultural technology and policy innovation whose principles could have a general global application.

#### 7.3.2.1 The Promise of Controlled-Environment Innovation

At 500 persons per square kilometer, Holland is a patchwork of urban centers interspersed primarily with farmland, such that there is urban and peri-urban land but almost no truly rural land. The way the Dutch not only feed themselves, but export food is truly remarkable. Holland is 1/270th the size of the USA, but exports more potatoes and onions than any other country in the world, and is the world's

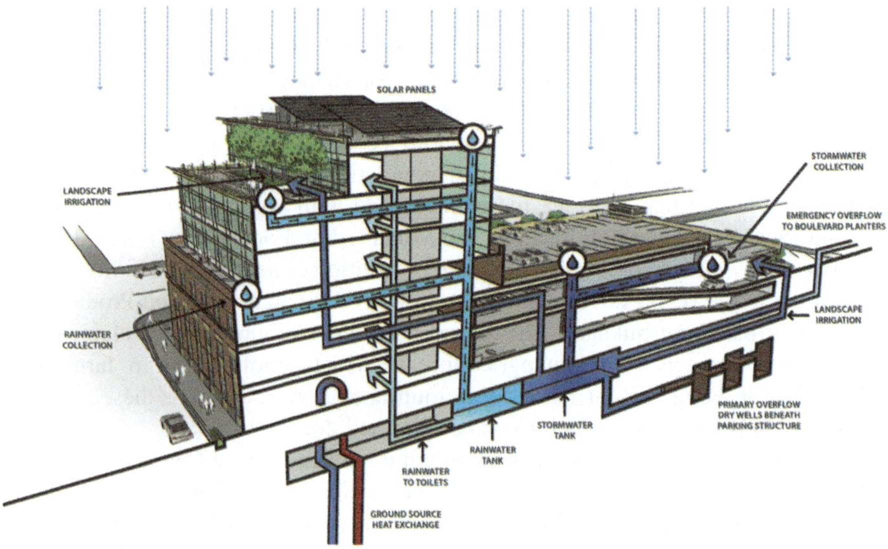


**Fig. 7.6** Building technologies needed for on-site irrigation: A building developer's guide to non-potable water collection, transport, and end uses. Urban agriculture would fit under the "Landscape Irrigation" category (William J. Worthen Foundation 2018)

second largest exporter in terms of value of horticultural vegetables (NG 2017). This is achieved through partnerships with foremost research institutions (Wageningen University, for example), clever symbioses with wastewater and waste heat and energy from powerplants and factories, integrated pest management, highly efficient lighting technologies, fine-tuned soil management, hydroponics, and greenhouses that comprise only 0.2% of the country's land mass. Inside the greenhouses, the Dutch have reduced water consumption by 90%, cut pesticide use almost entirely, and reduced fertilizer applications by 29%, all while increasing vegetable production by 28% (NG 2017). Outside the greenhouse, the Dutch have exported their knowledge and technology all over the world, especially to places that will experience the most urban growth, water stress, and food scarcity.

While recognizing the clear advances in agricultural yield, there is concern over the energy footprints of controlled-environment farming (Benke and Tomkins 2017; Al-Chalabi 2015; Clark and Tilman 2017). In order to be sustainable, and not trade one goal for the other (e.g., water for energy), these technologies often require adequate stormwater, wastewater, renewable energy, or waste heat to be viable. At the scale of individual buildings, numerous supporting utilities and infrastructure are required to use wastewater or stormwater for on-site irrigation of urban agriculture and landscaping (Figs. 7.6 and 7.7). Benke and Tomkins (2017) explored key performance indicators that weighed a host of advantages and disadvantages of conceptual urban agriculture technologies (Table 7.2). LCA is also useful to evaluate controlled-environment farming. Clark and Tilman (2017) compared produce grown in the open field with that grown in greenhouses and found mixed results in terms of

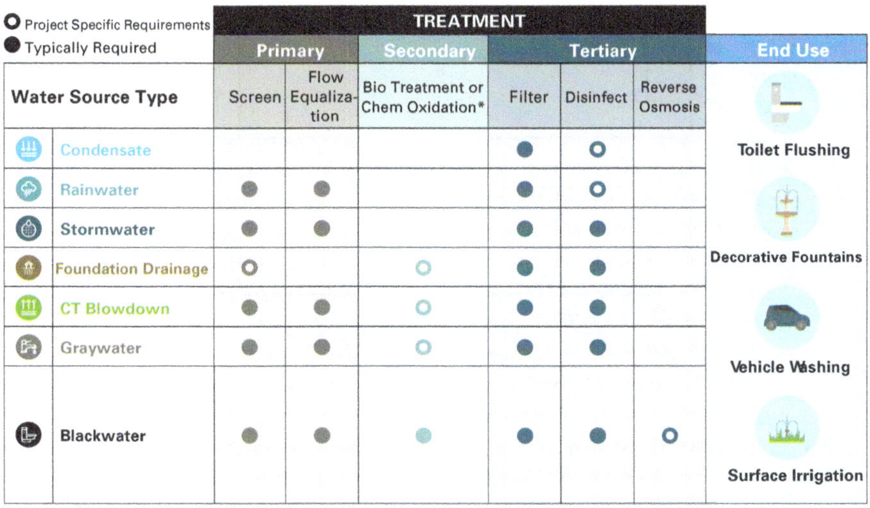


**Fig. 7.7** Wastewater treatment requirement for buildings from non-potable sources: A building developer's guide to non-potable water treatment for various end uses. Urban agriculture would fit under the "Surface Irrigation" category (William J. Worthen Foundation 2018)

**Table 7.2** Viability checklist for controlled-environment farming: Vertical farm—key performance indicators (KPIs) (Benke and Tomkins 2017)

| Key performance indicator | Satisfied | Partially satisfied | Not satisfied |
|---|---|---|---|
| Start-up costs | | | • |
| Energy consumption | | • | |
| Number of crop types | | • | |
| Production volume | | | • |
| Scaling-up issues | | • | |
| Venture capital | | • | |
| Skilled workforce for maintenance | | • | |
| Disruption to the rural sector | | • | |
| Transport savings | • | | |
| Clean, green, and gourmet food | • | | |

greenhouse gases, land use, eutrophication potential, acidification potential, and energy use (Fig. 7.8).

#### 7.3.2.2 The Promise of Low-Tech Innovation

Rwanda won its freedom from Belgium-administered United Nations' trusteeship on July 1, 1962. At that time, its capital city of Kigali had approximately 6000 residents (New Times 2011). Today, Kigali comprises well over one million people. Known to have one of the highest population densities in Africa at 6500 persons per square kilometer (World's Capital Cities 2019), its urban form is characterized by mostly low-rise commercial and informal residential housing. Kigali is not unlike other

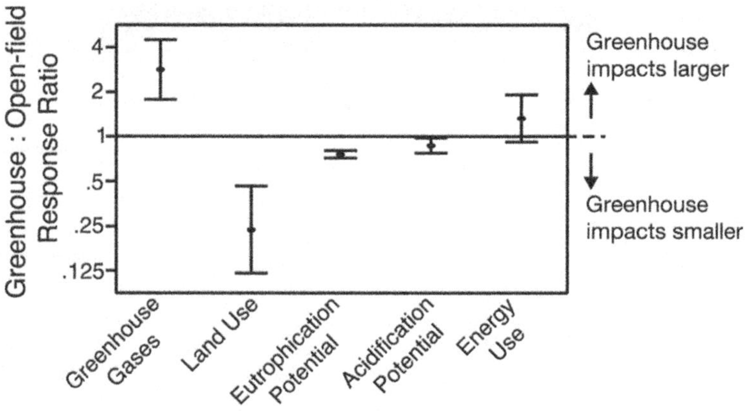


**Fig. 7.8** Comparison of environmental impacts of greenhouse production with open-field production: Response ratio of environmental impacts of greenhouse grown and open-field produce (Clark and Tilman 2017)

cities in the developing world, overcome by urban migration, facing limited food and water resources, and infrastructure stretched to the limit. In 2005, seeking to address the best and worst of urban growth, the city embarked on a conceptual urban master plan project that included existing, built-up areas extending to the outer city limits (City of Kigali 2008). The master plan process was not unlike that of the contemporary EcoDistrict protocol developed by EdoDistricts (Cassatella et al. 2018; DCDC 2015; EcoDistricts 2018), which aims at achieving multiple sustainability and infrastructure goals by defining boundaries, identifying needed zoning and deed reforms, including stakeholders, and identifying technologies and land uses to help achieve those goals. The EcoDistrict framework can be useful to identify the land use, zoning, and deed controls to provide consistent access to resources needed to implement technologies, such as co-location with other manufacturing processes and wastewater treatment plants.

Designers recognized the need to allow for future beneficial use of low-lying, fertile, agricultural areas dispersed throughout the city, identified in a detailed sub-watershed analysis (Fig. 7.9). The designers created a watershed land use pro-forma that was called an Environmental Treatment Zone (ETZ). Using gravity and, in some parts, existing land use patterns, the ETZ included all the components for organized solid and liquid waste recycling supporting small- to medium-sized businesses (Fig. 7.10). These became prime focus areas when urban agriculture, specifically, was incorporated into the Kigali Conceptual Master Plan (KCMP) in 2009, based on the FAO's recommendation that urban agriculture has the potential for commercial export of food, income generation, and increase a local, urban food supply (FAO 2008; Hamilton et al. 2014; Orsini et al. 2013; RUAF 2011; World Bank 2013).

**Fig. 7.9** Planning for urban agriculture taking advantage of urban watersheds: Each green dot represents the lowest elevation of a sub-watershed (not shown; beyond visible map boundary) that is suitable to manage and collect stormwater, wastewater, and compost other organic wastes in the proximity (City of Kigali 2008)

## 7.4 Conclusions and Recommendations

In this chapter, we have come to understand numerous aspects of water supply and use for food production in the global and local contexts. Water stress is common practically everywhere and so the question becomes whether a new form of use, water for urban agriculture, can be accommodated or provide more benefits than costs. Answers lie in many of the other chapters of this book, but also in this chapter,

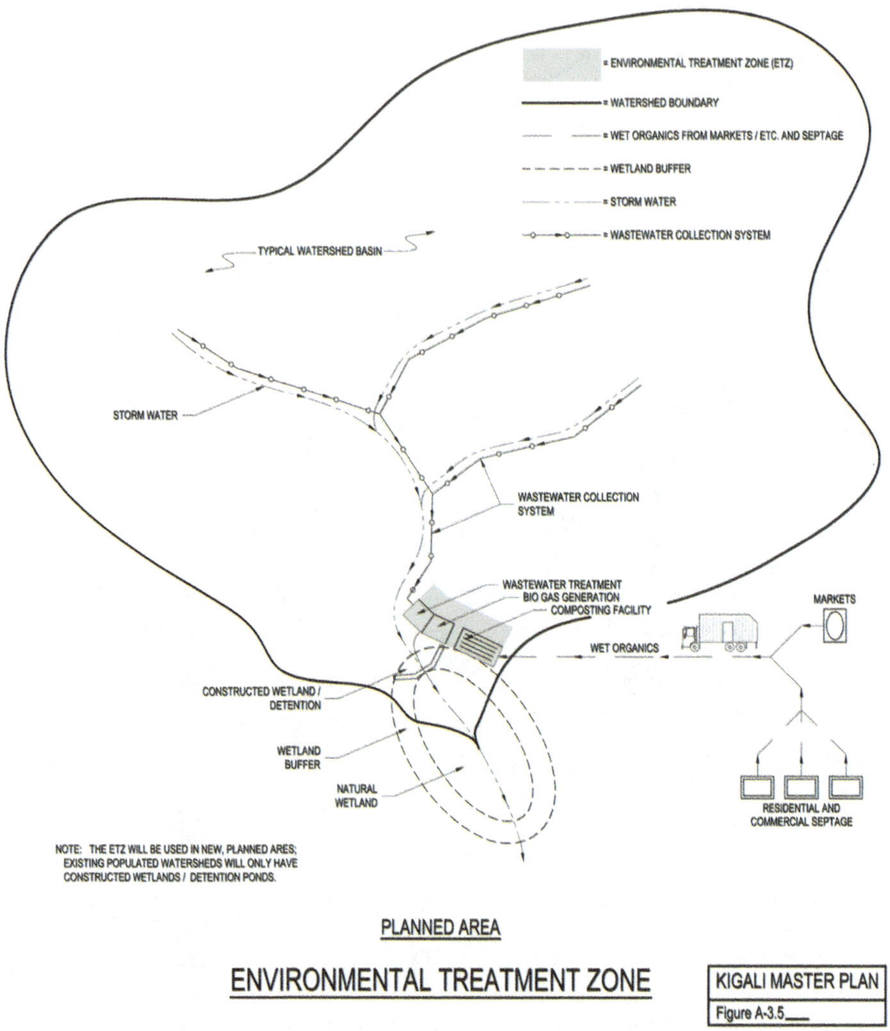


**Fig. 7.10** A watershed schematic plan that provides for urban agriculture: The pro-forma for an urban watershed shows flows of wastes and resources that can be used for urban agriculture. Urban agriculture would be co-located with the ETZ (City of Kigali 2008)

specifically, the way urban agriculture is irrigated. This involves water resources, water treatment, water conveyance, and production practices discussed in this chapter. Within the context of water for urban agriculture, I recommend actions to address the policy objects of feeding a hungry world, providing livelihoods, and increasing environmental sustainability. They are as follows:

1. Adopt frameworks that include end-user stakeholders. Are they poverty-stricken urban migrants, large commercial ventures, or building developers?
2. Increase research on life cycle assessment (including social life cycle assessment) on products, processes, and growing formats that are at least economically viable in order to inform further urban agriculture policy decisions.
3. Promote wastewater as a resource and educate on its safe use in all forms of growing urban food.
4. Establish a public health framework that not only takes into account morbidity and human health risk associated with wastewater for urban agriculture, but also adopts acceptable practices to mitigate risks rather than outright bans and prohibitions.

Planners must not neglect these engineering aspects of water for urban agriculture. The discussion of water use for new forms of urban agriculture requires related fields and cannot be discussed alone. We have seen that land use, social equity, growing technologies, and engineering issues, among others, are each necessary but insufficient for successful instances of urban agriculture.

## References


Al-Chalabi M (2015) Vertical farming: skyscraper sustainability? Sustain Cities Soc 18:74–77

Benke K, Tomkins B (2017) Future food-production systems: vertical farming and controlled-environment agriculture. Sustain Sci Pract Policy 13(1):13–26

Buechler S et al (2006) Wastewater use for urban and peri-urban agriculture. RUAF Foundation, Holland

Cassatella C et al (2018) Planning for the global urban agenda: shaping ecodistricts in Tokyo suburbs. Politecnico di Torino, Italy. http://www.biblio.polito.it/risorse/catalogo/isbn_del_politecnico_di_torino

City of Kigali (2008) Kigali City master plan report. http://www.masterplan2013.kigalicity.gov.rw/Downloads/

Clark M, Tilman D (2017) Comparative analysis of environmental impacts of agricultural production systems, agricultural input efficiency, and food choice. Environ Res Lett 12:064016

Dermody BJ et al (2014) A virtual water network of the Roman world. Hydrol Earth Syst Sci 18: 5025–5040

Despommier D (2010) The vertical farm: feeding the world in the 21st century. Thomas Dunne Books, New York, p 321

Detroit Collaborative Design Center (DCDC) (2015) Eastern market green stormwater management strategy ecodistrict best practices. http://www.dcdc-udm.org/projects/strategies/EMStormwater/

Drechsel P, Keraita B (eds) (2014) Irrigated urban vegetable production in Ghana – characteristics, benefits and risk mitigation, 2nd edn. International Water Management Institute, Colombo, Sri Lanka. http://www.iwmi.cgiar.org/Publications/Books/PDF/irrigated_urban_vegetable_production_in_ghana.pdf

EcoDistricts (2018) EcoDistrict Protocol version 1.3. https://ecodistricts.org/protocol/implementation-phases/

FAO (2008) Urban agriculture for sustainable poverty alleviation and food security. Position paper on urban and peri-urban agriculture, FAO Food for the cities multi-disciplinary group, October

FAO (2014) Aquastat: irrigated crops. http://www.fao.org/nr/water/aquastat/infographics/Irrigated_eng.pdf

FAO (2015) Aquastat: water withdrawal ratios by continent. http://www.fao.org/nr/water/aquastat/water_use/index.stm
FAO (2018) The role of cities in the transformation of food systems: sharing lessons from Milan pact cities. FAO, Rome
FAO (2019) Land, water and the SDGs. http://www.fao.org/land-water/overview/land-water-sdg/en/
Gössling S et al (2011) Food management in tourism - reducing tourism's carbon 'foodprint'. Tour Manag 32:534–543
Grewal S, Grewal PS (2012) Can cities become self-reliant in food? Cities 29:1–11
Hamilton A et al (2014) Give peas a chance? Urban agriculture in developing countries. A review. Agron Sustain Dev 34:45–73
Hayashi K et al (2006) Life cycle assessment of agricultural production systems. Proceedings of the international seminar on technology development for good agricultural practice (GAP) in Asia and Oceania
Heller MC, Keoleian GA (2003) Assessing the sustainability of the US food system. Agric Syst 76: 1007–1041
Jiménez B, Asano T (2001) Social, political, and scientific dilemmas for massive wastewater reuse in the world. In: Davis CK, McGinn RE (eds) Navigating rough waters. American Water Works Association, Washington, DC, pp 139–154
Kloppenburg J Jr et al (2000) Tasting food, tasting sustainability: defining the attributes of an alternative food system with competent, ordinary people. Hum Organ 59(2):177–186
Kulak M et al (2013) Reducing greenhouse gas emissions with urban agriculture: a life cycle assessment perspective. Landsc Urban Plan 111:68–78
Malik OA et al (2015) A global indicator of wastewater treatment to inform the Sustainable Development Goals (SDGs). Environ Sci Policy 48:172–185
McDonald R et al (2014) Water on an urban planet: urbanization and the reach of urban water infrastructure. Glob Environ Chang 27:96–105
Mogensen L et al (2008) Chapter 5 - Life cycle assessment across the food supply chain. In: Baldwin CJ (ed) Sustainability in the food industry. Wiley-Blackwell, Hoboken, NJ
National Geographic Society (NG) (2017) A tiny country feeds the world. National Geographic Magazine. https://www.nationalgeographic.com/magazine/2017/09/holland-agriculture-sustainable-farming/
New Times (2011) The history of city of Kigali. https://www.newtimes.co.rw/section/read/97496
Organisation for Economic Co-operation and Development (OECD) (2016) Agriculture and water. OECD Meeting of Agriculture Ministers, Background Note 5. http://www.oecd.org/agriculture/ministerial/
Orsini F et al (2013) Urban agriculture in the developing world: a review. Agron Sustain Dev 33: 695–720
Pfister S et al (2009) Assessing the environmental impacts of freshwater consumption in LCA. Environ Sci Technol 43(11):4098–4104
Resource Centres on Urban Agriculture and Food Security (RUAF) Foundation (2011) Urban agriculture: a sustainable solution to alleviating urban poverty, addressing the food crisis, and adapting to climate change. Case studies of the cities of Accra, Nairobi, Lima, and Bangalore undertaken by the RUAF Foundation, with financial support from the World Bank (project: CN# 43). Synthesis Report December
Sanyé-Mengual E et al (2015) Techniques and crops for efficient rooftop gardens in Bologna, Italy. Agron Sustain Dev 35:1477–1488
Stoessel F et al (2012) Life cycle inventory and carbon and water foodprint of fruits and vegetables: application to a Swiss retailer. Environ Sci Technol 46:3253–3262. https://doi.org/10.1021/es2030577
Thoma G (2018) Protected production of fruits and vegetables for nutrition security in urban and peri-urban environments. Presentation given at World Bank, July 18, 2018, Washington, DC

Thomaier S et al (2014) Farming in and on urban buildings: present practice and specific novelties of zero-acreage farming (ZFarming). Renew Agric Food Syst 30(1):43–54
U.S. Department of Agriculture (USDA) Economic Research Service (ERS). 2010. Vegetables and melons yearbook. https://usda.library.cornell.edu/concern/publications/3t945q78b?locale=en
U.S. Environmental Protection Agency (EPA) (2006) Life cycle assessment: principles and practice. EPA/600/R-06/060, May
UN World Health Organisation (WHO) (2006) WHO guidelines for the safe use of wastewater, excreta and greywater, Wastewater use in agriculture, vol II, WHO, Geneva
United Nations (UN) Food and Agriculture Organisation (FAO) (2004) In: Turner K, Georgiou S, Clark R, Brouwer R, Burke J (eds) FAO Water Reports 27. Economic valuation of water resources in agriculture - from the sectoral to a functional perspective of natural resource management. FAO, Rome. http://www.fao.org/docrep/007/y5582e/y5582e00.htm#Contents
Van der Hoek W et al (2002) Urban wastewater: a valuable resource for agriculture: a case study from Haroonabad, Pakistan. Research report 63. International Water Management Institute, Colombo, Sri Lanka
Van Veenhuizen R (2011) Productive use of urban water for urban agriculture. Final report WP 5.2. RUAF Foundation, Holland
Water Research Foundation (WRF) (2015) Methodology for evaluating water use in the commercial, institutional, and industrial sectors. Web report #4375. WRF, Denver, CO. https://www.waterrf.org/research/projects/methodology-evaluating-water-use-commercial-institutional-and-industrial-sectors
William J. Worthen Foundation (2018) Onsite non-potable water reuse practice guide. https://www.pankowfoundation.org/wp-content/uploads/WaterReuseGuide_FINAL.pdf
World Bank (2013) Urban agriculture: findings from four city case studies. Urban development series knowledge papers 80759. Urban Development & Resilience Unit. No. 18, July
World's Capital Cities (2019) Capital facts for Kigali, Rwanda. https://www.worldscapitalcities.com/capital-facts-for-kigali-rwanda/

**Stephen Fisher** is a licensed professional civil engineer with 30 years of experience in four main areas: environmental, infrastructure planning, land development, and international community development. Mr. Fisher has performed numerous studies and plans involving water resources, water facilities, community participation, erosion control, watershed management, and land use planning. His urban sustainability programming and design experience includes food systems and urban agriculture, systems theory, ethics, brownfields reuse, ecological footprinting, and resilience. Mr. Fisher participated in the creation of a state regulation of water reclamation and reuse. His international experience includes Argentina, Dominican Republic, Eritrea, Haiti, Honduras, Mexico, Peru, Rwanda, and Zimbabwe.

# 8 Evaluating Wastewater Reuse in Urban Agriculture from a Systems Perspective: Focus on Linkages with Water, Energy, and Health

Leslie Miller-Robbie and Anu Ramaswami

**Abstract**

When cities grow rapidly, they often displace surrounding agricultural lands and appropriate water previously used for irrigation. Sanitation infrastructure may struggle to contain flows and urban agriculture tends to move downstream of urban/riverine discharges. Irrigation of urban agriculture with domestic wastewater provides an opportunity for capturing valuable nutrients and water prior to release into nearby waterbodies. Cities invest capital and energy resources in wastewater treatment infrastructure in efforts to provide environmental and health benefits. Complex interactions in this food-energy-water-health (FEW-Health) nexus are location-specific; therefore, multiple impacts are explored in a site study in Hyderabad, India. Varying qualities of irrigation water (treated wastewater, untreated surface water, and groundwater) were evaluated, and the following impacts were quantified: water use, energy use and GHG emissions, nutrient uptake, and crop pathogen quality. Treatment plus reuse is shown to provide GHG mitigation when compared to the untreated case; however, land use needs are high to extract nutrients from dilute effluents. Also, harvesting practices and environmental factors contribute to crop pathogen content. Urban agriculture together with wastewater treatment and reuse is beneficial, but system-wide tradeoffs are complex. This chapter reveals key environmental, physical, and behavioral factors that constrain achievable benefits at the urban FEW-health nexus.

L. Miller-Robbie (✉)
University of Colorado Denver, Denver, CO, USA

A. Ramaswami
Princeton University, Princeton, NJ, USA



J. A. Diehl, H. Kaur (eds.), *New Forms of Urban Agriculture: An Urban Ecology Perspective*, https://doi.org/10.1007/978-981-16-3738-4_8

**Keywords**

Urban agriculture · Wastewater reuse · Wastewater treatment · Life-cycle · FEW-Health Nexus

## 8.1 Introduction

Food, energy, and water (FEW) systems are vital in providing materials to city residents. Environmental conditions both inside and outside city limits can affect the availability of FEW supplies, and the urban demand for FEW supplies impacts the local and surrounding environment (Ramaswami et al. 2017). Although cities generally cannot provision all FEW materials from within their boundaries, urban agriculture is one small piece of the larger, transboundary, urban food system that is local. Urban agriculture provides fresh produce to substitute for food grown elsewhere.

In many low- and middle-income cities, a large proportion of domestic wastewater is not treated, and nearby rivers receive the contaminated water (WWAP 2017). Urban agriculture can utilize nutrient-rich domestic wastewater, either treated or untreated including blackwater and greywater (Drechsel et al. 2010), as a source of irrigation water. In this way, water and nutrients are reutilized, and low-income urban households have greater access to fresh/healthy foods (Hanjra et al. 2015; Makoni et al. 2016). The fresh produce provided by urban agriculture is a valuable benefit in addressing food insecurity and undernourishment (Boyer and Ramaswami 2017). Urban agriculture also provides land treatment of wastewater, which affects emissions of greenhouse gases (GHGs). In cities where wastewater collection systems are not complete, wastewater treatment infrastructure is currently being implemented, which is expected to affect system-wide energy and GHG emissions; therefore, understanding linkages across sectors in the food, energy, water, and health (FEW-health) nexus in the context of such cities is important (Ramaswami et al. 2018). In high-income countries, wastewater effluent is not used in urban agriculture (biosolids are applied, but not effluent directly), so this situation does not arise.

As city populations grow, urban metabolism of FEW materials (resource consumption, energy use, and waste generation) also increases (Kennedy et al. 2007; Wolman 1965). Often in low-income nations, cities displace surrounding agricultural land and irrigation water, forcing agriculture downstream of urban riverine/wastewater discharges (Van Rooijen et al. 2005; Dutta 2012). Wastewater is a nutrient-rich resource that is valuable to farmers who are seeking a widely available and consistent source of irrigation water for their crops. Wastewater reuse in urban agriculture is not new or rare; in fact, it stems from ancient Greece, and today an estimated 200 million farmers irrigate at least 20 million hectares with raw or partially treated wastewater (Raschid-Sally and Jayakody 2008). This number accounts for approximately 8% of total worldwide irrigated land (263 million hectares in 1996), of which two-thirds lies in Asia (Howell 2001), and supports a population of farmers that represents approximately 15% of the total amount of people economically active in agriculture worldwide (FAOSTAT 2009). Wastewater reuse is employed to irrigate a variety of vegetable, fruit, and herb crops in cities in the Americas, Africa, and Asia (van der Hoek 2004). Because wastewater reuse in urban agriculture is widespread and legislation is difficult, the question is no longer if wastewater should be used for irrigation, but how it can be made more sustainable and safer (van Rooijen et al. 2005; Scott et al. 2004).

## 8.2 Advantages of Wastewater Reuse for Urban Agriculture

This practice has numerous advantages:

- Conservation of water: Water reused for urban agriculture means that less freshwater/groundwater is needed, which is important given increasing water scarcity (van der Hoek et al. 2002).
- Nutrient recycling: Wastewater contains nutrients, leading many farmers to prefer wastewater for irrigation because it is thought to increase productivity (Qadir et al. 2007).
- Avoided fertilizer (Asano 1998): Nutrients (nitrogen, phosphorus, potassium, and organic carbon) in wastewater could save the farmers money and have the indirect impact of saving energy and GHGs (Pitterle and Ramaswami 2009).
- Land treatment of wastewater: Without other treatment options, land application may provide some decrease in surface freshwater contamination (Raschid-Sally and Jayakody 2008) and GHG emission reductions.
- Spatial and temporal accessibility of irrigation water: Oftentimes, farmers have better access to wastewater as a source of irrigation water because it is in constant supply in urban and peri-urban areas, even in the dry season. This is because cities draw municipal drinking water from outside their boundaries and discharge it as wastewater after use (Qadir et al. 2007).
- Decreased need for expensive refrigerated transport or storage facilities: This is most valued in low-income countries with hot climates (Qadir et al. 2008).
- Nutrition: Urban agriculture, facilitated by wastewater reuse in many rapidly-urbanizing cities, provides both farmers and consumers with a local, fresh supply of vegetables (Qadir et al. 2008).
- Better livelihoods: Wastewater is an inexpensive source of water and nutrients allowing farming families to grow high-value and high-demand crops like vegetables (Kilelu 2004), which generates more income and raises living standards, including indirect benefits like education (Raschid-Sally and Jayakody 2008).

For these reasons, wastewater is considered a valuable resource for many. The articles/reports above are largely qualitative studies. Many of these benefits, along with savings in energy, greenhouse gas emissions, and water, are not well-understood quantitatively.

## 8.3 Disadvantages of Wastewater Reuse for Urban Agriculture

While there are many advantages, the practice of wastewater reuse in urban agriculture also poses public health and environmental problems as water, soil, and crops become increasingly contaminated.

- Contaminants: Wastewater contains a variety of pollutants including salts, metals, metalloids, pathogens, residual drugs, organic compounds, endocrine disruptor compounds, and active residues of personal care products (Qadir et al. 2007). Pathogens associated with wastewater irrigation include: hookworm, roundworm (*Ascaris lumbricoides*), *E. coli*, giardia (*Giardia lamblia*), hepatitis A virus, typhoid (*Salmonella typhi*), and cholera (*Vibrio cholerae*).
- Human health: Both acute and chronic diseases can result from exposure to contaminants in wastewater. The main threat to human health in the short term is pathogens, specifically intestinal nematode infections (Ensink et al. 2008).
- Soil and crop quality: Heavy metals and salts in wastewater adversely affect soil quality (Ganjegunte et al. 2018; Abd-Elwahed Mohammed 2018). Crop production is also hindered by high levels of heavy metals and soil salinity (Morugán-Coronado et al. 2011; Shahid et al. 2015).

Farmers in low-income countries often use water from a polluted stream, diluted wastewater, or untreated sewage directly on crops. Wastewater from any source is seldom treated before being applied to crops (Qadir et al. 2007).

## 8.4 Wastewater Treatment Plants for Water Reuse for Urban Agriculture

Domestic wastewater treatment plants (WWTPs) are large, centralized facilities that collect wastewater via piped systems that are connected to homes and businesses throughout a city. WWTPs utilize a variety of physical, chemical, and biological processes to remove contaminants from wastewater. They generally release the cleaner effluent water into a nearby surface water body. WWTPs are effective in removing pathogens and other harmful substances from water and have been shown to decrease health risks (Asano 1998). Rapidly-urbanizing cities that lack adequate collection and WWTP infrastructure face a large proportion of their sewage being released directly to the environment; therefore, they are implementing WWTP infrastructure to address this need for treatment of sewage-polluted water. With this infrastructure development, municipal energy use is expected to increase because WWTPs are energy intensive (Miller et al. 2013). However, energy investments are expected to offer various benefits in terms of pathogen reduction and may help in more sustainable wastewater reuse for agriculture. Also, overall reductions in carbon- and nitrogen-related GHG emissions may be achieved due to WWTP processes removing them from water, and via subsequent application of effluent to farmlands.

In this research, a systems approach was taken to explore linkages across sectors and outcomes in the FEW-Health nexus. Based on the above review, there are multiple and conflicting impacts: GHG emissions (energy- and non-energy related), economic benefits to farmer (food production), water reuse (water savings), monetary cost (infrastructure), and health benefits to society (pathogen risk reduction in food). In order to quantify these impacts, this chapter evaluates tradeoffs for three farm sites in a case study, irrigated by differing sources of water: groundwater, treated effluent from a WWTP, and untreated surface water representative of the sewage-contaminated riverine system.

## 8.5 Case Study in Hyderabad, India

Many location-specific factors affect the tradeoffs between GHG emissions, infrastructure costs, food production, pathogen risk reduction, and water savings; therefore, a case study approach was necessary. Hyderabad, India was chosen for the following reasons: centralized WWTP infrastructure is newly implemented (secondary treatment within the last 15 years), wastewater contamination of surface water is ubiquitous, and wastewater-polluted water is reused for urban agriculture.

For Hyderabad, 80% of the water supply is released as sewage (Ramachandraiah and Vedakumar 2007). According to a Ministry of Urban Development Report (2010), 40% of the produced wastewater in Hyderabad is collected and treated before discharge into the Musi River, which runs through the center of Hyderabad. This leaves an average of 175 million gallons of untreated wastewater entering the riverine system daily. For most of the year, which is dry season, the Musi River would not flow without the input of sewage water (van Rooijen et al. 2005; Ramachandraiah and Vedakumar 2007).

Downstream of Hyderabad, the Musi River is used extensively for irrigation, with nearly 40,000 hectares of farmland irrigated from the river (Hamilton et al. 2007). This has resulted in severe groundwater pollution (Foster et al. 2003) and an overall long-term decline in the productivity of untreated wastewater-irrigated lands by more than 50% (Devi et al. 2009). A few scientists have studied wastewater reuse in Hyderabad and the effect on the environment and the people (Gopal 2004; Srinivasan and Reddy 2009). Others have studied the role of Hyderabad's water supply network and sewage network in urban recharge of groundwater (Wakode et al. 2018), and the stresses on already-scarce surface and groundwater sources due to growing competition from the agriculture and urban-industrial sectors (van Rooijen et al. 2009; Celio et al. 2010; Venot et al. 2010a, b). The International Water Management Institute (IWMI) has pioneered much of the work in Hyderabad and throughout the world (Devi et al. 2009; Buechler and Devi 2003; Jacobi 2009; Amerasinghe et al. 2013). The Resource Centres on Urban Agriculture and Food Security (RUAF) are also active in Hyderabad and globally, with the primary aim to promote and institutionalize urban agriculture processes in cities (RUAF 2010).

There were four operating WWTPs in Hyderabad at the time of this case study, collecting and treating water in the south-east area of the city (Fig. 8.1a). The

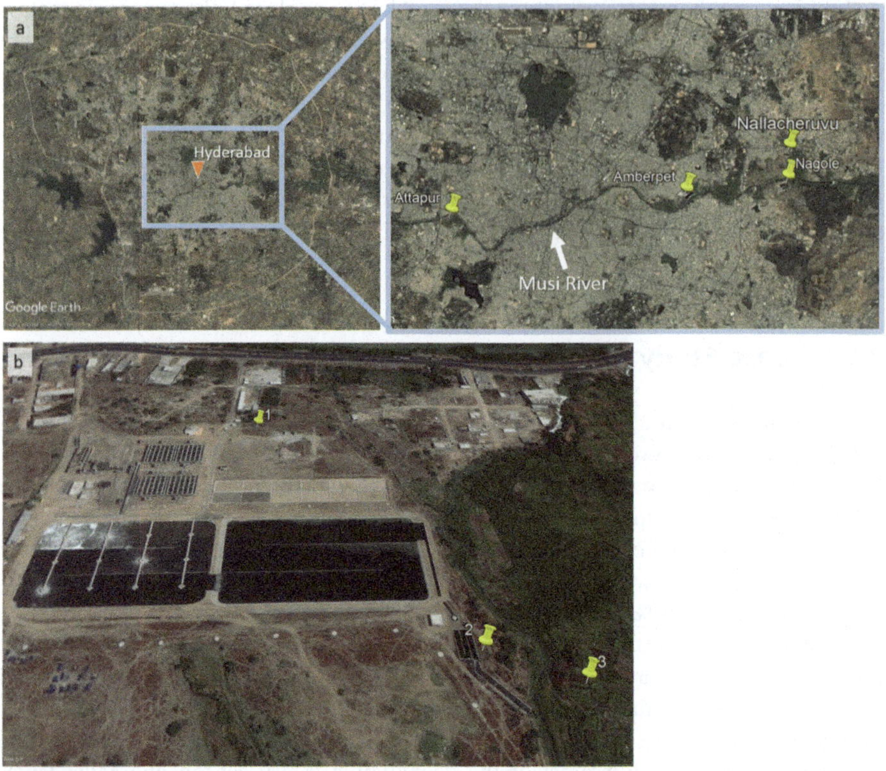


**Fig. 8.1** (**a**) Aerial view of Hyderabad, India showing the location of N-WWTP (Nallacheruvu) and the other three WWTPs (Attapur, Amberpet, and Nagole) near the Musi River, which flows from west to east (left to right); (**b**) Aerial view of N-WWTP showing co-location of urban agriculture plots (each 12 $m^2$) irrigated with: (1) groundwater from 50 ft deep; (2) N-WWTP effluent; (3) untreated surface water located on the other side of a stream from the WWTP effluent (retrieved from Google Earth Pro for years 2010 and 2011)

building of the Nallacheruvu WWTP (N-WWTP) in 2007 displaced urban farmers that had been farming in the area for up to 40 years (McCartney et al. 2008). Because the farmers used surface water to irrigate their crops, the area has a long history of wastewater contamination in both soil and groundwater. Today, adjacent to the N-WWTP, farmers grow crops such as spinach, coriander, mint, chilies, papaya, amaranth, fenugreek, fennel, and others.

The farming site at Nallacheruvu (Fig. 8.1b) was chosen for the following reasons: (1) its co-location of WWTP and urban agriculture, (2) ready access to three different qualities of water (groundwater, treated wastewater, and untreated surface water), (3) the availability of an experienced farmer, and (4) permission from the Hyderabad Metropolitan Water Supply and Sewerage Board for use of the study site and willingness to share data for N-WWTP. This field study took place during the dry season from March to May 2010, when water levels were at their lowest and stormwater would not dilute irrigation sources. Initial testing was done to choose plots that were similar in soil characteristics (physical texture and nutrient content) and distance, orientation, and slope to the nearby stream. The intent was to make all attributes between plots as similar as possible, with the exception of irrigation water quality. For the site study, the following parameters were measured during irrigation events throughout one crop growth cycle: irrigation water quality (pH, electrical conductivity (EC), total suspended solids (TSS), biochemical oxygen demand (BOD), chemical oxygen demand (COD), total organic carbon (TOC), nitrogen (N), phosphorus (P), potassium (K), *E. coli*, total coliform, Ascaris ova, Hookworm ova), irrigation water quantity (volume), soil quality (pH, EC, TOC, N, P, K, *E. coli*, total coliform, Ascaris ova, Hookworm ova), soil water quality (N, P, K, *E. coli*, total coliform), crop quality (N, P, K, *E. coli*, total coliform, Ascaris ova, Hookworm ova), and crop quantity (harvested bunches).

## 8.6 System-Wide Energy and Greenhouse Gas Impacts

System-wide energy and GHG emissions were evaluated for nearby streams, throughout the N-WWTP (Miller-Robbie et al. 2013), and for irrigating urban agriculture (Miller-Robbie et al. 2017). The values described below in the text are in terms of mg $CO_2e$ per liter water as opposed to metric tonnes $CO_2e$ per year (Fig. 8.2), as both are useful; the flow rate from March 2009 to March 2010 was 6570 million liters per year (MLY).

### 8.6.1 Untreated Wastewater in Streams

Uncontrolled release of untreated wastewater into streams results in the release of methane ($CH_4$) and nitrous oxide ($N_2O$), both potent greenhouse gases. Methane-related GHG emissions from wastewater were estimated using IPCC methods as the product of the maximum $CH_4$ producing capacity for domestic wastewater (0.25 kg $CH_4$ per kg COD (as measured via lab testing of water)) and a methane correction factor that was applied to represent the anoxic status of the receiving water body (Miller-Robbie et al. 2017; IPCC 2006). The estimation of $N_2O$ emissions from rivers was based on a meta-analysis of several stream $N_2O$ field studies (Beaulieu et al. 2011), which estimated 0.0075 kg $N_2O$ per kg dissolved inorganic nitrogen (DIN) (measured via lab testing of ammonia, nitrate, and nitrite) discharged to rivers is converted via denitrification and nitrification. Untreated wastewater contained an average of 514 mg/L COD and 84 mg/L DIN, resulting in 643 mg $CO_2e$/L attributed to $CH_4$, and 187 mg $CO_2e$/L attributed to $N_2O$ emissions (Miller-Robbie et al. 2017), for a total of 830 mg $CO_2e$/L untreated wastewater.

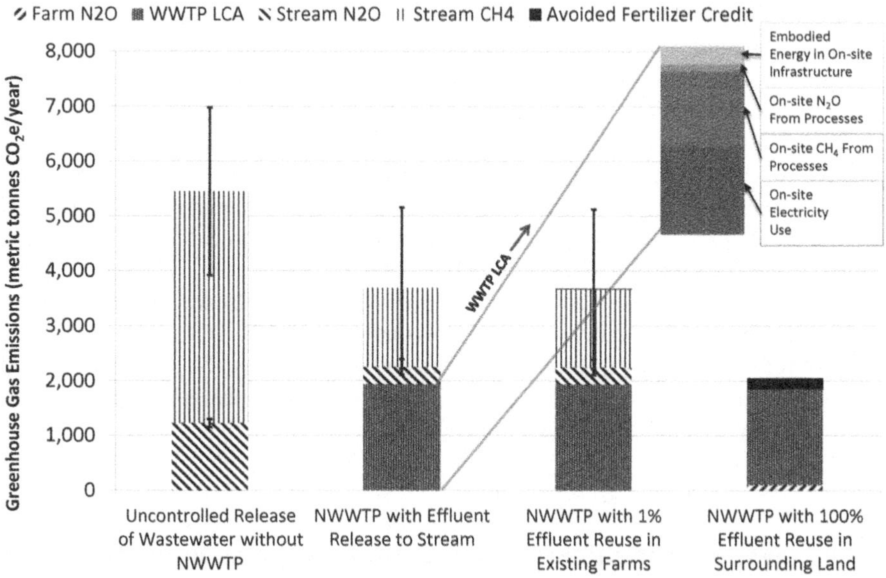


**Fig. 8.2** Results for the combination of wastewater treatment and reuse in agriculture, comparing the GHG emission impact from releasing untreated wastewater to various interventions. Error bars are based on the standard deviation in the concentration of COD or DIN in water. March 2009 to March 2010 flow rate of 6570 MLY (Miller-Robbie et al. 2017)

### 8.6.2 Wastewater Treatment Plant with Effluent Release to Stream

WWTP processes can be resource-intensive in terms of energy use and energy-related GHGs, and direct GHG emissions from the water surface. A life-cycle assessment (LCA) was employed to quantify energy consumed and GHGs emitted to achieve water quality improvements with WWTP infrastructure (Miller-Robbie et al. 2013). The four on-site components included were end-use energy in WWTP operations, embodied energy of infrastructure, process emissions of $CH_4$, and process emissions of $N_2O$. Total life-cycle GHG emissions were calculated as 295 mg $CO_2e$/L treated wastewater.

When treated effluent was released to the stream, it contained an average of 175 mg/L COD and 21 mg/L DIN (both measured in lab tests), resulting in 219 mg $CO_2e$/L attributed to $CH_4$, and 47 mg $CO_2e$/L attributed to $N_2O$ emissions (Miller-Robbie et al. 2017), for total life-cycle GHG emissions of 561 mg $CO_2e$/L treated wastewater. When compared to the emissions from untreated wastewater in streams in the previous section, a reduction of about 32% was estimated; the majority was due to the reduction in COD (and $CH_4$) and DIN (and $N_2O$) by WWTP operations. Contrary to expectations that the addition of a WWTP may increase system-wide GHG emissions, this study found that investing in energy and GHG emissions actually reduced overall GHG emissions because significant $CH_4$ and $N_2O$ were generated from untreated wastewater.

### 8.6.3 Wastewater Reuse for Urban Agriculture

GHG emissions from urban agriculture irrigated with treated wastewater were evaluated using the DAYCENT model, developed by the Natural Resource Ecology Laboratory at Colorado State University. DAYCENT is well-documented and widely used to estimate GHG emissions from cropped fields, usually with major crops such as corn, soybean, wheat, alfalfa, and cotton in the USA (Del Grosso et al. 2005, 2009; Jarecki et al. 2007; USEPA 2011). This study utilized DAYCENT for wastewater irrigation of vegetables in the context of India. The DAYCENT model utilizes multiple parameters for input data: local weather, historical data on land use, physical and chemical soil characteristics, irrigation events, crop characteristics, nitrogen, phosphorus, and organic matter addition events, carbon/nitrogen ratio, and relative concentrations of nitrogen species. An $N_2O$ emission factor was the model result of interest because it is the only GHG produced from agriculture under non-flooded conditions; the aerobic environment of agriculture does not facilitate COD (or BOD) conversion to $CH_4$, so $CH_4$ is negligible in this case. Based on the treated effluent plot, the DAYCENT model estimated an emission factor of 0.00070 $gN_2O$-N flux/g DIN applied to agriculture—about tenfold less than the river emission factor of 0.0075 $gN_2O$-N/g DIN (Beaulieu et al. 2011). If all of the treated wastewater was reused for irrigation, the emissions would be only 23 mg $CO_2e$/L, attributed to $N_2O$ emissions from cropped fields (Miller-Robbie et al. 2017), for total life-cycle GHG emissions of 318 mg $CO_2e$/L treated wastewater. Thus, in general, the DAYCENT model shows that urban agriculture would be effective in further reducing the production of GHGs as compared to the release of treated wastewater to the stream. This is an important and counter-intuitive result which indicates that both water and GHG benefits can arise due to applying WWTP-treated wastewater to urban agriculture.

## 8.7 Practical Constraints of Treated Wastewater Reuse in Urban Agriculture

WWTPs are commonly placed at a low elevation near a river at the outflow from a city. Therefore, the potential to irrigate urban agriculture with treated wastewater is limited by terrain, in the absence of additional piping and pumping infrastructure. Approximately 562,000 $m^2$ of available land is adjacent to the flow between the outlet of N-WWTP boundary and inflow to the Musi River; however, farmers employ gravity-driven irrigation with surface water and the actual land under farming that is readily gravity-fed from the effluent channel was estimated to be only 1% (approximately 5500 $m^2$). While nutrients in the water suffice for the crops (according to lab test results and the success of the crops in the absence of additional fertilizer), the limiting factor is the topography; since only 1% of water can be readily diverted by gravity to urban agriculture in this case study, the impact of urban agriculture on nutrient cycling and GHG mitigation is relatively small (Fig. 8.2). In the event that 100% of N-WWTP effluent could be reused in agriculture, the hypothetical reduction in system-wide GHG is ~66%; however, the additional energy associated with diverting irrigation water is not included in the model. For this particular site, extensive infrastructure and energy would be required to pump water above the stream banks to irrigate land, illustrating practical constraints.

## 8.8 Environmental and Behavioral Causes of Crop Contamination

The water quality of the three irrigation waters (groundwater, N-WWTP effluent, and untreated surface water) at the three different sites differed consistently throughout the study, as determined by lab testing of composite water samples taken during irrigation events. For example, average total nitrogen measured was at 3, 37, and 48 mg/L for groundwater, treated effluent, and untreated surface water, respectively. Nitrogen levels were relatively high in the treated effluent because nitrogen is not one of the primary treatment targets of N-WWTP; the treatment is focused on meeting the Indian disposal standards of 5-day BOD below 30 mg/L and fecal coliforms below 10,000 MPN/100 mL, among other parameters (Miller-Robbie et al. 2013). The higher nutrient and organic matter content of the irrigation water was beneficial to crops (Miller-Robbie et al. 2017), with the treated effluent and untreated surface water plots producing the highest crop yields; the groundwater plot yielded only 12% of the sellable bundles as compared to the other two plots at harvest.

Although the water quality improved by several orders of magnitude due to WWTP treatment (Fig. 8.3a), crop quality did not improve significantly, as measured by indicator organisms, *E. coli* and nematode ova (Miller-Robbie et al. 2017). As seen in the crop *E. coli* results (Fig. 8.3b), there were clear differences of at least two orders of magnitude, on average, between the *E. coli* content of the three irrigation waters throughout the study. However, the *E. coli* content on the spinach at harvest was not as different as in the irrigation water; at harvest the crop samples were within one order of magnitude of each other when crops were harvested by the farmer using his usual harvesting practices. Even the spinach grown with relatively clean groundwater was not significantly different from that grown with treated effluent ($p > 0.1$), which had a much higher irrigation water *E. coli* content. However, the spinach grown with WWTP effluent had significantly lower *E. coli* content than that grown with untreated water ($p < 0.025$). Similar results were seen for Ascaris and hookworm content of water, soil, and crops (Miller-Robbie et al. 2017).

Several behavioral and environmental factors were explored to identify reasons why the *E. coli* on spinach were not dissimilar across the three farm plots, even though irrigation water quality differed by orders of magnitude. First, the researcher observed farmer handling at the time of mid-point crop sampling, and noticed the farmer-harvested spinach with great speed, resulting in frequent contact between the leaves and the soil, which contained high levels of *E. coli* in all three plots. The farmer also placed the harvest under a pre-moistened (wastewater-soaked) gunnysack to prevent wilting in the heat. The researcher collected samples at final harvest

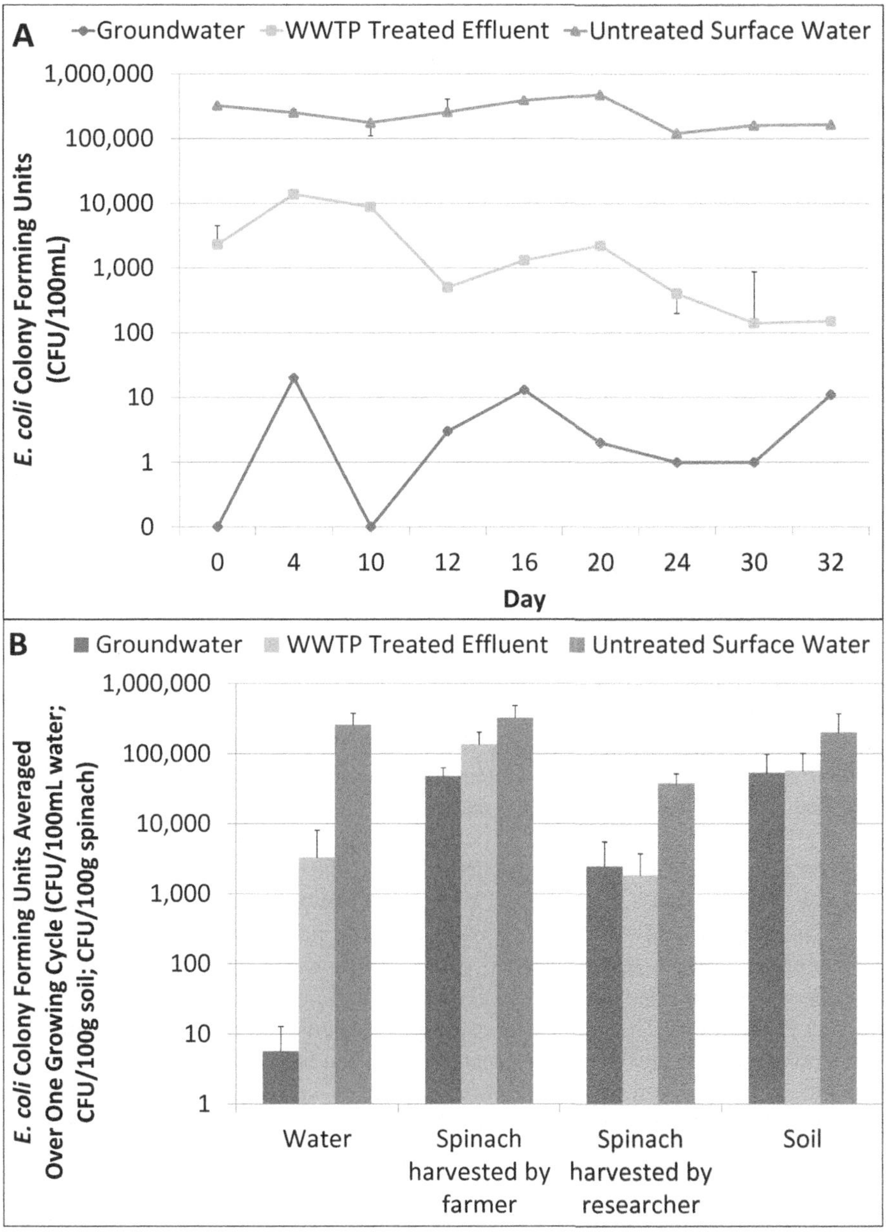


**Fig. 8.3** (**a**) Daily average *E. coli* in irrigation water over the course of the study. (**b**) *E. coli* averaged over one growing cycle for all media: irrigation water, spinach harvested by farmer and by researcher, and soil. *CFU* colony forming units (Miller-Robbie et al. 2017)

in an effort to minimize recontamination by using hand sanitizer between samples and taking care to not touch anything except the crop and sterile sample bag.

The researcher-harvest yielded crop samples were at least one order of magnitude less for *E. coli* content than the farmer-harvested samples (statistically significant at $p < 0.1$) (Fig. 8.3b). However, as seen with farmer harvesting, there was not a significant difference ($p > 0.1$) between groundwater and treated effluent irrigated crops, but both were considerably different from crops irrigated with untreated surface water (Fig. 8.3b). Thus, the data show that the WWTP did reduce microorganism concentration on crops, but not as dramatically as in the irrigation water.

Other factors such as extreme summer heat (soil temperatures as high as 58 °C in direct sunlight), wind-blown dust, soil, and aerosol particles from the WWTP could also be important. Therefore, this field study demonstrates that energy investments in WWTP reduce *E. coli* in water by several orders of magnitude, but have a significantly smaller effect for crops produced from urban agriculture due to a combination of environmental and behavioral factors.

## 8.9 Determining Health Risks Associated with Crop Microorganism Content

To determine the health risk due to ingesting pathogens on leafy vegetables irrigated with treated and untreated wastewater, a basic quantitative microbial risk analysis model was used (Mara 2008), in accordance with World Health Organization 2006 Guidelines. The measured *E. coli* content of the farmer-harvested spinach (Fig. 8.3b) was used as an indicator bacterium to estimate rotavirus concentration. Assumptions included: 0.1 to 1 rotavirus expected per $10^5$ *E. coli*; $10^{-2}$ to $10^{-3}$ rotavirus die-off between last irrigation and consumption (Mara 2008), and that this lettuce-based model was appropriate for spinach. Consumption of wastewater-irrigated crops was the focus of this study and farmer exposure was not quantified. In addition to consumption, farmers are exposed to pathogens in wastewater through their skin (e.g., hookworm species) and orally (aerosols and via unwashed hands/other items) (van der Hoek 2004). The probability of infection calculation considers consumption of uncooked wastewater-irrigated spinach, which can be considered as a worse-case scenario in comparison to consumption of cooked spinach. There are education programs to encourage farmers to grow crops that are more suitable for irrigation with wastewater, i.e., trees, shrubs, flowers, livestock fodder, and crops that are not eaten raw (RUAF 2020).

To estimate the probability of infection due to one dose (100 g) of spinach, the β-Poisson dose-response model was used:

$$\mathrm{PI}(d) = 1 - \left[1 + \left(\frac{d}{\mathrm{N50}}\right)\left(\frac{21}{\alpha} - 1\right)\right] - \alpha \qquad (8.1)$$

where

$PI(d)$ = probability of infection in an individual due to ingestion of a single dose, $d$
N50 = median infective dose; 6.17 for rotavirus (Mara 2008).
$\alpha$ = pathogen infectivity constants; 0.253 for rotavirus (Mara 2008).

One dose of spinach (100 g) irrigated with groundwater, WWTP effluent, and untreated surface water was estimated to contain 0.4, 1.2, and 2.9 rotaviruses, respectively, due to the *E. coli* indicator concentration; using Eq. (8.1), the probability of infection from these single doses was 0.16, 0.29, and 0.41, respectively. When compared with the tolerable infection risk for rotavirus in developing countries of $7.7 \times 10^{-4}$ per person per year (given by WHO 2006 guidelines), which equates to a dose per exposure event of $3.9 \times 10^{-5}$ rotaviruses (Mara 2008), the amounts contained in one dose from this study were orders of magnitude larger; therefore, the health risks are exceedingly high for all three farm plots.

## 8.10 Assessment of System-Wide Tradeoffs

System-wide tradeoffs, between energy use/GHG emissions, food production, groundwater use, infrastructure monetary cost, and health risk reduction, were assessed and relative comparisons were made between the three farm sites (Table 8.1). This study found that the urban agriculture groundwater scenario was the least beneficial for food production and groundwater use categories, and had a minimal impact on energy use/GHG emissions and infrastructure monetary cost, and the lowest spinach pathogen indicator (*E. coli*) content, although enough to pose a health risk. Use of treated effluent and untreated surface water for urban agriculture were more similar for some categories; they yielded higher food productivities, while avoiding groundwater extraction. Despite the added embodied energy and GHG emissions in WWTP infrastructure, the treated effluent case did emit fewer GHGs overall than the untreated surface water case due to reduced COD and DIN in the effluent water when released to streams (Fig. 8.2), and did have less crop *E. coli* content; however, the health risk was still significant.

**Table 8.1** Relative system-wide positive benefits and negative costs for relevant tradeoffs

| | Energy use/GHG emissions | Food produced | Groundwater used | Infrastructure cost | Pathogen indicator on crop |
|---|---|---|---|---|---|
| Groundwater | + | – | – | + | ± |
| Treated effluent | ± | + | + | – | ± |
| Untreated surface water | – | + | + | + | – |

## 8.11 Key Findings and Future Recommendations

As cities grow and domestic wastewater is either released to the environment without treatment or WWTPs are built and wastewater is treated, this study strives to quantify the holistic impacts of wastewater use for urban agriculture. The key findings are as follows:

1. Contrary to expectations, investments of energy and GHG emissions, in terms of constructing, operating, and maintaining WWTP infrastructure, actually reduce system-wide GHG emissions. This is because significant $CH_4$ and $N_2O$ are generated from untreated wastewater in streams. Urban agriculture further reduces system-wide GHG emissions because $CH_4$ emissions are negligible when wastewater is reused as irrigation water.
2. Because the nutrients in wastewater effluent are dilute, a very large amount of urban agricultural land is needed to capture the water and nutrients. This limits the potential for wastewater reuse for irrigation water within city limits where large amounts of land are less available; however, peri-urban areas are often nearby and more open. Pumping and piping infrastructure would likely be needed to maximize the amount of land used.
3. Although the water quality in this study improved by several orders of magnitude due to WWTP treatment, crop quality did not improve when irrigated with higher-quality water. Both behavioral and environmental causes were found to contribute to contamination.
4. Although water was treated via the WWTP and subsequently utilized for crop irrigation, the treated water still posed a health risk to consumers. Therefore, precautions and education programs are important.

Overall, quantitative analysis of urban water contamination shows that investing in WWTP infrastructure offers the most benefits in the FEW-Health nexus; however, key environmental/behavioral factors need to be considered when evaluating wastewater reuse in urban agriculture. While the purpose of WWTP implementation is not specifically to provide irrigation water to urban farmers, farmers can benefit from WWTP-treated water for use on their crops. There is little guidance due to few published, quantitative studies on appropriate water quality standards in low-income countries for urban agriculture. Therefore, field studies that measure pathogens on crops in many locations, climates, and seasons could help to inform these parameters.

Benefits to urban agriculture may be better realized from other methods of wastewater treatment. Assessing the potential of natural treatment/vegetative buffer strips for megacities where the majority of wastewater is untreated, or alternatives to flush toilets leading to centralized WWTPs, could be more favorable from the perspective of water reuse for urban agriculture.

## References

Abd-Elwahed Mohammed S (2018) Influence of long-term wastewater irrigation on soil quality and its spatial distribution. Ann Agric Sci 63(2):191–199
Amerasinghe P, Bhardwaj RM, Scott C et al (2013) Urban wastewater and agricultural reuse challenges in India. International Water Management Institute Research Report 147. Colombo, Sri Lanka
Asano T (1998) Wastewater reclamation and reuse, vol 10. CRC Press, Boca Raton, FL
Beaulieu JJ, Tank JL, Hamilton SK et al (2011) Nitrous oxide emission from denitrification in stream and river networks. Proc Natl Acad Sci 108(1):214–219
Boyer D, Ramaswami A (2017) What is the contribution of city-scale actions to the overall food system's environmental impacts?: assessing water, greenhouse gas, and land impacts of future urban food scenarios. Environ Sci Technol 51:12035–12045
Buechler S, Devi G (2003) Household food security and wastewater-dependent livelihood activities along the Musi River in Andhra Pradesh, India. International Water Management Institute and the World Health Organization, Geneva, Switzerland
Celio M, Scott CA, Giordano M (2010) Urban–agricultural water appropriation: the Hyderabad, India case. Geogr J 176(1):39–57
Del Grosso SJ, Mosier AR, Parton WJ et al (2005) DAYCENT model analysis of past and contemporary soil $N_2O$ and net greenhouse gas flux for major crops in the USA. Soil Tillage Res 83(1):9–24
Del Grosso SJ, Ojima DS, Parton WJ et al (2009) Global scale DAYCENT model analysis of greenhouse gas emissions and mitigation strategies for cropped soils. Glob Planet Change 67(1–2):44–50
Devi MG et al (2009) Valuing a clean river: a case study of Musi River, Hyderabad, India. Proceedings of the 53rd Australian Agricultural and Resource Economics Society conference, Cairns, Australia, 11–13 Feb 2009
Drechsel P, Scott CA, Raschid-Sally L et al (eds) (2010) Wastewater irrigation and health: assessing and mitigating risk in low-income countries. Earthscan with the International Development Research Centre and the International Water Management Institute, London, UK
Dutta V (2012) Land use dynamics and peri-urban growth characteristics reflections on master plan and urban suitability from a sprawling north Indian city. Environ Urban Asia 3(2):277–301
Ensink JHJ, Blumenthal UJ, Brooker S (2008) Wastewater quality and the risk of intestinal nematode infection in sewage farming families in Hyderabad, India. Am J Trop Med Hyg 79(4):561–567
FAOSTAT (Food and Agriculture Organization Corporate Statistical Database) (2009) FAO statistical yearbook 2007–2008
Foster S, Garduño H, Tuinhof A et al (2003) Urban wastewater as groundwater recharge: evaluating and managing the risks and benefits. In: Sustainable groundwater management: concepts and tools, Global water partnerships briefing note series, vol 12. World Bank, Geneva, Switzerland, pp 1–6
Ganjegunte G, Ulery A, Niu G et al (2018) Organic carbon, nutrient, and salt dynamics in saline soil and switchgrass (Panicum virgatum L.) irrigated with treated municipal wastewater. Land Degrad Dev 29(1):80–90
Gopal SMG (2004) Water: issues faced in Hyderabad metropolitan. Hyderabad Metropolitan Water Supply and Sewerage Board, Hyderabad, India
Hamilton AJ, Stagnitti F, Xiong X et al (2007) Wastewater irrigation: the state of play. Vadose Zone J 6(4):823–840
Hanjra MA, Drechsel P, Wichelns D et al (2015) Transforming urban wastewater into an economic asset: opportunities and challenges. In: Wastewater: economic asset in an urbanizing world. Springer, Dordrecht, pp 271–278
Howell TA (2001) Enhancing water use efficiency in irrigated agriculture. Agron J 93:281–289
IPCC (Intergovernmental Panel on Climate Change) (2006) In: Eggleston S et al (eds) 2006 IPCC guidelines for National greenhouse gas inventories. IGES, Japan

Jacobi J (2009) Agricultural biodiversity as a livelihood strategy?: the case of wastewater-irrigated vegetable cultivation along the Musi River in Peri-urban Hyderabad, India. Masters of Science, Institut für Physische Geographie

Jarecki MK, Parkin TB, Chan ASK et al (2007) Validation of DAYCENT model based on daily measurements of nitrous oxide emission from a corn field of nitrous oxide emission from a corn field. 4th USDA greenhouse gas conference, USDA ARS GraceNet Project, Baltimore, USA

Kennedy C, Cuddihy J, Engel-Yan J (2007) The changing metabolism of cities. J Ind Ecol 11(2): 43–59

Kilelu CW (2004) Wastewater irrigation, farmers' perceptions of health risks and institutional perspectives: a case study in Maili Saba, Nairobi. Cities feeding people report within the International Development Research Centre

Makoni FS, Thekisoe OMM, Mbati PA (2016) Urban wastewater for sustainable urban agriculture and water management in developing countries. In: Sustainable water management in urban environments, Handbook of environmental chemistry, vol 47. Springer, Cham, pp 265–293

Mara DD (2008) A numerical guide to volume 2 of the guidelines and practical advice on how to transpose them into national standards. In: Using human waste safely for livelihoods, food production and health: information kit on the 3rd edition of the guidelines for the safe use of wastewater, excreta and greywater in agriculture and aquaculture. World Health Organization, Geneva, Switzerland

McCartney M, Scott C, Ensink J et al (2008) Salinity implications of wastewater irrigation in the Musi River catchment in India. Ceylon J Sci 37(1):49–59

Miller L, Ramaswami A, Ranjan R (2013) Contribution of water and wastewater infrastructures to urban energy metabolism and greenhouse gas emissions in cities in India. J Environ Eng 139: 738–745

Miller-Robbie L, Ramaswami A, Kumar P (2013) Life cycle energy use and greenhouse gas emission analysis for a wastewater treatment plant in India. Water Environ Res 85(7):621–631

Miller-Robbie L, Ramaswami A, Amerasinghe P (2017) Wastewater treatment and reuse in urban agriculture: exploring the food, energy, water, and health nexus in Hyderabad, India. Environ Res Lett 12(7):075005

Ministry of Urban Development (2010) Improving service outcomes 2008–09. In: Service level benchmarking databook. Government of India, New Delhi

Morugán-Coronado A, García-Orenes F, Mataix-Solera J (2011) Short-term effects of treated wastewater irrigation on Mediterranean calcareous soil. Soil Tillage Res 112:18–26

Pitterle M, Ramaswami A (2009) Hybrid life cycle assessment of energy, economics and greenhouse gas emissions at us waste wastewater treatment plants. Dissertation, University of Colorado Denver

Qadir M, Wichelns D, Raschid-Sally L et al (2007) Agricultural use of marginal-quality water - opportunities and challenges. In: Molden D (ed) Water for food, water for life: a comprehensive assessment of water management in agriculture. Earthscan, London, UK, pp 425–457

Qadir M, Wichelns D, Raschid-Sally L et al (2008) The challenges of wastewater irrigation in developing countries. Agric Water Manag 97(4):561–568

Ramachandraiah C, Vedakumar M (2007) Hyderabad's water issues and the Musi River need for integrated solutions. Paper presented at international water conference, Berlin, Germany 12–14 Sept 2007

Ramaswami A, Boyer D, Nagpure AS et al (2017) An urban systems framework to assess the trans-boundary food-energy-water nexus: implementation in Delhi, India. Environ Res Lett 12: 025008

Ramaswami A, Bettencourt LM, Clarens A et al (2018) Sustainable urban systems: articulating a long-term convergence research agenda. A report by The Advisory Committee for Environmental Research and Education. National Science Foundation, Washington, DC

Raschid-Sally L, Jayakody P (2008) Drivers and characteristics of wastewater agriculture in developing countries: results from a global assessment. International Water Management Institute, Colombo, Sri Lanka, pp 1–39

RUAF (Resource Centres on Urban Agriculture and Food Security) (2010) http://www.ruaf.org/. Accessed 19 Dec 2019
RUAF (Resource Centres on Urban Agriculture and Food Security) (2020) Urban agriculture: what and why? https://www.ruaf.org/urban-agriculture-what-and-why. Accessed 9 Jan 2020
Scott C, Faruqui N, Raschid-Sally L (eds) (2004) Waste water use in irrigated agriculture: confronting the livelihood and environmental realities. International Water Management Institute, Colombo, Sri Lanka
Shahid M, Khalid S, Abbas G et al (2015) Heavy metal stress and crop productivity. In: Crop production and global environmental issues. Springer, Berlin, Germany, pp 1–25
Srinivasan JT, Reddy VR (2009) Impact of irrigation water quality on human health: a case study in India. Ecol Econ 68(11):2800–2807
USEPA (United States Environmental Protection Agency) (2011) Inventory of U.S. greenhouse gas emissions and sinks: 1990–2009. Office of Atmospheric Programs, Washington, DC
van der Hoek W (2004) A framework for a global assessment of the extent of wastewater irrigation: the need for a common wastewater typology. In: Scott CA, Faruqui NI, Raschid-Sally L (eds) Wastewater use in irrigated agriculture: confronting the livelihood and environmental issues. International Water Management Institute, Colombo, Sri Lanka, pp 11–24
van der Hoek W, Ul Hassan M, Ensink JHJ et al (2002) Urban wastewater: a valuable resource for agriculture: a case study from Haroonabad, Pakistan. International Water Management Institute, Colombo, Sri Lanka
van Rooijen DJ, Turral H, Biggs TW (2005) Sponge city: water balance of mega-city water use and wastewater use in Hyderabad, India. Irrig Drain 54:S81–S91
van Rooijen DJ, Turral H, Biggs TW (2009) Urban and industrial water use in the Krishna Basin, India. Irrig Drain 58(4):406–428
Venot J-P, Bharati L, Giordano M et al (2010a) Beyond water, beyond boundaries: spaces of water management in the Krishna river basin, South India. Geogr J 177(2):160–170
Venot J-P, Reddy VR, Umapathy D (2010b) Coping with drought in irrigated South India: farmers' adjustments in Nagarjuna Sagar. Agric Water Manag 97(10):1434–1442
Wakode HB, Baier K, Jha R et al (2018) Impact of urbanization on groundwater recharge and urban water balance for the city of Hyderabad, India. J Soil Water Conserv 6(1):51–62
Wolman A (1965) The metabolism of cities. Sci Am 213(3):179–190
WWAP (United Nations World Water Assessment Programme) (2017) Wastewater, the untapped resource. In: United Nations World Water Development Report 2017. UNESCO, Paris, France

**Leslie Miller-Robbie PhD** is a researcher in environmental engineering at the food-energy-water-health nexus and an expert in life cycle assessment. She has taught at the University of Colorado Denver, Metropolitan State University of Denver, and Humboldt State University. She holds a PhD in Civil Engineering Systems, for which she focused on water/wastewater infrastructure, urban metabolism, and water reuse in urban agriculture. For her post-doctoral fellowship at Colorado School of Mines, she evaluated the environmental impacts of water and energy resource recovery systems. Her interdisciplinary background includes molecular, micro-, and environmental biology.

**Anu Ramaswami PhD** is a professor at Princeton University in India studies, civil and environmental engineering, and affiliated with the High Meadows Environmental Institute. She is an interdisciplinary environmental engineer recognized as a pioneer and leader on sustainable urban infrastructure systems. Her work explores how seven key sectors—water, energy, food, buildings, mobility, connectivity, waste management, and green/public spaces—shape human and environmental well-being. She is inaugural director of the M.S. Chadha Center for Global India at Princeton University, lead principal investigator and director of the National Science Foundation (NSF)-supported Sustainable Healthy Cities Network, and serves on United Nations Environment's International Resource Panel and NSF's Advisory Committee for Environmental Research and Education. She received her BS in Chemical Engineering from the Indian Institute of Technology Madras in Chennai and her PhD in Civil and Environmental Engineering from Carnegie Mellon University.

# Frontier Agriculture: Climate-Smart and Water-Saving Agriculture Technologies for Livelihoods and Food Security

9

Dorte Verner, Saleema Vellani, Elisha Goodman, and David C. Love

**Abstract**

Many refugee and host populations are food insecure and poor. At the end of 2020, nearly 82.4 million people worldwide were forcibly displaced, and the United Nations Refugee Agency reported that more than 26 million people were living in refugee-like situations. The rapid and large influx of refugees adds additional pressure to host countries' water and public resources, which amplifies the need for more climate-smart and sustainable food production. There is an urgency to engage with and support refugee livelihoods.

This chapter shows that Frontier Agriculture, which comprises climate-smart and water-saving agriculture technologies, such as hydroponics and aquaponics, can contribute to improve overall well-being and nutritional status for farmers and groups of people that are less integrated into the labor market. Frontier agriculture can leverage scarce resources, such as water and arable land, and promote inclusive economic activities that increase access to nutritious food, improve livelihoods, create jobs, promote entrepreneurship, enhance skills, and build social cohesion. It can also assist with building communities and help recover from the loss of assets and from trauma of fleeing from conflicts. Previous experiences suggest that small-scale hydroponic and aquaponic projects targeting vulnerable populations can be implemented rather quickly and produce meaningful results within a short timeframe.

D. Verner (✉) · S. Vellani
The World Bank, Washington, DC, USA
e-mail: dverner@worldbank.org

E. Goodman
Massachusetts Institute of Technology, Boston, MA, USA

D. C. Love
Johns Hopkins University, Baltimore, MD, USA



J. A. Diehl, H. Kaur (eds.), *New Forms of Urban Agriculture: An Urban Ecology Perspective*, https://doi.org/10.1007/978-981-16-3738-4_9

**Keywords**

Hydroponics · Aquaponics · Circular economy · Climate change · Food security · Urban agriculture

## 9.1 Introduction

The Middle East and North Africa (MENA) is experiencing unprecedented levels of forced displacement. As many as 82.4 million people worldwide were forcibly displaced at the end of 2020, according to UNHCR.[1] The water crisis, coupled with fragility,[2] may fuel more migration and place more pressure on scarce water resources and land (World Bank 2017b). Migration can act as risk multipliers in fragile contexts.[3] The rapid and large influx of people adds additional pressure to host countries' water and land resources, amplifying the need for more climate-smart and sustainable food production.[4] Currently, agriculture uses nearly 85% of the water in MENA. While many farmers have implemented drip irrigation and other water-saving technologies in recent decades (World Bank 2017d), more innovation is needed to increase the production of and access to nutritious food using approaches that have not been exploited in the past.

Climate-smart and sustainable agriculture is important to achieve nutritious food security[5] and increase income-generating activities. Many refugee and host populations are food insecure and poor. Creating livelihoods and engaging in economic activities in the new environment are a challenge for those who have been displaced. Besides contributing to food security, water-saving agriculture technologies and innovations can provide jobs and livelihoods along with skills

---

[1] UNHCR (2021), https://www.unhcr.org/flagship-reports/globaltrends/.

[2] A fragile situation is defined as having either: (a) a composite World Bank, African Development Bank and Asian Development Bank Country Policy and Institutional Assessment rating of 3.2 or less; or (b) the presence of a United Nations and/or regional peace-keeping or peace-building mission (e.g., African Union, European Union, NATO), with the exclusion of border monitoring operations, during the past 3 years (World Bank 2019).

[3] Source: World Bank (2017b), Water Management in Fragile Systems, *Building Resilience to Shocks and Protracted Crises in the Middle East and North Africa.*

[4] According to FAO, climate-smart agriculture (CSA) is an approach that helps to guide actions needed to transform and reorient agricultural systems to effectively support development and ensure food security in a changing climate. CSA aims to tackle three main objectives: sustainably increasing agricultural productivity and incomes; adapting and building resilience to climate change; and reducing and/or removing greenhouse gas emissions, where possible. CSA is an approach for developing agricultural strategies to secure sustainable food security under climate change. CSA provides the means to help stakeholders from local to national and international levels identify agricultural strategies suitable to their local conditions. CSA is one of the 11 Corporate Areas for Resource Mobilization under the FAO's Strategic Objectives. It is in line with FAO's vision for Sustainable Food and Agriculture and supports FAO's goal to make agriculture, forestry, and fisheries more productive and more sustainable.

[5] Food security is a measure of (1) availability and (2) access of food (World Bank 2017c).

and human capital upgrading for both host communities and forcibly displaced populations.

Traditional growing methods are not effective solutions given the context of forcibly displaced populations and their lack of access to resources. Growing in the urban context would enable a shorter supply chain, less transport, less packaging, conservation, and labor needed, leading to substantial decreases of resources and energy use (e.g., up to 79% of the retail price in US conventional food distribution) (Wohlgenant 2014). Shortening and simplifying food supply chains can drastically diminish their environmental impacts, while providing cities and rural areas with fresh, highly nutritious produce.

Water and agriculture are key to stabilization and ultimately to peacebuilding through producing, processing, and selling food, generating income and employment, rebuilding household-level food security, and rebuilding social cohesion and institutions from the bottom up. Building resilience in water and agricultural systems in fragile and conflict-affected areas requires both the short- and long term to be considered in planning, bridging the humanitarian-development divide. World Bank (2017b) found that "when water quality and quantity are reduced, water for irrigation may be curtailed, leading to conditions that can breed fragility, such as rural unemployment, rural-urban migration, job competition and price inflation in urban areas, and consequent instability." Regions where a large portion of employment and livelihoods depend on irrigated agriculture are particularly exposed to these types of risks.

The rationale for mapping and analyzing the potential of "matching" Frontier Agriculture (FA) technologies with the needs of refugees and host populations in MENA is fourfold:

1. The food-water-energy nexus is important for recovery and stabilization of countries and communities. The core of this nexus is the need to establish food security for all individuals. These emerging agriculture technologies can potentially make an important contribution to reduce water use in agriculture (more crops per drop) and to increase well-being, food security, and resilience of vulnerable people, while also reducing multiple-dimension poverty. Moreover, agriculture is the first sector to recover from conflict situations because production inputs can be rapidly mobilized, including seeds, tools, and water.
2. Poor refugees and host communities are economically insecure and spend a large amount of time trying to meet basic needs. The return to skills from prior occupations and education is often low due to lack of jobs and economic opportunities. Policy simulations show that typical development policies that invest in skills, education, and employability are unlikely to succeed in improving welfare unless they are accompanied by more comprehensive measures aimed at creating adequate economic opportunities.
3. There is an urgent need to bridge the humanitarian-development divide and assist displaced populations to rebuild their active lives through concerted development efforts while also supporting host communities. When forcibly displaced populations do not have access to economic opportunities, their human and social

capital deplete, and they survive on short-term coping strategies, which include putting children to work, marrying off girls at a very young age, and disposing of their few assets.

4. The humanitarian system is under pressure and underfunded. The large-scale emergencies continue to drive increases in humanitarian assistance needs. Multi-sector requirements in UN appeals have increased 13-fold between 2005 and 2015. The pace of growth slowed between 2015 and 2016, with a 6% increase and appeals reaching US$27.3 billion in 2016, of which the UN-coordinated appeals accounted for US$20.5 billion and 40% of the requested amount remained unfunded. Several sectors are particularly underfunded, such as agriculture, education, and security. Moreover, there is a need to advance not only the social side, but also the economic and productive sectors, while shifting from providing humanitarian assistance to development assistance.

Within the broader frameworks above, this chapter analyses the potential of FA, and more specifically hydroponics for innovation and for development engagement that have a positive impact on the lives of refugees and host communities in the MENA region. This chapter begins by exploring the potential of FA (Box 9.1) to contribute to an improvement in well-being of displaced populations, including nutritional status, and analyzes which FA technologies are appropriate for different settings. Additionally, we further analyze how new and improved livelihoods, increased economic integration, and expanding markets can potentially reduce the burden of hosting a large number of refugees on host communities and countries while simultaneously providing opportunities for vulnerable host communities, including youth.[6] There is evidence from ongoing initiatives of hydroponics that this activity not only contributes to increasing skills and knowledge and improving livelihoods but can also assist in building communities recovering from the loss of assets and trauma of fleeing from conflict.[7]

[6] MENA is facing a youth bulge, and strategies are needed for integrating youth in the economy. Youth shares in MENA countries are typically higher than global averages, both as share among the total population and the working-age population. Yemen and Palestinian Authority have much higher youth shares in the latter category with more than 26% of the working-age population between ages of 15–24, compared to a global average of 19%. Moreover, as of 2016 only approximately one fifth of females over 15-years old is in the labor force, implying that women's labor force inclusion in MENA is the lowest in the world (source: World Development Indicators, 2016 World Bank).

[7] See https://www.weforum.org/agenda/authors/dorte-verner and www.enosh.org.il on an example of community building through hydroponics. World Bank (2017a): Forcibly Displaced. Towards a Development Approach Supporting Refugees, the Internally Displaced, and their Hosts. Overview, *notes (p1), "development actors should help reduce—even eliminate vulnerabilities. The forcibly displaced have often acquired vulnerabilities that are specific to them, such as catastrophic losses of assets and trauma.*" The issue is further discussed on p. 8 of the aforementioned report.

**Box 9.1 Frontier Agriculture**

"Frontier agriculture" is a term that encompasses climate-smart and water-saving agriculture technologies and comprises horticulture production applying hydroponic systems, hence growing vegetables with significantly reduced water usage (80–95%), minimal land area, and fewer inputs compared to traditional farming.

There are different types of hydroponic systems ranging from low to high-tech, including open and closed circulation systems. The most common systems are water culture, drip system, and nutrient film technique (NFT). Hydroponics can be installed in urban, peri-urban, and rural locations. The systems can be small, portable, and easy to manage and can be installed in homes, on roofs, and in other private and public spaces. People that have limited or no access to land and who cannot use traditional farming methods can be provided with opportunities to produce food with hydroponics.

Currently, hydroponics is mainly used to grow tomatoes, cucumbers, peppers, leafy greens, and a variety of specialty herbs and crops. Plants use equal amounts of water in hydroponics and conventional soil methods; however, a hydroponic system delivers water more efficiently to plant roots so overall water use is significantly reduced. Since the systems support production of fresh vegetables and herbs, they have the potential to positively impact household members' nutrition and household incomes through sales of fresh produce.

Source: https://www.weforum.org/agenda/authors/dorte-verner

The majority of refugees in the MENA region are facing poverty, food insecurity, and malnutrition. About 88% of refugees in Jordan are poor or vulnerable to poverty. Comparatively, in Lebanon, 71% of refugees live in poverty, with some districts reaching poverty rates of 80%. In both places, refugees are younger and there is a larger share of female household heads than in the host country. A significant share of Syrian refugees in Jordan and Lebanon are not only food insecure but have experienced deterioration of their dietary quality and reduction in the number of daily meals. In Jordan, 48.7% of the PAs[8] (Principal Applicants) reported reducing food quantity, 45.7% skip meals, 42.5% limit size at meals, 27.2% purchase food on debts, and 17.9% borrow food or rely on help from friends and/or neighbors. Refugees in Lebanon are also subject to significant food insecurity, and the data show that 89% experience reduced food quality, 61.4% skip meals, 46.2% reduced food quantity, 38.7% borrow food or rely on help from friends and/or neighbors, and 6.1% reported that women in the household eat less than men (Verner et al. 2017).

Different nutritious food items are not consumed by refugees on a regular basis due to shortages of nutrient-rich foods such as fruits and vegetables, eggs, and meat.

[8]"Principal Applicant" is defined as the primary applicant on a petition, in this case on a petition for refugee status.

The average number of days per week that refugees in Jordan reported going without access to specific food items was as follows: deprived of oils and fats about 5 out 7 days and deprived of eggs, dairy, cereal, pasta, canned food, and vegetables about 3 out of 7 days. Notably, there was no difference based on the PA's occupational background. Lack of sufficient nutritious food can affect refugees' health outcomes—and is especially critical for children as it affects brain and general development.

The main objective of this chapter is to increase the knowledge of Frontier Agriculture—water-saving, soilless, climate-smart agriculture technologies—that can increase nutrition and food security, economic engagement, and livelihoods and skills for disadvantaged refugee groups and their host communities. The analyses focus on situations in the MENA region. This chapter introduces two forms of soilless climate-smart agriculture: hydroponics and aquaponics and discusses the different types and adaptability and the requirements of the technology for different environments. It ends with a summary of the potential for Frontier Agriculture technologies to contribute to increased livelihoods and incomes of refugees and host communities.

## 9.2 Frontier Agriculture Technologies

In MENA, a shift from immediate, reactive responses to a balanced, long-term development approach is necessary to address the water and fragility challenges (World Bank 2017b). There is a vicious cycle of water and fragility due to their compounding nature. Water scarcity challenges are becoming worse with climate change, rising demands, inter-sectoral competition, and urbanization. FA water-saving technologies not only help address food security and meet other basic needs, but also reduce water scarcity issues and conflicts by leveraging the opportunities and productive potential of water.

Several initiatives have been launched to address challenges of limited arable land and water resources through soil-based farming methods, such as small plots, community gardens, and drip irrigation gardens. This chapter goes beyond addressing limited arable land and water resources through soil-based farming methods, such as small plots, community gardens, and drip irrigation gardens. Hydroponics and aquaponics technologies require less water, no soil, and minimal use of land. Hydroponics may be a valid alternative to produce nutritious food while increasing livelihoods in a natural resource-constrained environment.

The following sections describe water-saving agriculture technologies, namely different types of hydroponic and aquaponic systems, ranging from simple, low-tech to more advanced techniques. It covers the inputs and outputs, different aspects of production, and the costs and labor involved in these technologies. There is a special emphasis on simplified hydroponic and aquaponic systems.

### 9.2.1 Hydroponics

Hydroponics is a method of growing plants using a nutrient solution, which is a mixture of water and nutrient salts, without the presence of soil (Gericke 1940, 1945; Hoagland and Arnon 1950). Hydroponics is not a new phenomenon; early examples of hydroponic growing include the hanging gardens of Babylon and the floating gardens of the Aztecs of Mexico (Resh 1995). In traditional farming, soil is the main input to store the various nutrients required for plant growth. When water saturates the soil, it picks up these nutrients, so they can more readily interact with the plant roots (Campbell and Reece 2002). In hydroponics, soil is replaced with the use of a nutrient solution.

Hydroponic systems use 80–95% less water than open field agriculture, with more advanced systems using less water than simplified systems (Despommier 2010). Hydroponic techniques range from simple systems that do not need electricity and deliver water to buckets using only gravity, to sophisticated systems, that are stacked vertically in tall buildings requiring a power source to pump and circulate water, such as aeroponics. Hydroponic farming is possible across diverse climates and agro-ecological zones, including arid areas (Heredia 2014). Growing in more extreme environments can be done by farming indoors or in greenhouses in a controlled environment. These farming methods separate the production area from the ecosystem and greatly reduce the land area required for agricultural production, thereby lessening the impacts on ecosystem services (e.g., biodiversity, habitat, carbon sequestration, building soil, water purification, etc.) compared to traditional agriculture. Thus, hydroponic food production can have a positive impact on the environment and on natural resource management.

Hydroponic farming is being established in both urban and rural areas as consumer demands increase for fresh produce with high nutritional value. Since hydroponic systems do not depend on external conditions, they can be set up almost anywhere, including unused or recycled spaces such as parking lots, building rooftops, warehouses, and shipping containers. Producing in urban areas minimizes the distance between the food producer and consumer (Bellows et al. 2004).

### 9.2.2 Types of Hydroponic Systems

There is a continuum of at least seven types of hydroponic systems, from the most simplified to the most sophisticated types. From simple to advanced, systems include wick systems, deep water culture, ebb and flow, drip method, nutrient film technique (NFT), aeroponics, and aquaponics. The type of system chosen depends primarily on the type of plant as well as any limitations of the grower and/or growing space (Jensen 1997). While these system types may share many features, including design, they fundamentally differ in how they manage the nutrient solution. The most popular are water culture, drip system, and NFT (Resh 1995). Figure 9.1 compares the different types of systems and their advantages and disadvantages.

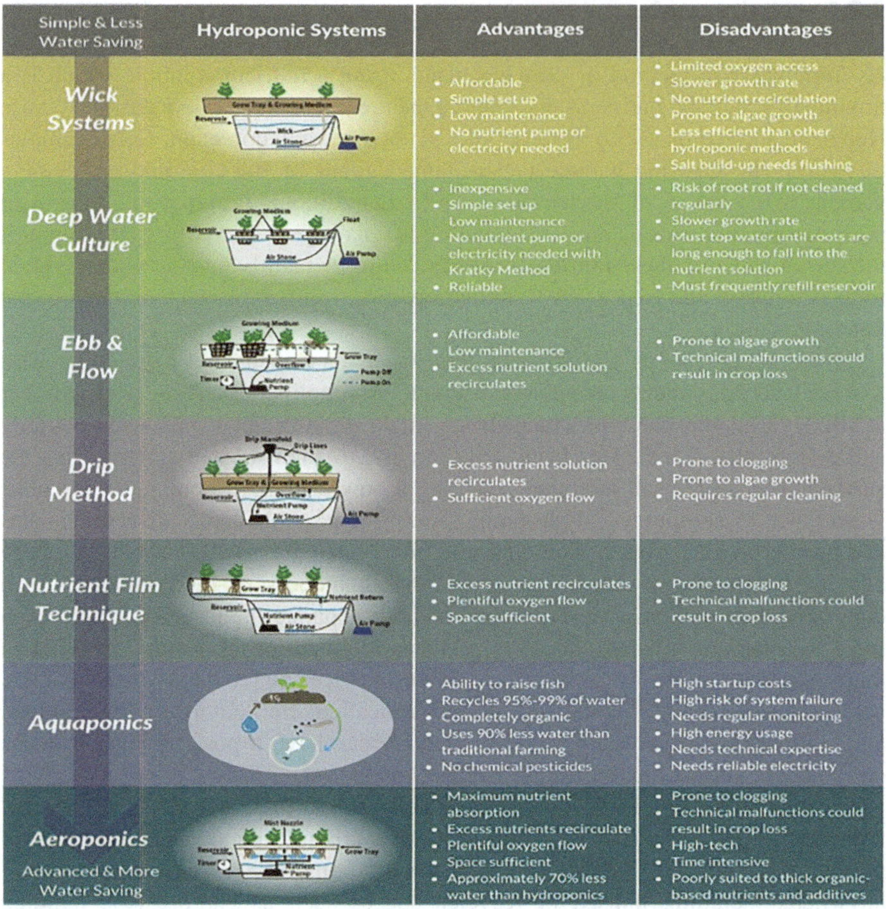


**Fig. 9.1** Types of hydroponic systems and advantages and disadvantages. (Source: Authors)

Hydroponic systems can generally be delineated into open and closed systems (Abd-Elmoniem et al. 2006; Jensen 1997). Open systems, also known as "run to waste systems," do not employ water reuse measures; and the nutrient solution flows through the system only once and is discarded (Jensen 1997; Nederhoff and Stanghellini 2010). On the other hand, closed systems reuse the nutrient solution via recirculation for an unspecified length of time (Lykas et al. 2006). Open systems provide two primary advantages: they eliminate the need for nutrient solution maintenance and reduce the risk of plant pathogens and infection (Jones Jr. 2005). Despite these advantages, open systems are known to waste a large amount of water and nutrients (Nederhoff and Stanghellini 2010) and may not be appropriate for arid regions. In a closed system, more water and nutrients are added to top-up instead of replace the entire solution (Jensen 1997; Nederhoff and Stanghellini 2010). The nutrient solution is regularly monitored and adjusted to maintain proper nutrient ratios. As a result, closed hydroponic systems use 20–40% less water and nutrients than open hydroponic systems. However, they are more difficult to monitor and maintain, which arises from ion accumulation as the nutrient solution recirculates (Lykas et al. 2006). Also, recirculation requires an infrastructure of reservoirs and pumping systems that have to be monitored and maintained in order to perform optimally, which can make them more susceptible to failure if not managed well (Nederhoff and Stanghellini 2010).

### 9.2.3 Hydroponic and Aquaponic System Inputs

The main inputs to hydroponics are the seeds or seedlings, nutrient solution, water, growing medium, container, and, in some cases, electricity and lighting. Aquaponics will have fish fry and fish food in place of a nutrient solution. A more thorough demonstration of the inputs and outputs of hydroponics and aquaponics is shown in Fig. 9.2.

#### 9.2.3.1 Wick Systems

The most simple, low-tech hydroponic systems are wick systems, which are non-circulating systems comprising raised garden beds that have a water reservoir below the plant roots. Water is supplied through a pipe to the water reservoir and the water is drawn upward into the root zone by capillary action, enabling the plants to absorb the amount of water they need. Therefore, there is no need for overhead watering and a lot less water is lost through evaporation. The roots growing in the moist soil have a continuous supply of water, oxygen, and nutrients.

The wick system technique works well in dry, water-scarce environments with limited and unreliable technical inputs, such as electricity, and where technical assistance is not readily available. For example, this technique is currently being used by women in the Palestinian Territories despite the challenges faced by the territories to access materials and inputs (Box 9.2, Fig. 9.3).

**Box 9.2 Wicking Bed Systems in the Palestinian Territories**

In March 2012, the Applied Research Institute Jerusalem (ARIJ), in partnership with the Polish Center for International Aid, piloted a project to adopt Nutrient Film Technique (NFT) and wicking bed production to increase food security, nutrition, women empowerment, income generation, and the competitiveness of the agricultural cooperatives sector in Palestine. Thirty-five NFT units and 52 wicking bed units were established to benefit marginalized and underprivileged families in remote areas of Bethlehem and Hebron governorates. Education models were also established at Al-Arroub Agricultural School to train students in these technologies. The student's families consume the food produced and the surplus is marketed to generate income

(continued)

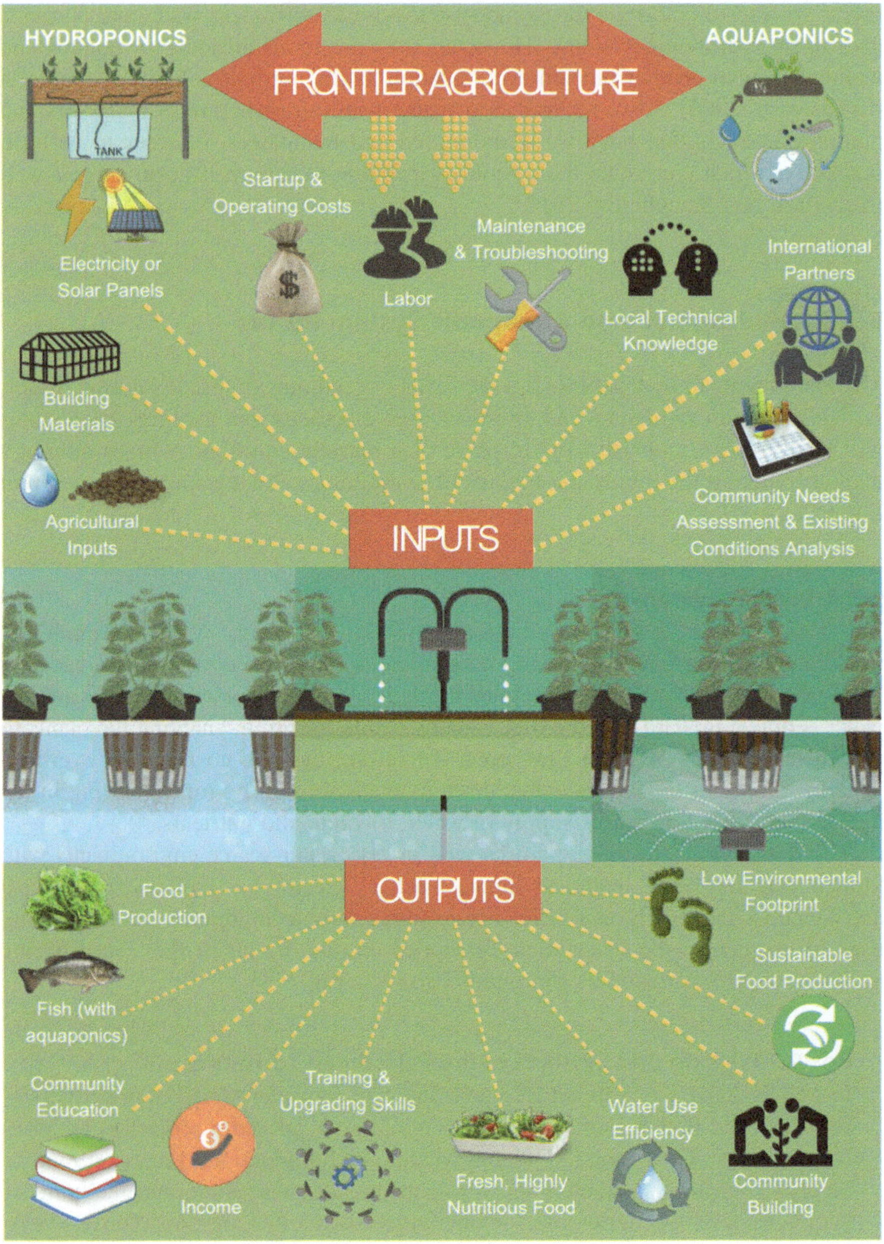


**Fig. 9.2** Inputs and outputs of hydro/aquaponics. (Source: Authors)

**Box 9.2** (continued)
and assist collaborating cooperatives in sustaining their social and humanitarian missions.

Each wicking bed unit kit costs US$820, which contains four beds with a total area of 4-square meters and can plant 200 seedlings per season for three to four seasons per year. Six hundred to 800 plants can be grown to produce 450–600 kg of food per year; thus, one hydroponic unit can produce food with a market value of US$400 to US$550 per year. Each wicking bed unit comprises four separate units that can be replanted three to four times a year, yielding the same production capacity as the NFT system. The project beneficiaries, who are mostly women, were trained to manage the units, which comprise of water pumps, a pumping regulator, and various fittings, such as pipes and other simple equipment. ARIJ provides training, technical support, and follow-up services to the beneficiary families.

Both NFT and wicking bed systems, which are portable, are suitable for urban and rural areas, particularly water-scarce environments, and reduce the usage of irrigation water by 50%. In addition to being safe and pesticide-free, the systems are conducive to family participation in planting and caring for the plants. These systems enhance food security at the household level for vulnerable populations and there is potential to transform these pilot units into a means of income generation.

#### 9.2.3.2 Deep Water Culture

Deep Water Culture (DWC) is another simple, non-circulating hydroponic technique suitable for areas with little to no electricity. The system requirements include a water reservoir to supply nutrients to the plants and a polystyrene platform to float the plants on top of the nutrient solution. Water culture systems are highly desirable to grow leafy greens, such as lettuce, because these plants grow fast and consume large amounts of water.

The Kratky Method (KM) is a type of water culture where the farmer builds or uses a watertight container as a water reservoir, such as five-gallon plastic storage containers or trash bins, filled with the nutrient solution (Fig. 9.4). The KM is simple to operate, requires little to no maintenance, is inexpensive, and is suitable for inexperienced farmers. Plants are grown in net pots on top of the tank cover and are continuously watered since the entire growing medium becomes moistened by capillary action. The roots of the plants are only partially submerged in the water and the top of the plant roots has access to oxygen, creating a moist air. Aside from planting or transplanting, no additional labor is required until harvesting. Electricity and pumps are not needed, so the additional production costs and complexities associated with aeration and circulation in many other hydroponic systems are avoided by this method (Kratky 2009).

**Fig. 9.3** Wick system. Photographs © Applied Research Institute–Jerusalem. Used with permission from Applied Research Institute–Jerusalem

### 9.2.3.3 Ebb and Flow

The ebb and flow method, also called flood and drain, is the classic hydroponic method that is widely used due to its inexpensive cost, dependability, and simplicity—although it does require a power source. This method feeds plants by flooding the plant site with a nutrient solution and allowing that solution to drain back into the reservoir. It uses pots filled with inert media, placed inside a tray or container. During the growing cycle, the tray or container is filled automatically several times a day by a pump that uses a timer.

### 9.2.3.4 Drip Method

The drip system is another widely used hydroponic technique with water circulating through the system using drip emitters. The drip emitters drip water rather than spray or run it, and a dripper runs to every plant placed in a growing medium. After the water passes through the cup holding each plant, it goes back into the water reservoir

**Fig. 9.4** The Kratky Method bucket system. Photograph © Eyal Barkan/FARM-IT. Used with permission from Eyal Barkan

and gets recycled through the system again. Plants can be grown in buckets or trays. The system requires electricity to power a submersible pump to disperse the water and an air stone to mix the water in the reservoir.

#### 9.2.3.5 Nutrient Film Technique

The Nutrient Film Technique (NFT) is a method that places plants in long plastic grow trays with water circulating through the system. Plants are then supported in smaller plastic net cups filled with a growing medium. A water level is set in the tube depending on the maturity of the plants. When the plants are younger, the water level is set higher allowing the roots to reach the water. Once the plant roots mature, the water level is lowered to promote root growth. With this technique, the nutrient solution is pumped past the plant roots allowing the plants to meet their water and nutrient requirements. The drawback of this system is that it is susceptible to power outages and pump failures. Once a failure occurs, the plant roots dry out very rapidly.

#### 9.2.3.6 Aeroponics

Aeroponics is a relatively new method for growing edible plants and the most advanced method of hydroponics. Aeroponics utilizes a fine mist of nutrient-laden water created by its passage through a pressurized nozzle that is then directed toward the enclosed root system of the plants. Aeroponics uses approximately 70% less water than hydroponics, while delivering the same amount of nutrients to the roots. Recent advances in nozzle design have improved the reliability of the system for creating the spray by eliminating clogging due to nutrient build-up, a major issue in earlier models. As a result, more vertical farms are adopting aeroponics as their main growing strategy. See Box 9.3 for more information on vertical farming.

**Box 9.3 Vertical Farming**

Vertical farming allows produce to be grown in vertically stacked layers, on vertically inclined surfaces, or integrated in other structures, increasing productivity in terms of the amount of produce grown per square meter (Christie 2014). Vertical farms are a space-saving form of controlled environment agriculture and different types of hydroponic systems can be used to grow produce vertically. Vertical farms can range from simple NFT hydroponic systems using recycled materials to multistory buildings using aeroponics, such as AeroFarms in the USA (Fig. 9.3), containing an environment conducive to the growing of fruits, vegetables, and nonedible plants for biofuels, drugs, and vaccines. Today, the leading countries employing vertical farms include Japan, Singapore, Taiwan, China, the USA, and a few countries in Western Europe.

Vertical farming works well for small spaces including in urban settings such as rooftops and unused spaces. It allows for a higher yield to be obtained per square foot and uses less land than traditional farming. Vertical farming is more of a structural technique than an actual agricultural system since both hydroponics and other types of water and soil-saving techniques can be set up vertically. For example, vertical hydroponic gardens in Israel utilized NFT with a snake-like structure (Fig. 9.5), and a vertical triangular structure for a hydroponic garden in Palestine utilized NFT, but with a lower-tech set up and lower-cost materials (Fig. 9.5). Another vertical garden system in the Palestinian territories was constructed of recycled large water bottles and soil for production of small plants and herbs (Fig. 9.5).

## 9.2.4 Aquaponics

Aquaponics is a method for producing food that combines recirculating aquaculture, raising fish and aquatic animals in land-based tanks with hydroponics, cultivating plants in water. Aquaponics recycles 95–99% of the water introduced in the system and it is this recycling of water that distributes nutrients throughout the system. Aquaponics systems produce the same types of plants as hydroponic systems, but also provide protein by growing various types of fish and aquatic animals as demonstrated in Fig. 9.6. Depending on the type of system, aquaponics often uses less water than hydroponic systems, with the exception of aeroponics. Since fish plays a key role in the system and would die with the use of chemical pesticides, aquaponics grows chemical-free, all-natural produce.

Aquaponics is scalable and adaptable to many different uses: it can be used as a small or large-scale commercial farm, a recreational or hobby activity, for

**Fig. 9.5** Examples of vertical farming approaches. Top left and top center: hydroponic example in Israel. Bottom left and bottom center: hydroponic example in Palestinian Territories. Right: hydroponic example at AeroFarms. (Source: Authors (left four images) and AeroFarms (right image))

community-based projects, as a hands-on teaching tool in the classroom, and can be incorporated with school curricula. The size of an aquaponic operation ranges in scale from a table-top fish tank to a large warehouse. The three main types of aquaponic systems incorporate deep water raft culture, media beds, and NFT. Deep water culture is most popular among commercial producers while media beds are most popular among home gardeners. Box 9.4 describes how aquaponics is used in the Palestinian Territories to increase food consumption, increase livelihoods, empower women, and address food security.

**Box 9.4 Aquaponics in the Gaza Strip, Palestinian Territories**

In response to the crisis in Gaza and given the high number of food insecure female-headed households in urban areas, the FAO has been implementing several small-scale aquaponics projects in partnership with European donors since 2010. In the first phase of their initial project, 119 food insecure female-headed households were provided with innovative vertical rooftop units connected to fish tanks. With little daily physical effort and in the comfort of carrying out these activities in their own homes, all the beneficiaries increased their household food consumption (FAO 2016). Aquaponics enabled women

(continued)

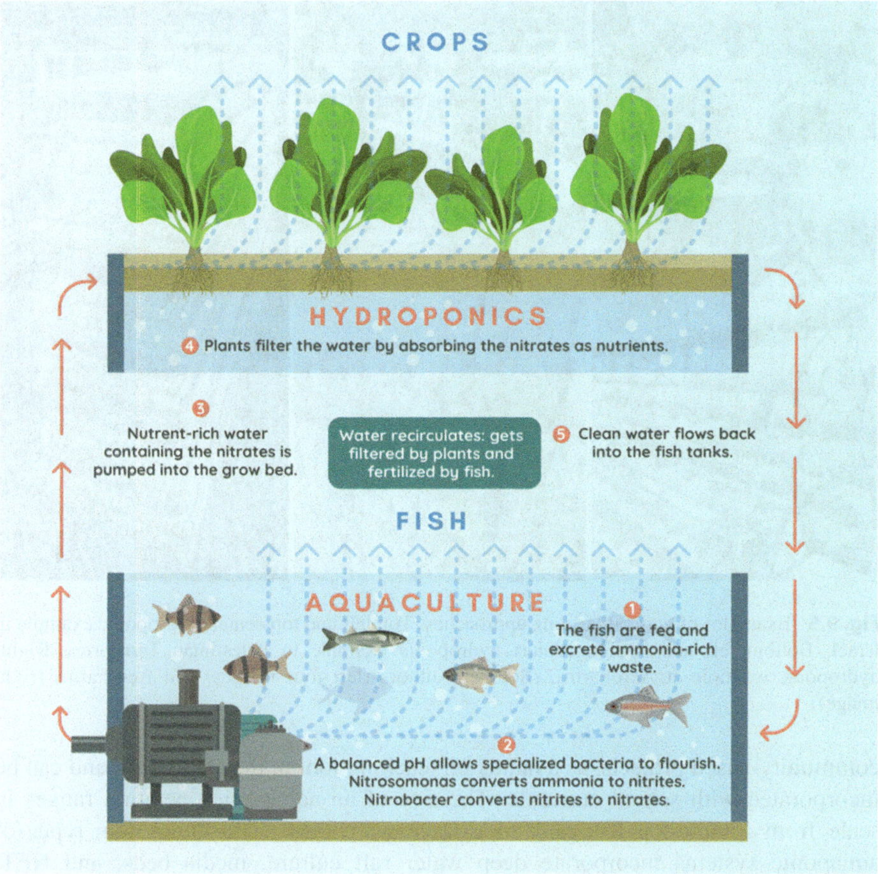


**Fig. 9.6** The aquaponics cycle. (Source: Authors)

**Box 9.4** (continued)

to simultaneously improve their food security and income while caring for their homes and children.

An aquaculturalist and supplier of fish to the FAO based in Beit Lahiya expanded his small aquaculture farm by integrating a semi-commercial sized plant production component to his operation. He effectively transformed his livelihood and created the largest aquaponics unit in Gaza (FAO 2016). The FAO is closely monitoring his progress and providing technical support when necessary as this initiative sheds light on the potential for vulnerable farmers in Gaza to generate income in semi-commercial aquaponic systems (FAO 2016).

#### 9.2.4.1 Types of Crops

The ability to produce certain crops depends on the size of the growing system. In smaller spaces, such as for domestic use, crops such as leafy greens and herbs are common. Such plants grow quickly, can be continuously harvested, and do not require much space to expand. In larger spaces, such as a greenhouse, garage, or patio, a more advanced system may be used, and voluminous plants can be grown that require trellises and deep root support.

In commercial hydroponics, some crops do better than others. Tomatoes, lettuce, bell peppers, and cucumbers do very well in large-scale greenhouse facilities. Herbs and leafy greens do well in warehouse facilities that are vertically oriented. The quality and number of crops that can be produced will largely determine the viability of the commercial hydroponic operation.

Currently, hydroponics is mainly used to grow leafy greens, tomatoes, cucumbers, peppers, herbs, and several other crops (Spensley et al. 1978; Brentlinger 1997; Jensen 1999). These crops have demonstrated the revenues required to make a hydroponic operation profitable (Jensen 1999). Vegetables with both a vegetative state (leaf, root production) and a generative state (fruit production) were found to grow much more efficiently in soilless culture.

#### 9.2.4.2 Nutrient Solution

The main input in hydroponics, the nutrient solution, is a combination of water and nutrient salts mixed to specific concentrations to meet plant requirements and ensure healthy plants (Hoagland and Arnon 1950; Graves 1980; Jones Jr. 2005; Resh 2013). The nutrient solution is fully controllable and can be delivered to plants on an as-needed basis. This makes hydroponics capable of high yields while minimizing water usage and nutrient consumption.

Maintaining an optimum pH range between 5 and 7 is essential since there is no soil to act as a pH buffer. Nutrient solution pH is a common parameter used in hydroponic growing. The pH of the root zone effectively determines which nutrients are available to the plant, as plants can only uptake certain ions within a specific pH range. Nutrient solution is not used in aquaponics, rather, the fish effluent provides fertility to the plants. Balancing pH for fish health is also critical.

#### 9.2.4.3 Water

By volume, water is the primary ingredient in a nutrient solution and, therefore, the single most important factor to growth (Graves 1983). Plants consume equal amounts of water in hydroponics and conventional soil methods; however, the hydroponic system delivers the water more efficiently (Sanchez 2014). Hydroponic farming, in closed systems, uses 80–99% less water than conventional irrigated farming since the plants only consume the water they need while recycling the unused water back to the reservoir.

In rain fed agriculture, however, these comparisons become more difficult because rainwater cannot be "wasted" or "saved"—it is merely part of the hydrological cycle (e.g., evaporation, condensation, and precipitation). On the other hand, when plants are grown indoors or in greenhouses, water is not lost to deep percolation, runoff, and evaporation (Heredia 2014). Other variations and more advanced forms of hydroponics, such as aquaponics and aeroponics, use less water than simpler hydroponic systems (Pantanella et al. 2010).

#### 9.2.4.4 Growing Mediums

The most widely used growing medium is rockwool, which is a melted balsamic rock spun into fibers. However, once used for growing vegetable crops, the disposal cost in landfills can be excessive as rockwool is biologically non-degradable. A replacement to rockwool that is becoming increasingly common is coconut coir, the husk of coconuts, which is found between the hard, internal shell that contains the coconut meat, and the outer coat of a ripe coconut. It is a renewable resource unlike peat moss, which is considered a non-renewable resource. In warmer regions of the world, peat moss biologically breaks down rapidly while coco coir is slow to decompose. Other popular options for growing media are perlite and/or vermiculite, often mixed with peat moss, as a growing medium in the production of greenhouse vegetable crops, especially in the production of vegetable transplants. Sand, gravel, and sawdust are also artificial media options. Sand is quite popular in arid/semi-arid regions of the world and sawdust in the forested regions of northern latitudes.

#### 9.2.4.5 Energy Use

Addressing energy needs is one of the key challenges facing the hydroponics and aquaponics industry, particularly in northern latitudes. High-tech hydroponic systems tend to have high energy use due to incorporation of lighting, pumping, and air moderation systems. But, energy use for hydroponics can be part of a renewable energy use strategy for cutting down carbon emissions. Electricity can be sourced from wind or solar systems with a commercial battery to store excess renewable energy when needed.

When farming indoors and in greenhouses, most of the energy use in hydroponic farming can be attributed to the heating and cooling loads as well as supplemental artificial lighting. For example, heating is a major component of operational costs for greenhouses in northern Europe and other countries with cold winters. Greenhouses located in more moderate climates, such as climates closer to the greenhouse set point temperature, will experience a lower energy demand. In fact, in certain climates, heating and cooling systems may not be required, but instead replaced by a passive ventilation system, thus reducing the overall energy demand considerably. The feasibility of hydroponic systems is heavily reliant on the climate of farming locations (Barbosa et al. 2015). Though lighting increases the energy use, artificial light in indoor environments can make hydroponics feasible in areas with unreliable access to sunlight due to seasonal conditions or the surroundings.

For aquaponics systems, depending on the climate, type of system, and species of fish, there can be additional energy requirements to heat, aerate, and pump the water in fish tanks. Air stones and sprayer bars can help to aerate the water, and systems in warmer climates may not need to heat the water. Additionally, biodigestors may provide a more sustainable source of heat.

### 9.2.5 Production

Hydroponics allows for continuous production year-round in many areas and, on average, more annual growing seasons and shorter harvest cycles than soil-based farming methods. Hydroponic farmers have learned to adopt new growing methods and shifted away from traditional cultivars to achieve higher yields (Christie 2014). The productivity and hence economics of hydroponic food production continue to be main drivers for expansion. Today, commercial hydroponic farms can produce three to four times the yields compared to soil production while using significantly less water (Ly 2011). These higher yields result from the controlled environmental conditions maintained within the greenhouse or indoor farm, which allow for continuous production year-round. The controlled environment promotes a reduction in the number of days required for each harvest cycle, allowing for multiple crops per year. Also, plants grown hydroponically are generally less stressed than soil-grown plants since the plants are in their optimum growing conditions all the time, and in turn create less waste than conventional farming (Treftz and Omaye 2015). In aquaponic systems, fish stocking densities and thus production levels are constrained by the oxygenation levels of the tanks. One conservative estimate of the amount of fish that can be produced in a year is seven tilapia fish per 38 l of water (Goodman 2011).

#### 9.2.5.1 Nutrition

The amounts of key nutrients in hydroponic produce are the same as in conventionally grown produce and are sometimes even higher reference. In conventional farming, plants obtain nutrients from soil, whereas in hydroponics, plants obtain nutrients from a solution instead. Plants generate their own vitamins; therefore, vitamin levels tend to be similar whether a vegetable is grown in soil or hydroponically. However, the mineral content can vary in hydroponic crops, which depends on the fertilizer used. The nutrient levels of a plant can be enhanced by simply adding nutrients to the solution, such as calcium, magnesium, or minor elements such as zinc or iron. Nevertheless, the nutrients and phytochemicals slightly vary for different crops in general, regardless of the growing method. The nutritional profile of each crop depends on the crop variety, the season it is harvested, the length of time between harvest and consumption, and how the crop is handled and stored during that time. These minor variations in nutrient levels are unlikely to have a significant impact on overall consumer health.

#### 9.2.5.2 Pest Management and Plant Survival

Hydroponically grown plants, though not immune, are usually more pest resistant than plants grown using soil and may not need application of herbicides or pesticides. Plants grown in hydroponics are generally stronger and healthier than their soil-grown counterparts since they are fed precise nutritional requirements in a carefully controlled environment. In addition, natural preventative measures against infestations are implemented in most hydroponics systems. For example, companion planting is one method commonly used in hydroponics where crops are intermixed with plants that act as pest deterrents for the primary crop. Biological controls such as beneficial insects may also be used.

According to a study that compared hydroponic and soil systems for growing strawberries in a greenhouse, the hydroponic plants had a higher survival rate at 80% compared to the soil-grown strawberries, of which less than 50% survived (Treftz and Omaye 2015). The lower plant survival rates from soil-based farming are attributed to increased pest infections. Although both growing systems received identical integrated pest management treatments, the in-soil plants suffered more, and pests thrived in the soil-grown strawberries, particularly aphids and spider mites. This is due to increased beneficial bacteria and microbes that pests thrive on in soil conditions (Resh and Howard 2012). Although the hydroponic plants were affected by pests, to a lesser extent, the pests were not able to thrive in hydroponic conditions (Treftz and Omaye 2015).

#### 9.2.5.3 Cost and Labor

Hydro/aquaponic systems vary in terms of cost and labor depending on the system and the materials used, as well as local factors such as climate conditions and energy costs. The startup costs of a hydro/aquaponics system are usually higher than the cost to set up a soil-growing operation. Some of the startup costs of a hydro/aquaponics system can be offset with reduced operating costs due to the system's efficiency in the use of labor, water, fertilizers and pesticides. Also, there tends to be less waste with hydro/aquaponics compared to soil-growing operations.

Startup costs of a hydroponic greenhouse can range anywhere from 2 to 20 times more than soil agriculture (Mathias 2014). When including the hydroponic growing system, estimated costs for greenhouses range from US$52 to US$140 per square meter in research conducted in the 1990s in the USA (Jensen and Malter 1995). Commercial operations may also require a warehouse or other building or structure (Pantanella et al. 2010), which may create added startup costs. Some commercial hydroponic operations require controllers, computer systems, large-scale lighting fixtures, ventilation and heat recovery systems, irrigation and rainwater harvesting, as well as skilled labor (Pantanella et al. 2010). Electricity and utility costs can also be extremely high.

Non-commercial and simpler systems that use existing local materials can cut startup costs considerably, and can be beneficial where imports are expensive or specific technology and materials are unavailable. One wicking bed system in the Palestinian Territories costs $820 for four beds, growing materials, and a simple grow structure of shadow nets with iron skeleton and plastic sheeting for weather and sun protection. Lower-cost solutions, such as simplified, lower-tech variations of the technology are increasingly being implemented in developing countries, particularly those with an arid landscape and water scarcity, such as Jordan.

Labor costs for hydro/aquaponic systems vary depending on the complexity of the system chosen, the amount of trained labor required, and local technical knowledge. Despite this variability, labor costs for a hydro/aquaponic system represent a much larger share than traditional farm labor, which is estimated by the USDA to vary from 17% to 40% of total operating costs in labor-intensive farm production (Daly and Fink 2013), and for hydro/aquaponics systems could be has high as 56–70% (Goodman 2011). In a commercial system, in order to lower labor costs and keep them at an even rate, automated technology may be beneficial.

In terms of labor requirements, the more sophisticated the system, the more technical expertise is needed to monitor and troubleshoot when problems arise. Hydroponic growers must know technical details about the species being produced, plant health problems and how to fix them, symptoms of nutrient deficiency and toxicity, management of nutrient solution, anticipation of possible power outages, and the consequent lack of water circulation in the channels. Aquaponics systems have additional labor requirements related to animal husbandry, water quality, and simple plumbing concepts. Examples from Jordan, Palestinian territories, and the UAE show that the required skills and techniques can be rapidly acquired for people with little formal education.

#### 9.2.5.4 Income and Profitability

When planning a hydro/aquaponics system, it is challenging to estimate production levels and income as production varies greatly from system to system and climate to climate. Further, income is influenced by the type of crops that can be grown, local demand, and food prices. Local food prices may shift dramatically over time, allowing producers to charge higher or lower prices for their product (Goodman 2011). In an aquaponics system, it is less possible to quickly change the aquaculture component to adjust to market conditions; the systems can be finicky, can fail, and often are not profitable enterprises. It is advisable to use conservative estimates to account for mishaps with the system and unsold produce (Engle 2015). Notwithstanding these challenges, there is a path for profitability, which can involve careful siting, development of knowledge and skills, and use of alternative revenue streams (Love et al. 2015). Moreover, economies of scale, alternative business models, and other creative ways to increase income and reduce expenses such as producing inputs on-site or procuring items for free can affect cash flow and help an aquaponics operation be profitable (Goodman 2011). Economic advantages of soilless culture systems include a potentially fast and flexible soilless cropping period, which allows growers to quickly change production to take advantage of market conditions. It is also important to note that the hydroponic system would last through multiple seasons without the need to amend the soil with fertilizer or organic matter.

## 9.3 Matching Needs of Refugees and Hosts with Frontier Agriculture Technologies

The MENA region faces two large challenges. First, the increasingly water-scarce region applies 85% of its water in agriculture and second, the recent escalation of the global refugee crisis which, to a large extent, is a MENA crisis. There is a need for increased intake of nutritious food, livelihoods, and jobs for a large share of the more than 18 million adult and youth population living in refugee-like situations in

MENA. It is necessary for the protracted situation to be addressed through the development lens to provide solutions that reactivate the lives and skills of the displaced populations. Moreover, the humanitarian system is under pressure and cannot provide enough resources to meet the needs of forcibly displaced people in the MENA region and beyond. The remainder of this report attempts to merge two agendas: food insecurity among the refugees in a water-scarce region and find solutions through innovative technologies.

Given that water and arable land are scarce in MENA, one way of increasing food production is through frontier agriculture. Hydroponics and aquaponics are climate-smart, innovative, and effective technologies that can produce nutritious food with less water (at least 80%) without requiring arable land. Hydroponic systems are easy to operate and can be installed for small-scale use in homes and community cooperatives to large-scale, commercial farms. Due to the adaptability and flexibility of hydroponics and aquaponics to most environments, and their ability to provide additional nutritious food and marketable produce beyond the capacity of arable land, these technologies are being employed in some of the most challenging areas in MENA, such as in the Palestinian Territories. The selection of the type of hydro/aquaponic system depends on the access to inputs and the level of creativity to produce, reuse, or upcycle inputs. Since the technology is flexible and adaptable to local conditions, the simplest system can jump-start or supplement existing food production. It is a solution that can be introduced in places that previously had no or very limited food production.

The basic inputs to hydro/aquaponics are available or acquirable in all countries in MENA. Hydroponics systems provide high-cost savings on water, land, fossil fuels, and chemical purchases compared to traditional farming. The startup and operating costs entirely depend on the type of system chosen and level of complexity. The more advanced and complex the system, the higher the startup and operating costs. There also tends to be less waste with hydroponics and overall better resource management. This system allows for more crop cycles in a year than traditional farming and more high-value crops in some areas.

We developed a flexible decision matrix that can be used as a tool to determine which type of system would be suitable depending on the local conditions of the growing site. The decision matrix in Table 9.1 is a guide to systematically identify, analyze, prioritize, and compare different systems under consideration for implementation in frontier agriculture. The decision matrix presents the technologies discussed in this chapter and ranks them using a Likert-type scale on a variety of attributes: water use, energy use, technological complexity, maintenance, startup costs, financial sustainability, and mobility. Given that each situation requires a different set of social, ecological, and economic considerations, there may not be one single most effective technology for all applications, but hybrids can be constructed to meet specific needs of people, enterprises, and communities.

While advanced hydro/aquaponic systems may be appropriate for some regions, simplified hydroponic systems that are feasible with minimal training and a small initial investment are preferable for refugees and host communities in MENA. Though the yields from simplified systems are lower than advanced systems,

**Table 9.1** Decision matrix for water-saving technologies (Source: Authors' elaboration)

| Technology | Food | Water use[a] | Energy use | Technological complexity | Maintenance | Start-up costs | Financially self-sustaining | Mobility |
|---|---|---|---|---|---|---|---|---|
| Wick systems | Crops | Low | None | Simple | High | Low | High | Low-high |
| Deep water culture | Crops | Low | Medium | Medium | Low | Med-high | Medium | Low |
| EBB & flow | Crops | Low | Low-high | Complex | High | Med-high | Low | Low |
| Drip method | Crops | Low | High | Complex | Low | Med-high | Low | Low |
| Nutrient film technique | Crops | Low | High | Complex | Med-high | High | Medium | Low |
| Aquaponics | Crops, fish | Low | Low-high[b] | Complex | High | Med-high | Low | Low |
| Aeroponics | Crops | Low | High | Complex | High | High | High | Low |

[a]Open systems recirculate water, closed systems do not recirculate water
[b]Depending upon pump size and heating requirements. Aquaponics requires a constant electrical source or backup energy (battery, generator)

low-tech systems outperform conventional farming methods and use at least 80% less water. Initially, a needs assessment should be conducted at the local community or individual level to identify and rank the priorities and objectives, which can be used to select an adequate hydroponic system or to design the appropriate system. Regardless of the system chosen, this technology can provide important social, economic, and nutritional benefits.

We propose a three-pronged approach using hydroponic farming systems to address some of the existing needs by providing opportunities for those forcibly displaced, particularly those most poor and/or vulnerable, and their hosts. There are groups with more needs than others, these include refugee women in Lebanon and Jordan that previously worked in agriculture and as housekeepers. They were the most food insecure, had the lowest cash incomes, and the majority were not engaged in paid work. Social barriers, education, skill matches, and household responsibilities seem to prevent many women from participating in the labor markets. Women and girl refugees, and women and girls in host communities in the case of Djibouti, face low education levels, health constraints, and limited access to economic opportunities.

Besides contributing to food security, water-saving agriculture technologies and innovations are ways to improve livelihoods, provide jobs and economic integration with skills, and human capital upgrading for both host and forcibly displaced populations in MENA and those most in need (see above). The three prongs focus on:

- Increasing access to nutritious food
- Improving livelihoods, providing jobs, and supporting economic integration and entrepreneurship
- Enhancing skills

First, increase access to nutritious food: most refugees are food insecure and have a Vitamin A deficiency. Less than 10% of the refugee population in Lebanon and Jordan are food secure. Moreover, different nutritious food items (fruits, vegetables, eggs, meat) are not consumed by refugees on a regular basis. In Djibouti, both refugees and rural host communities are food insecure. The WFP and UNHCR assessment in the refugee camps of Holl Holl and Ali Addeh reports figures of 66% and 44%, respectively. Also, the host communities around the two camps are food insecure; 62% and 44%, respectively (see more in Verner et al. 2017).

The simplest systems in hydroponics, such as the deep water culture Kratky Method and wick bed systems, do not require electricity or land and need a fraction of the water required in open field agriculture. Hydroponic systems can grow a wide variety of fruits and vegetables, especially leafy greens—which grow fast and provide leaves within a few weeks—that help address Vitamin A deficiency. If the primary priority is to address food insecurity among refugees and host communities, households can be trained to maintain the simplest hydroponic systems using basic materials such as buckets and local rocks. Conversely, if the overarching goal is to increase economic activity among refugees and increase incomes, a large NFT system can be constructed at the community level, in which case households can consume from the production and the surplus can be sold in the local market or beyond.

Second, improve livelihoods, providing jobs and supporting economic integration and entrepreneurship: job opportunities need to be created for both displaced populations and host community populations to reduce rampant poverty and vulnerability. Many forcibly displaced and host populations lack jobs and income, which is one of the main reasons as to why they face nutritious food insecurity and poverty as mentioned above. Hydroponics provides different types of employment. Based on field observations, the wicking bed systems used by women in Palestine provided one part-time job per unit for self-consumption and the surplus produce is sold to the market. These women only need to work 2–3 h a day maintaining the system and 2–3 days per week. For larger-scale commercial operations, it is difficult to obtain data on employment, costs, etc. as it is private information, however, evidence shows that using a DWC or NFT system to grow leafy greens on one acre of land provides approximately 18–22 full-time jobs on average.

Hydroponics provides an opportunity to promote entrepreneurship. There is also potential for production that exceeds individual needs, which could lead to the creation of local markets for such produce and additional jobs. The revenue generated by selling excess production could turn into an important source of income for refugees and allow them to meet other basic needs. Other entrepreneurial opportunities not directly related to hydroponics may arise, especially when the refugees can combine other skills with their training on these systems. For example, they can contribute to a higher level in the value chain, such as producing dried blueberries or essential oils, or create inputs to hydroponics such as upcycling materials or creating hydroponic fertilizer. There are also opportunities for refugees to collaborate with host communities. For example, based on field observations in the Palestinian Territories, a group of entrepreneurs in a village near Ramallah secured contracts with prospective restaurant clients in advance of constructing a hydroponics farm.

And third, enhance skills: skills are a key to increase economic integration and expand the private sector. Training and knowledge acquired in hydroponic operations are a way to upgrade human capital, which is transferable to other locations, including the home country after conflict recedes and reconstruction begins. Refugees who return to their origin communities or relocate to other countries will bring the practical knowledge with them and potentially start new hydroponics operations. The training process and increase in human capital may empower refugees to find or create employment or other income-generating opportunities, potentially related to hydroponics. For example, some may choose to work in education in a related field to hydroponic farming and others may choose to work in another part of the value chain, such as producing hydroponic fodder for livestock. Training can also provide social capital to create social enterprises.

## 9.4 Conclusion

In this chapter we discussed how Frontier agriculture that comprises climate-smart and water-saving agricultural technologies such as hydroponics and aquaponics can improve the livelihoods and well-being of refugee communities and other vulnerable communities, including their host communities, who are often equally food insecure and poor. Frontier agriculture reduces the pressures that host communities experience on their water and other resources due to influxes of refugees. Frontier agriculture can leverage scarce resources, such as water and arable land, and promote inclusive economic activities that increase access to nutritious food, improve livelihoods, create jobs, promote entrepreneurship, enhance skills, and build social cohesion. Frontier agriculture can also contribute to improved overall well-being and nutritional status of people, assist in building community, and support recovery from the loss of assets and from trauma related to fragility and conflict.

While advanced hydro/aquaponic systems may be appropriate in some locations, simplified hydroponic systems that are feasible to implement with minimal training and a small initial investment may be a solution for starting up a food production system for refugees and host communities. Experience suggests that small-scale hydroponic and aquaponic projects targeting vulnerable populations can be implemented rather quickly and produce meaningful results within a short timeframe.

The impacts and benefits of frontier agriculture on food security and livelihoods may vary based on local growing conditions, local market factors, and type of growing system(s) employed. Further research could analyze in more detail crop-yields feasibility and economics in different geographical and local contexts, including growing conditions, labor requirements, input prices, and crop prices. This research can provide a valuable resource to farmers, including refugees, interested in hydro/aquaponics food production.

## References


Abd-Elmoniem EM, Abdrabbo MA, Farag AA, Medany MA (2006) Hydroponics for food production: comparison of open and closed systems on yield and consumption of water and nutrients. In: Second international conference on water resources and arid environments. King Saud University, Riyadh, Saudi Arabia, pp 1–8

Barbosa GL, Gadelha FDA, Kublik N, Proctor A, Reichelm L, Weissinger E, Wohlleb GM, Halden RU (2015) Comparison of land, water, and energy requirements of lettuce grown using hydroponic vs. conventional agricultural methods. Int J Environ Res Public Health 12 (6):6879–6891

Bellows AC, Brown K, Smit J (2004) Health benefits of urban agriculture. In: Community Food Security Coalition's North American Initiative on Urban Agriculture, Portland, OR. http://www.co.fresno.ca.us/uploadedfiles/departments/behavioral_health/mhsa/health%20benefits%20of%20urban%20agriculture%20(1-8).pdf

Brentlinger D (1997) Status of the Commercial Hydroponic Industry in the United States of America. Acta Hortic 481:731–734

Campbell NA, Reece JB (2002) Biology, 6th edn. Benjamin Cummings, San Francisco, CA

Christie E (2014) Water and nutrient reuse within closed hydroponic systems. Georgia Southern University, Georgia

Daly W, Fink J (2013) Economic assessment of hydroponic lettuce production at Goucher college. Goucher College, Baltimore, MD

Despommier D (2010) The vertical farm: feeding the world in the 21st century. St. Martin's Press, New York

Development Initiatives: Global Humanitarian Assistance Report (2016)

Engle CR (2015) Economics of aquaponics. SRAC Publication No. 5006. Southern Regional Aquaculture Center

FAO (2016) Urban agriculture in the Gaza strip through vertical gardens and aquaponics. Food and Agriculture Organization of the United Nations, Rome. fao.org/fileadmin/templates/FCIT/PDF/Fact_sheet_on_aquaponics_Final.pd

Gericke WF (1940) Soilless gardening. Prentice Hall, New York

Gericke WF (1945) The meaning of hydroponics. Science 101(2615):142–143

Goodman E (2011) Aquaponics: community and economic development. Massachusetts Institute of Technology, Cambridge, MA

Graves CJ (1980) The nutrient film technique. Hortic Rev 5:1–44

Graves CJ (1983) Chapter 1: The nutrient film technique. In: Janick J (ed) Horticultural reviews. Avi Publishing Company, Inc. https://doi.org/10.1002/9781118060728

Heredia NA (2014) Design, construction, and evaluation of a vertical hydroponic tower. California Polytechnic State University, San Luis Obispo

Hoagland DR, Arnon DI (1950) The water culture method for growing plants without soil, vol C347. College of Agriculture, University of California, Berkeley, CA

Jensen MH (1997) Hydroponics. HortScience 32(6):1018–1021

Jensen MH (1999) Hydroponics worldwide. Acta Hortic 481(87):719–729

Jensen MH, Malter A (1995) Protected agriculture: a global review. Technical paper 253. World Bank, Washington, DC

Jones JB Jr (2005) Hydroponics: a practical guide for the soilless grower, 2nd edn. CRC Press, Boca Raton, FL

Kratky BA (2009) Three non-circulating hydroponic methods for growing lettuce. In: Proceedings of the international symposium on soilless culture and hydroponics. Acta Horticulturae, vol 843, pp 65–72

Love DC, Fry JP, Li X, Hill ES, Genello L, Semmens K, Thompson RE (2015) Commercial aquaponics production and profitability: findings from an international survey. Aquaculture 435:67–74

Ly HM (2011) Converting soil grown production methods to hydroponics in protected cropping. Nuffield International. No 1014
Lykas C, Katsoulas N, Giaglaras P, Kittas C (2006) Electrical conductivity and pH prediction in a recirculated nutrient solution of a greenhouse soilless rose crop. J Plant Nutr 29:1585–1599
Mathias MC (2014) Emerging hydroponics industry. Pract Hydroponics Greenhouses:18–21
Nederhoff E, Stanghellini C (2010) Water use efficiency of tomatoes in greenhouses and hydroponics. Pract Hydroponics Greenhouses 115:52–59
Pantanella E, Cardarelli M, Colla G, Rea E, Marcucci A (2010) Aquaponics vs. hydroponics: production and quality of lettuce crop, vol 927. Università della Tuscia, Italy, pp 887–893
Resh MH (1995) Hydroponic food production: a definitive guidebook of soilless food-growing methods, 5th edn. Woodbridge Press Publishing Company, Santa Barbara, CA
Resh HM (2013) Hydroponic food production: a definitive guidebook of soilless food-growing methods, 7th edn. CRC Press, Santa Barbara, CA
Resh HM, Howard M (2012) Hydroponic food production: a definitive guidebook for the advanced home gardener and the commercial hydroponic grower. CRC Press, Santa Barbara, CA
Sanchez SV (2014) Avaliação de alface crespa produzidas em hidropônia tipo NFT em dois ambientes protegidos em Ribeirão Preto (SP). Paulista State University, College of Agricultural and Veterinary Science. http://www.fcav.unesp.br/download/pgtrabs/pv/m/2802.pdf. Accessed 9 Sept 2014
Spensley K, Winsor GW, Cooper AJ (1978) Nutrient film technique - crop culture in flowing nutrient solution. Outlook Agric 9(6):299–305
Treftz C, Omaye ST (2015) Comparison between hydroponic and soil systems for growing strawberries in a greenhouse. Int J Agric Ext 3(3):195–200
UNHCR (2021) Global trends: forced displacement in 2016, Geneva
Verner D, Vellani S, Klausen A, Tebaldi E (2017) Frontier agriculture for improving refugee livelihoods: unleashing climate-smart and water-saving agriculture technologies in MENA. The World Bank, Geneva
Wohlgenant MK (2014) Chapter 16: Marketing margins: empirical analysis. Handb Agric Econ 2001(1):933–970
World Bank (2017a) Forcibly displaced. Towards a development approach supporting refugees, the internally displaced, and their hosts. World Bank, Geneva
World Bank (2017b) Water management in fragile systems. Building resilience to shocks and protracted crises in the Middle East and North Africa. Background Paper
World Bank (2017c) Beyond scarcity. Water security in the Middle East and North Africa. World Bank, Geneva
World Bank (2017d) Refugees and water saving agriculture technologies workshop, 17 Aug 2017
World Bank (2019) World Bank Group. 2019. World Bank Group strategy for fragility, conflict, and violence 2020–2025. World Bank Group, Washington, DC

**Dorte Verner** is a Lead Agriculture Economist at the World Bank. Dorte has published extensively in peer-reviewed journals and produced books and reports on climate change, frontier agriculture, value chains and jobs, and poverty reduction. Previously, Dorte was the Principal Economist at the IADB and led the evaluation of IADB's climate interventions. She also served as the Climate Coordinator and developed investment and policy operations for the World Bank. She has won three World Bank Sustainable Development awards for her projects and research. Before joining the World Bank, Dorte worked at the OECD, the European University Institute, and Paris I (Sorbonne). Dorte holds a PhD in Economics.

**Saleema Vellani** is an Innovation Strategist at the World Bank Group. Saleema's work has been focused on Latin America and the Caribbean, the Middle East, and Africa, where she has co-led knowledge products and published papers in the areas of agriculture innovation, food security, and economic development. Before she joined the World Bank, she co-authored several approach papers and evaluations at the Inter-American Development Bank. Saleema currently teaches Design Thinking and Entrepreneurship at Johns Hopkins University. She holds a Master's degree in International Relations and International Economics from the Johns Hopkins University School of Advanced International Studies. You can connect with her at www.saleemavellani.com.

**Elisha Goodman** is an organic farmer at heart, transplanted to the city where she works as a strategist on sustainability, food, waste, innovation, and economic and community development. She is co-founder of Maa-Bara, a social enterprise that took aquaponics overseas to Africa. Elisha holds a Master's in City Planning from MIT and is a certified Permaculturist. She has previously written on aquaponics in her MIT Master's thesis titled, "Aquaponics: Community and Economic Development," and while consulting for the World Bank. Elisha has taught courses at MIT D-Lab on innovation and human-centered design and has conducted international development work in Colombia, Nicaragua, and Kenya. You can connect with her at www.elishagoodman.com.

**David C. Love PhD, MSPH** is an Associate Scientist at the Johns Hopkins Bloomberg School of Public Health, Department of Environmental Health and Engineering and Center for a Livable Future. Dave's works focus on the fisheries and aquaculture sectors using food systems, environmental science, and public health research methods and tools. You can follow Dave's work on Twitter (@davelove1).

# Food Security Achieved Through Utilizing Waste Materials in Part of Durban and Rural Surrounds, South Africa

# 10

M. G. Leech

**Abstract**

Food security for all citizens is provided at Constitutional Level and is further entrenched in other legislation in South Africa. Where citizens are unable to provide these basic needs for themselves, social assistance (in the form of food) should be provided equally to all to ensure a health life for all who live in South Africa. Town Planning and Health legislation and regulations are in place to ensure that the built environment in which we live and work is healthy and safe for all. Yet, food provision by urban agriculturists is ensnared in legislation, ordinances, and regulations, making it difficult to ensure that the safety of food is guaranteed and is fit for human consumption such as the food brought into the city from outside the city. The land that urban agriculturists are using could be a health hazard source and should be checked before use, as should the water sources which are known to be polluted. Care must be taken to ensure that water is treated or filtered before use in agriculture to ensure safe food is produced. The aim of this paper was to establish how safe and feasible it is for poor residents to produce sufficient quality food for their needs. Questions asked of the officials and urban agriculturists revealed that conflict exists between the regulations and the needs, and therefore there is a need to negotiate to ensure that each gets what he or she needs. Sustainability is also an essential target to be strived for through re-use and recycling materials as part of food provisioning. With limited funding available to undertake this work it is imperative that attempts be made to utilize suitable materials readily available to reach the goals and stretch funding to meet food security needs.

M. G. Leech (✉)
IERM, Sandton, South Africa

SACLAP, Gauteng, South Africa



J. A. Diehl, H. Kaur (eds.), *New Forms of Urban Agriculture: An Urban Ecology Perspective*, https://doi.org/10.1007/978-981-16-3738-4_10

**Keywords**

Food security · Town planning · Health regulations · Water · Local authority

## 10.1 Introduction

The migration of thousands of South Africans to urban areas in search of a better life and job opportunities has placed a heavy burden on health and nutrition endeavours (eThekwini: Integrated Development Plan: 145 & 222). In response, many smallholder gardeners and entrepreneurs have attempted to establish vegetable gardens in urban and peri-urban areas to meet the increasing demand for fresh and healthy food. However, various challenges related to town planning compliance, health regulations, safe water, organic waste, and fencing to protect crops from man and animal thefts have been experienced, and the sustainability of these farms and gardens is under threat. Against this background, a mandate of local and provincial governments to ensure food and health security among all communities was acknowledged, and several food security projects were launched among members of the local farming community in various selected areas in and around Durban to encourage the cultivation of nutritious and cost-effective crops of a large variety of vegetables. Participants in the projects formed Agriculture Hubs in their respective areas. This chapter reports on one particular Hub in which we formed a local farming enterprise in Mariannridge and surroundings (Western Area). Our aim was to identify urban food gardeners and provide servicing/guiding and training for them. We also provided organic manure, organic liquid fertilizer, and organic pest control preparations. This chapter illuminates the legislative background that guided this project and the challenges that were experienced. We end with a discussion of the various strategies that were initiated to assist the participating smallholder farmers.

## 10.2 Applicable Essential Legislation in South Africa

Various legislations in South Africa attempt to ensure that the health and well-being of the country's citizens are of a high standard. For example, section 27(1)b of the Bill of Rights, which is entrenched in the Constitution of the Republic of South Africa Act No. 108 of 1996, states that "sufficient food and water is a right" of all South Africans. Furthermore, section 27(2) of the Bill mandates the State to ensure the "progressive realization of this right," be it vested in provincial or local government structures. Section 28 1(c) of the Bill of Rights also states that children have a right to food and water (RSA 1996). Moreover, under the Public Administration section of the same Act, section 195(1), it is stated that food provisioning services as well as all other services need to be provided "impartially, fairly, suitably and without bias" to all residents. Chapter 1 of the National Environmental Management Act No. 107 of 1998 also states that all South African citizens are entitled to "benefits and services that meet "basic human needs" and it urges that humans' "well-being must be pursued" (RSA 1998a, b).

Considering the above, the Municipal Systems Act No. 32 of 2000 states that consultation must take place between the Council and residents to determine what the residents' wants and needs are. According to section 73(i) of this latter Act, attention must be given to the effects of the provisions of the Constitution and, in terms of section 73(a), priority must be given to the basic needs of members of all communities. These needs must be reflected in a municipal Integrated Development Plan as is demanded by sections 2 and 3 of the Municipal Systems Act (RSA 2000).

Historically, prior to the National Elections of 1994, poor communities in South Africa have devised various measures to ensure they had enough food for their survival. One such endeavour has been agricultural activities in rural, peri-urban, and urban environments, including planting on road verges outside their dwelling's (personal observation). This food growing practice was taking place despite regulations issued by the Provincial Town and Regional Planning Commission of the KwaZulu-Natal Land Use Management Scheme (KNZ 2001, 2004) preventing smallholder farmers from practising agriculture in built environments in KwaZulu-Natal, which includes Durban metro (KZN 2001, 2004). Fortunately, more recent national legislation in the form of the Land Use Management Act No. 16 of 2013 has allowed various changes to earlier land use legislations "in the public interest" and, more specifically, "in the interest of a disadvantaged community" (RSA 2013). The urban agriculturists believe that the changes laid down in the legislation should be put in practice but unfortunately the officials are not in a hurry to implement this legislation.

According to this legislation, provision must be made for an Integrated Development Plan by municipal councils to ensure the development of communities. Councils must also provide the necessary funding to address the identified needs of society, one of which is undeniably ready access to affordable fresh produce which has resulted in a growing demand for urban agriculture space (RSA 2000). However, the results of a recent survey (Leech 2014) showed that no proper consultation had been initiated by the Durban Metro Council or officials, and thus no provision could be made in the Integrated Development Plan (IDP) of the Council for the development of urban agricultural endeavours in this region, albeit that such developments are both viable and necessary.

The viability of the development of urban agriculture in the Durban area is undeniable, as numerous undeveloped pieces of public open spaces exist in all residential areas (Leech 2000). There are also open spaces available on school properties that are ear-marked for the development of sport fields, but that are currently undeveloped due to a lack of funds. These sites, that currently lie fallow, could be partially developed as community gardens if residents/parents of pupils can grow vegetables and fruits here as recreational, horticultural, and even educational endeavours until funding becomes available to develop these sites. The soil, which is a sandy loam, could be greatly improved whilst these pieces of land are used for horticultural/agricultural purposes. However, horticultural practices on such available pieces of land have not been allowed under town planning regulations, even though it would save the gardeners the cost of leasing other properties as these properties are classified as Public Open Space and belong to the Council. It would also save the Council the expense of cutting the rank grass and undergrowth on such public open space properties, as demanded by health by-laws when owners fail to respond to notices served on them to clear the overgrown vegetation. Well-devised and supported agricultural activities could also protect valuable trees and other vegetation in these areas, especially near water courses.

## 10.3 Nutrition and Health: Food Safety

Environmental Health Practitioners are tasked in terms of the Health Act No. 61 of 2003 to ensure the whole food production cycle from soil to table is safe and healthy. Unfortunately, the food produced within the urban built environment by passes the checks and balances introduced to ensure safe and healthy food. Therefore, even though consumers should be protected by regular inspections are not happening with urban grown food.

The National Environmental Health Norms and Standards Notice 1225, 2015 highlights nine important check points for food and nutritional health. Five of the check points are applicable to food production, demanding water quality monitoring, food control, health surveillance of premises, environmental pollution control, and chemical safety associated with food. However, the results of a recent survey questionnaire revealed that the necessary attention had not been given to these important check points by the environmental health practitioners (Leech 2014). As a result, the onus is on the appointed urban agriculturists to attend to these check points, whilst rolling out the new food security projects. This point is further confirmed by the Annual Inspection Report of the Health Sector (2016/2017).

One contentious area in the food security programme is that the Health Act, 2015 does not state whether the Provincial or the Municipal (i.e., the local) health departments are responsible for urban pollution control. This important control aspect has not yet been resolved, as was confirmed by a survey that assessed the impact of the quality of water available for human consumption and food production in the survey area (Western Area of eThekwini) (Leech 2014). The findings from the survey established the need for those rolling out the food security program to monitor essential food security check points. Unfortunately, due to previous limitations in the land use scheme (as described above), locally grown food was produced in conflict with town planning regulations and was not checked to the same rigorous control levels that food brought into a city or town adheres to.

## 10.4 Water Security and Garden Irrigation

Water is a critical component in food production but is often in short supply in the greater Durban Metro, especially during autumn and winter (May to September). Water use is governed by the National Water Act No. 36 of 1998. According to schedule (1)(a) of this Act, water flowing in urban and peri-urban areas may be sourced for reasonable domestic use, whilst section 1(b)(ii) allows the use of such water for small gardening, e.g., 4 square metres, but not for commercial purposes (RSA 1998a, b). For larger market-oriented gardens, e.g., 1 ha, a water permit must be obtained and paid for by the gardener/agriculturist. In the Durban Metro, such gardeners/agriculturists are in most instances unemployed and have no fixed income (Leech 2014).

Because water is scarce during the dry season, farmers rely on roof run-off water that is captured in tanks and sumps during the rains September to April. Tank inlets are fitted with first flush filters to remove dirt and leaves before being stored in above ground tanks or underground sumps. Water may also be sourced using modified storm water catch-pits which was an initiative devised by the author in 2013 (Leech 2013). The modified road storm water catch-pit consists of a sump on one side—usually the top—and built-in filter screens to collect road surface dirt. The sumps collect and retain about 500 l of water. A storm water pipe is connected to the outlet side to lead the excess water away. However, because most streams and rivers in the study region run through inhabited areas, the water is known to be polluted (eThekwini Aquatic Biomonitoring Report, July 2010). Thus, the storm water collected in these sumps cannot be used safely for human consumption or for the growing of vegetables. One solution to safeguard small quantities of water, e.g., 20 l units, is to stir a cup of liquid bleach into the water and allow it to settle for several hours before use.

Another method used for open-air larger tanks or weirs is to leave receptacles open to the elements. These receptacles are left in the sun for several days before the water is used. Larger quantities of water can be gravity filtered through a mobile wetland that consists of a $4 \times 2 \times \frac{1}{2}$ m fish tank on stands. These tanks are planted with wetland type plants that are stacked so that water will flow through them and overflow into a tank below. The flow is repeated until the water reaches a final tank where it is safe to use. Gardeners typically use pedal water pumps fitted with non-return valves in the pipeline to pump the water out of a stream and into the mobile wetland.

Drip irrigation with gravity-fed tanks is another method used by smallholder farmers to filter storm water (Leech—personal experience; Fig. 10.1).

Some farmers also use water retaining crystals that are added to soil being planted with food seedlings. The use of crystals dramatically reduces seedling losses (personal experience). Unfortunately, water retaining crystals have a limited lifespan and must be replaced bi-annually.

Whilst water is readily available in streams and rivers and can appear clean and safe for use within the home environment for watering, washing, and drinking, this is not the case within the built environment of eThekwini. The Environmmental Health Practitioners in terms of their duties and responsibilities laid down in legislation they are required to ensure that the water is safe for use (RSA 2003) through tests and inspections. Any other use of water is controlled by Part V of the Environmental Conservation Act No. 73 of 1989.

**Fig. 10.1** A gravity feed tank and drip irrigation in use during the winter period when no rain is expected. (Source: author)

## 10.5 Implementing Urban Agriculture Support

Zulu citizens were traditional rural farmers who grew food close to their homes, which were mainly attended by the women of the house. With the family migration to the cities there was still the need for producing food whilst they looked for paid work.

### 10.5.1 Amending the Soil

The initiative to support urban farmers by the municipality in the study area commenced in 2013. The program framework was to educate the urban agriculturists on how to grow enough food bio-intensively, firstly, to feed their immediate family, thereafter, to feed those in need. Critical to this process was the fact that the same piece of land was to be used so that it was producing food on a continuous basis by aerating and enriching the soil. In the roll-out phase of the project we had to contend with viable vegetable growing areas that had previously been planted and harvested and then left fallow. With limited expertise in checking the safety pros and cons of an urban agriculture area (soil had not been checked to see if it was safe to use), we had to progress with the roll out of food growing expertise and related materials to ensure that enough food would be produced eventually.

One important focus was to ensure fertility of soil. Soil enrichment using compost or manure can be expensive and, therefore, due to a limited budget, alternative organic materials had to be sourced in this case. Fortunately, horse manure was available free of charge from racing stables in the area—the manures only had to be transported to the patches of land before being used for food cultivation. Care was taken to source manure only from stables where medication had not been given to the horses on a regular basis. Such medication could potentially leach into the manure and be taken up by plant roots to be stored in edible parts of the plant. In addition to concerns about medication bioaccumulation, the fresh hay and saw dust found in the manure also had to be treated by composting before use. To do this, the manure was left until the dry material had broken down, at which time the manure was ready for application. The manure was stacked in windrows with air vents (i.e., creosoted poles were spaced 3–5 m apart). The poles were removed once a windrow was at the correct height and air could flow through to decompose the piles of manure. Water was also added to hasten the decomposition process.

Soil was also enriched using mulch in the form of bark and wood chips, twigs, and small branches. Mulch was readily available from the Municipality as a result of workers who regularly pruned and chipped street trees. Mulch was only applied when needed, and other material, e.g., Vetiver grass and Napier fodder, was more commonly found in damp/moist areas, this was then shredded and used as mulch among the plants.

To further enrich the soil, a minimum quantity of 100 l per garden of various types of liquid fertilizer per month had to be manufactured for large scale distribution. Liquid fertilizers, including manure tea, compost tea, and green tea, the respective materials needed, were placed in cheesecloth in containers and aerated. Three sizes were mixed: when 22 l containers were used, the material (manure, compost, or green leaves) was augmented with 1.5 kg mushroom compost or comfrey/borage leaves or horse manure, plus 25 ml molasses, 25 ml seaweed emulsion, and 5 ml citric acid. Using 500 l containers, a mixture of 3.4 kg of material was placed in cheese cloth bags, and 585 ml (2.3 cups) molasses, 585 ml seaweed emulsion, and 115 ml citric acid (half a cup) was added. Quantities were doubled when 1000 l tanks were used. A compressor was used to pump air through solid pipes to the bottom of the tank where the air could bubble up for 48 h. The mixture was decanted into previously used, but clean, 5 l containers (donated by the office staff) for distribution during site visits or at monthly meetings attended by the farmers. The liquid fertilizer material was also pumped into a water tanker before pumping any rain/storm water into it. The water was then delivered to the gardens where it was needed. Farmers in outlying areas were also provided with 20 l containers, fish tank air pumps, and enough piping, air stones, and molasses to manufacture their own liquid fertilizer. Small quantities of the liquid fertilizer could be kept in a refrigerator for up to 21 days should any farmer need more. After that period, it would need to be discarded on a compost heap because the micro-organisms had a short shelf-life.

Another form of compost, known as worm compost, was made from left-over plant material from the liquid fertilizer process, left-over vegetables from the gardens, and discarded vegetables collected from vegetable retailers. Worms breed out in such discarded material and work their way through from the bottom to the top, making rich worm castes and worm "tea" which is better than normal compost and liquid fertilizer (Personal Training Received in Seattle in 2009). Unfortunately, only a limited quantity could be produced due to the limited vegetative material available.

### 10.5.2 Containing and Securing Urban Gardens

Fencing food gardens is critical because of the loss of valuable food. A common and cheap fencing type known as "Bonnox" was decided against using around food gardens because pigs, chickens, goats, and other animals can squeeze through the openings and destroy the vegetables and seedlings. Instead, packing pallets were procured, after ensuring that the planks were still well attached. The wood was sealed, and the pallets were erected side by side and attached to pieces of discarded electrical wooden poles cut to the correct length and buried half a metre in the ground. Recycled fencing wire was attached to the pallets so that food/fruit bearing creepers such as granadillas could be planted and trained up the wires. Pallets could also be joined and used as inverted V frames to train beans and peas to grow up them. We also used thorny plants such as the indigenous Kei Apple (*Dovyalis caffra*) which produces edible fruit and *Agave americana* along the fence-lines when other fencing material was not available.

Due to the topography—in some cases a slope of one in 50 m was found—terraces were constructed along the natural curves of the land and sloped downhill. This ensured that any rain from regular thunderstorms would collect on the terraces and percolate into the soil. Terraces were made 1.8 m wide to accommodate a 0.6 m path and a 1.2 m bed. Culinary herbs were grown on the slope above the path and food was grown in the bed next to the path.

The beds were first covered with suitable compost and then dug using a double dig board according to the John Jeavon's method (Jeavons 2001). This process takes the compost down to a depth of two spade lengths into the soil whilst aerating it so that the maximum amount of food can be grown in the least space. The beds allowed people to kneel on the edges to weed and plant, without standing in the beds thereby pressing out the air needed for ideal water penetration. The double dig fork (see Fig. 10.2a) was used in alternate years to ensure the beds were aerate properly to two spade lengths for root penetration when new plantings were done.

The paths around the gardens were constructed using rejected concrete fence panels laid end to end and patched where there were breaks or cracks. The paths allowed easy access of wheelbarrows to bring in compost and mulch for use in the beds.

Planting templates were made using off-cut hardboard and old street signs (Fig. 10.2b). The different planting distances were marked on the boards in permanent ink and the templates were used to indicate planting spots. Planting distances ranged from 5 cm for peas and radishes to 35 cm for cabbages and kale.

a b

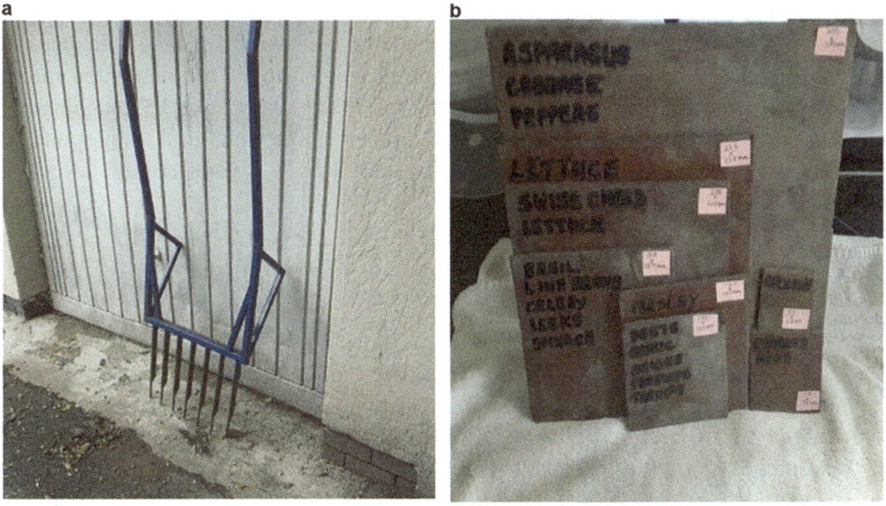

**Fig. 10.2** Critical tools for the double dig process (**a**) Double Dig Fork—Made from Spring Steel (**b**) Vegetable Planting templates made from off-cut hardboard. (Source: author)

### 10.5.3 Pest Control

Pest control was applied when necessary and care was taken to ensure that gardeners did not purchase and use chemical pest control products, which require a waiting period after application and before harvesting due to potential negative impact on food consumers' health. In the case that chemical pest control products were used, recommended waiting periods were implemented where crops were sprayed before the crops were harvested.

A variety of non-chemical options were made available for dealing with pests. A selection of organic pest control materials in liquid form was provided with returnable handheld spray units for use. For instance, to control ant infestation, an extract of orange skin oil in water with an added wetting agent was provided. This mixture was obtained by placing five large orange skins in 2 l of water for several days. The oil leaches out into the water and, by adding two teaspoons of liquid soap, an effective ant repellent is created. During the rainy season, mildew was combatted by using the water rinsed from milk containers with the necessary wetting agent, which helped it stick to sprayed surfaces. A combination of onions, chillies, and garlic was also used as a successful pest control agent. A mixture was made by boiling ingredients in water and allowing the mixture to cool before adding the wetting agent. Onions, chillies, and garlic were also dried and kept in airtight packets at the garden sites. When required, boiling water was added to the contents of the packets and allowed to cool. The wetting agent was then added before spraying. Great success was also achieved when five large leaves of rhubarb were boiled in water. When cooled, a wetting agent was added, and the mixture was as used as a pest repellent. To control moles, the green seed pods and leaves of *Datura stramonium* were placed in the mole runs, which successfully curbed their activities. In areas where carrots and turnips were planted, the beds were trenched, and chicken wire was laid down before pacing the planting soil on top. This prevented moles from consuming crops from underneath. And, birds were kept away by attaching old DVD disks to pieces of string, hung on trees and, when the wind blew, they spun and reflected the rays of the sun in different directions. This was a surprisingly effective bird repellent.

### 10.5.4 Nutrition

Due to the high levels of malnutrition reported by clinic staff in the case study area (personal communication with nursing staff in the area), the farmers were encouraged to grow high calorie vegetables (Table 10.1; Fig. 10.3) in preference to the usual crops such as carrots, spinach, and cabbages. Traditional harvests were not discouraged, however, but an effort was made to educate farmers about food that would provide a more balanced diet. Laminated flyers showing the nutrient value of vegetables were disseminated. These flyers reflected the protein, calorie, and calcium content of vegetable varieties that could be planted. A list of companion plants was also provided.

## 10.6 Future Development Plans

With the use of the bio-intensive methods of food production, large quantities of food can and have been produced through the Mariannridge agricultural hub region. These excesses can be distributed through NGOs and Churches to the poor, or through to the school feeding schemes in the area. Plans were in the making to

**Table 10.1** List of vegetables and their nutrient value

| Variety | Protein content (g) | Calorie content | Calcium content |
|---|---|---|---|
| Artichoke | 5.3–7.2 | 213–345 | 44–93 |
| Beans, Lima bush | 92.5 | 1533 | 327 |
| Beans, white | 102 | 1510 | 499 |
| Beans, mung | 109.8 | 1574 | 535 |
| Peas, bush | 109.4 | 1542 | 612 |
| Peanuts | 117 | 2572 | 313 |
| Carrots | 4.1 | 195 | 134 |
| Mealies | 8.7 | 400 | 7 |
| Spinach | 10 | 86 | 367 |
| Onions | 6.2 | 172 | 111 |
| Eggplant/Brinjal | 4.4 | 118 | 44 |
| Tomato | 5 | 95 | 59 |

Source: Jeavons (2001)

**Fig. 10.3** Artichoke Plants grown in the Mariannridge Depot for consumption and training. (Source: author)

establish deep freeze units and pantries within the community halls located in the nearby suburbs. Excess food could then be converted into deep freeze produce, bottled in brine (salt-water solution), or dried for later distribution and consumption. Value-added products could also be made and bottled for sale or later use, for example, in one case, an excess of beetroot was converted into beetroot chutney which was sold before the produce was bottled.

## 10.7 Conclusion

Malnutrition and diarrhoea are very prevalent amongst the children in South Africa and the cause of the high mortality rate in this age group (eThekwini 2017). Malnutrition is also prevalent in the adult population, especially those suffering from HIV and TB who need food sustenance to help them recover and lead a reasonably normal life (eThekwini 2017). The food safety focus by officials is limited at the retail level and misses other food sources that as a result by-pass the inspections being done. But this inspection oversight needs to change to ensure food and water from all sources is safe for consumption. Critical attention also needs to be given to ensuring that provision is made for the food and water needs of all the community equally in terms of legislation. With limited funding available, alternative sources of material that can be used to ensure that a sustainable source of safe food and water must be found and utilized. This will enable the budget to stretch as far as possible. And, it will protect the rights of all the citizens for safe food and water are met.

## References

eThekwini Municipality (2010) The state of health and integrity of rivers within the eThekwini Municipal area. eThekwini Aquatic Report, July, Internal Document

eThekwini Municipality (2017) Integrated development plan 2017–2022. Internal Document

Jeavons JC (2001) Biointensive sustainable mini-farming: II. Perspective, principles, techniques and history. J Sustain Agric 19(2):65–76

Kwa-Zulu-Natal. Provincial Planning and Development Commission (2001) Land use management system: guidelines for the preparation and implementation of schemes (updated). Kwa-Zulu Natal Provincial Planning and Development Commission, Pietermaritzburg

Kwa-Zulu-Natal. Provincial Planning and Development Commission (2004) Land use management system: guidelines for the preparation and implementation of schemes (updated). Kwa-Zulu Natal Provincial Planning and Development Commission, Pietermaritzburg

Leech MG (2000) Development strategy for playground provision. M. Dip. Thesis, Tshwane University of Technology

Leech MG (2013) Plan of water capture catch pit (Not published)

Leech MG (2014) Strategy for a viable, sustainable urban agriculture in a dynamic, urbanising society. Doctoral Thesis, UNISA

South Africa (1996) Constitution of the Republic of South Africa. Act 108 of 1996. Government Printers, Pretoria

South Africa (1998a) National Environmental Management Act No. 107 of 1998 [Online] www.info.gov.za/view/downloadFileAction?id=7064. Accessed 3 Dec 2012

South Africa (1998b) National Water Act-Act 36 of 1998. www.info.gov.za/viewDownloadFileAction?id=70693. Accessed 3 Dec 2012

South Africa (2000) Local government: Municipal Systems Act 32 of 2000 [Online]. https://cisp.cachefly.net/assets/articles/attachments/03552_locgovmunsystact32.pdf. Accessed 18 May 2020

South Africa (2003) National Health Act 61 of 2003 [Online]. www.info.gov.za/viewDownloadfileAction?id=68039. Accessed 3 Dec 2012

South Africa (2013) Spatial Planning and Land Use Management Act 16 of 2013 [Online]. http://www.sacplan.org.za/documents/SpatialPlanningandLandUseManagementAct2013Act16of2013.pdf. Accessed 23 Sept 2013

**M. G. Leech** was born in Windhoek, Namibia, and schooled in Pretoria and Boy's Town South Africa. He was trained as a horticulturist in Pretoria and Pinetown and obtained three national technical certificates as well as two national diplomas and a higher diploma in horticulture. He branched out and obtained a Master's in Parks and Recreation Management. He learned about Urban Agriculture through an American Community Garden Scholarship and applied what he learned when he returned. He commenced a research thesis on Urban Agriculture in his region earning him a doctorate in Environmental Management. He has branched out further into the field of arboriculture, organic pest control, and the manufacture and use of organic fertilizer.

# Part III

# Labor

# Contextualizing Urban Agriculture in Quito, Ecuador: A Look at Urban Production and Producer Traits

# 11

Kate Oviatt

**Abstract**

Research on urban agriculture has identified a great number of benefits, including, but not limited to, improved food security and increased economic well-being. While such outcomes provide strong reasons for engaging in urban agriculture, it is important to recognize that these benefits are not experienced uniformly among all who participate in urban agriculture. Rather, the benefits must be understood in relationship to the characteristics of urban producers. The characteristics of urban producers will heavily influence who engages in urban agriculture, the reasons they have for engaging in it, and the type of benefits that they realize from engagement. This chapter uses findings from a case study on the AGRUPAR urban agriculture program in Quito, Ecuador to explore how the practice of urban agriculture differs among producers based on three primary characteristics: migration history, age, and gender. The findings from this case study demonstrate how the personal characteristics of producers can influence how urban agriculture manifests and the benefits associated with it, underscoring the importance of taking producer traits into consideration when studying urban agriculture.

**Keywords**

Urban agriculture · Gender · Migration · Age · Livelihood · Ecuador

K. Oviatt (✉)
University of Colorado Denver, Denver, CO, USA
e-mail: katie.oviatt@cuanschutz.edu



J. A. Diehl, H. Kaur (eds.), *New Forms of Urban Agriculture: An Urban Ecology Perspective*, https://doi.org/10.1007/978-981-16-3738-4_11

## 11.1 Urban Agriculture and Its Producers: A Case Study

The practice of urban agriculture is a worldwide phenomenon; the Food and Agriculture Organization of the United Nations (FAO 2019) estimates that approximately 800 million people engage in urban agriculture in some form. As urban populations continue to grow, interest in urban agriculture and its potential benefits for urban residents has increased, with cities around the world integrating it into city plans, policies, and community development efforts as a way to address growing issues of poverty, health, and well-being.

An example of one such effort is the urban agricultural project, AGRUPAR, in Quito, Ecuador. AGRUPAR (*Agricultura Urbana Participativa*: Participatory Urban Agriculture) is a long-standing program through the municipal government of Quito that trains residents in urban agricultural production and provides extension services for those who complete the program, including access to inputs, infrastructure, and help from professional agronomists. AGRUPAR supports thousands of producers throughout the city, who range from hobbyist gardeners with patio-sized container gardens, to larger-scale producers who depend on intensive production for their livelihood. The program targets low-income communities, using urban agriculture as a tool for addressing poverty and food insecurity, but membership is open to anyone interested in joining.

The potential benefits of urban agriculture, in general, and of the AGRUPAR program, in particular, are myriad, ranging from pragmatic to transformative. Urban agriculture has been used as a means for addressing both urban poverty and urban food insecurity; food production in the city gives urban households increased access to healthy food (Corrigan 2011; Litt et al. 2011; Zezza and Tasciotti 2010), while also providing economic benefits through the sale of garden products and saving money on food (Bryld 2003; Cook et al. 2014; van Veenhuizen 2006). It also has been associated with environmental gains (Ackerman et al. 2014; Brown and Jameton 2000; Galluzzi et al. 2010), improvements in physical and mental health (Brown and Jameton 2000; Hale et al. 2011; WHO 2016), as well as individual and social benefits such as increases in self-esteem, confidence, gains in social capital, and greater community engagement (Battersby and Marshak 2013; Bradley and Galt 2014; Brown and Jameton 2000; Olivier and Heinecken 2017; Pudup 2008; Teig et al. 2009; Webber et al. 2015).

To understand how these potential benefits manifest in the real lives of urban agricultural producers, research was undertaken with participants in the AGRUPAR program in Quito. A critical insight that emerged from researching the AGRUPAR program was that, while the program provided participants with the same training and extension services, there was significant variability in terms of how participation in the program affected the lives of participants. Notably, participants' engagement in urban agriculture was mediated through personal characteristics that situated them differentially in relationship to urban agriculture, such that how they engaged in it and the effects it had in their lives varied.

The findings of this research indicate that the characteristics of producers must be considered in order to understand both the variations in how urban agriculture is practiced, as well as the benefits associated with it. This chapter will explore the case study of AGRUPAR, focusing in particular, how the practice of urban agriculture was found to be influenced by three producer characteristics: migration history, age, and gender.

## 11.2 Researching AGRUPAR

The AGRUPAR program has been active in Quito since 2000 when it was implemented in partnership with the International Development Research Centre as part of an effort to understand how municipal governments could facilitate urban agriculture. Since then, the program has become an important part of the city's effort to address both food insecurity and un/under-employment among city residents. The program's mission is to "work to fight against poverty and to improve the living conditions of vulnerable groups by producing healthy food, creating employment opportunities and improving income, while also encouraging environmental stewardship, conserving indigenous knowledge, and promoting unity and solidarity among participants" (CONQUITO 2015).

In an effort to meet these goals, AGRUPAR has developed a multi-pronged approach to encourage and support urban agriculture among its participants. All participants complete a comprehensive training program that provides them with the knowledge and skills they need to become urban agricultural producers (Fig. 11.1a). As part of its commitment to foster both health and environmental stewardship, AGRUPAR's training is based on organic, agro-ecological methods, an approach that avoids the use of petrochemical inputs. In addition to this training, a key part of AGRUPAR's success is due to the comprehensive support that it gives to producers across the chain of production. Some of the key ways in which the program supports its producers include, but are not limited to, the provision of ongoing extension services from professional agronomists (Fig. 11.1b), the provision of free or discounted materials for irrigation systems, greenhouses (Fig. 11.1c), and organic seeds and inputs, access to the city's specialized markets to sell agricultural products (Fig. 11.1d), and assistance in becoming officially certified organic producer. This comprehensive approach provides participants with extensive support, making it far more likely that they will be able to successfully engage in urban agricultural production.

To understand how participants' lives had changed since joining the AGRUPAR program as urban agricultural producers, fieldwork was conducted in Quito, Ecuador from 2014 to 2015. A mixed methods approach was applied, utilizing both quantitative and qualitative data collection methods to explore changes within economic, social, health, environmental, and personal domains of participants' lives.

The first stage of research was the administration of a survey. Two hundred gardens registered with AGRUPAR were randomly selected (representing approximately 29% of the gardens in the program), and a survey was conducted with a participant associated with the garden. A total of 192 surveys were included in the final analysis. Surveys were administered in the comfort of participants' homes or

**Fig. 11.1** Images of the AGRUPAR program: (**a**) AGRUPAR group training; (**b**) Agronomist visit; (**c**) Greenhouse in neighborhood; (**d**) Producers at AGRUPAR's *bioferia*. (Source: K. Oviatt)

gardens and were designed to capture their perspective on how their lives had changed since they started their garden. The data from the surveys provided a macro-level understanding of how participants' lives had changed since they joined AGRUPAR, specifically considering changes in economic well-being, social engagement, health status, environmental behavior, and participant's sense of agency.

The survey was followed by in-depth, semi-structured interviews with a sub-sample of participants. Within the sample of survey participants, a quota sampling approach was used to ensure that participants with diverse traits were represented in the interviews, selecting participants by gender, age, and selling status (whether or not they sold their garden products), for a total of 18 interviews. The interviews were organized around the same domains as the survey and, while the questions were the same for each participant, the responses were open-ended, allowing participants to freely express themselves. A semi-structured format ensured that responses were comparable across interviews and able to gain participants' perspectives about the changes they had experienced in each domain.

Findings from both methods were integrated to develop an understanding of urban agriculture participation, practice, and effects in participants' lives. Data from the 192 surveys were analyzed using descriptive techniques including frequency distributions and comparison of means. Preliminary findings from the survey were used to inform the development of the interview guide. Data from the 18 interviews were coded using a combination of coding strategies to capture themes that emerged from the data (open-coding) as well as theoretical constructions defined prior to fieldwork (a priori coding). This mixed methods approach provided a fuller understanding of the changes participants had experienced than either method could on its own. The quantitative data provided breadth, giving an understanding of how participation effected change among participants as a whole and made it possible to compare variations among different groups of producers. Conversely, qualitative data from the interviews provided great depth, giving insight into the details of how participants' lives had changed and what mattered most to participants themselves.

The relevancy of different producer traits in understanding urban agriculture emerged from this analysis. The survey data revealed that the reported changes within the five primary domains were not experienced uniformly among all participants. Rather, they were experienced differentially based on certain producer traits, most notably, migration history, age, and gender. Interview data supported this with participants from these different backgrounds expressing unique experiences, priorities, and perspectives. The remainder of the chapter will be dedicated to exploring how these particular characteristics influenced the practice of urban agriculture and the benefits associated with it for AGRUPAR participants.

## 11.3 Findings: Comparing Differences in Urban Agricultural Producers

### 11.3.1 Migration History

The first characteristic found to have some effect on the practice of urban agriculture was the migration history of producers. While there were many similarities between people who had migrated to Quito and those who were from the city (Quiteños),

**Table 11.1** Migration history summary

| | Quiteños | | All migrants | | Long-term migrants | | Recent migrants | | Total | |
|---|---|---|---|---|---|---|---|---|---|---|
| | % | *n* | % | *n* | % | *n* | % | *n* | % | *n* |
| | 58 | 118 | 37 | 74 | 34 | 65 | 5 | 9 | 100 | 192 |
| Sell products | 81 | 95 | 65 | 48 | 63 | 41 | 55 | 5 | 74 | 143 |
| Sell at *bioferias* | 33 | 39 | 24 | 18 | 25 | 16 | 10 | 1 | 29 | 57 |
| Bottom quartile of earners[a] | 16 | 15 | 31 | 15 | 32 | 13 | 22 | 2 | 34 | 48 |

[a]Percentage calculated based on total number of producers who sell products, not total number of producers

there were some notable differences. For the purposes of this research, participants who moved to the city as infants (age 2 or younger) were considered as being from Quito. Participants who had moved to the city were more likely to have a background in agriculture. In interviews, many described how as a child they had lived in rural provinces of Ecuador and had parents who depended on agriculture for their livelihood. This background was part of what inspired them to become a part of AGRUPAR; agriculture was an activity they enjoyed, not just a means to increased food security or economic savings.

The most significant aspect in which migrants differed from Quiteños was in terms of economics (Table 11.1). Migrants were significantly less likely to sell their garden products compared to producers from Quito. The vast majority (81%) of Quiteños sold at least some of their products, compared to 65% of migrants. When considering the most recent migrants (people who moved to Quito in the last 15 years), the number who engaged in sales dropped to 55%. Similarly, producers from Quito were more likely to be a part of AGRUPAR's *bioferias*, which were markets organized by the municipality specifically for AGRUPAR producers to help them sell their products: 33% of producers from Quito sold at the *bioferias*, compared to 24% of all migrants, and just 10% of more recent migrants. Additionally, migrants who sold their products reported lower revenue from sales than producers from Quito: 31% of migrants that sold their products were in the bottom quartile of earners (earning less than $50 USD a month in sales), while just 16% of Quiteño sellers were among the bottom earners.

### 11.3.2 Age

The second characteristic found to influence how people engaged in urban agriculture was age (Table 11.2). Producers were categorized into three age groups: younger (18–34 years), middle aged (35–54 years), and older (55+ years). First, the age of producers had some effect on the economic aspects of urban agriculture. Producers engaged in selling in distinct ways depending on their age: a substantial majority (76%) of middle aged and older producers sold their garden products, compared to just over half (56%) of younger producers. They are also more likely

**Table 11.2** Age summary

| | 18–34 years | | 35–54 years | | 55+ years | | Total | |
|---|---|---|---|---|---|---|---|---|
| | % | *n* | % | *n* | % | *n* | % | *n* |
| | 8 | 16 | 52 | 100 | 39 | 75 | 100 | 191 |
| Sell products | 56 | 9 | 76 | 76 | 76 | 57 | 73 | 142 |
| Sell in *bioferias* | 19 | 3 | 30 | 30 | 32 | 24 | 29 | 57 |
| Bottom half of earners[a] | 78 | 7 | 50 | 38 | 42 | 24 | 49 | 69 |
| Primary benefit of UA is providing food to family | 47 | 7 | 54 | 53 | 47 | 35 | 50 | 95 |
| Primary benefit of UA is helping the environment | 33 | 5 | 10 | 10 | 9 | 7 | 12 | 22 |
| Primary benefit of UA is enjoyment of activity | 7 | 1 | 16 | 16 | 23 | 17 | 18 | 34 |
| Much more active in community | 56 | 9 | 60 | 59 | 40 | 29 | 51 | 97 |
| Know community much better | 50 | 8 | 54 | 53 | 27 | 20 | 43 | 81 |
| Much more confident with others | 75 | 12 | 75 | 74 | 53 | 39 | 66 | 125 |
| Much more self-confident | 93 | 15 | 80 | 80 | 65 | 49 | 75 | 144 |

[a]Percentage calculated based on total number of producers who sell products, not total number of producers

to sell in the *bioferias*: just over 30% of middle aged and older producers were *bioferia* sellers compared to 19% of younger producers. Additionally, when they did engage in selling, younger producers tended to earn less: 78% of young producers were in the bottom half of earners compared to about half of middle aged and older producers.

Younger and older producers also varied in what they most valued about practicing urban agriculture. The benefit that was most valued by producers of all age groups was that they were able to produce food for their families (between 47% and 54% across age groups). But beyond this shared value, different age groups of producers diverged in other valued benefits. For example, younger producers were more likely to say that helping the environment through organic, agro-ecological production methods was the primary benefit they valued about having a garden (33% compared to 9–10% of older and middle-aged producers). As producers increased in age, they were more likely to cite the enjoyment of working in a garden as the primary benefit: 23% of older producers as compared to 16% of middle-aged producers and only 7% of younger producers.

There was also variation in terms of social changes that producers experienced as a result of being a part of AGRUPAR. Younger and middle-aged producers were more likely to say that they were more active in their neighborhoods, with 56–60% saying they were much more active, compared to 40% of older producers. Participants self-defined what they meant by active, but examples include talking more with neighbors, leaving their house more, or working with others in the community towards a common goal. The same age groups were also more likely to say they felt they knew their neighborhood better after having participated in the program: 50–54% of younger and middle-aged producers said they knew their neighborhood much better compared to 27% of older producers.

On a more personal level, younger and middle-aged producers also experienced greater changes in their feelings of confidence. When asked if they had more confidence when talking with their family and friends as a result of their participation in AGRUPAR, 75% of younger and middle-aged producers said they had gained a lot more, compared to slightly more than half (53%) of older producers. Similarly, younger and middle-aged producers also experienced greater increase in their self-confidence as a result of participation in the program: 93% of younger producers and 80% of middle-aged producers said they had gained a lot more self-confidence, compared to 65% of older producers. While all age groups made gains in confidence, younger and middle-aged producers gave much more positive answers and appear to have experienced greater gains.

### 11.3.3 Gender

Gender was found to be a third characteristic with a substantial influence on how people engaged with urban agriculture. Men and women varied significantly as urban agricultural producers in terms of how they practiced and the benefits they experienced. Participation in the AGRUPAR program was dominated by women: at the time of research, the program was 67% female. The predominance of women was most likely due to the fact that many of the women in the program were housewives, a traditional and common role for women in Ecuador, and urban agriculture was an activity that could easily be incorporated into their domestic roles. In contrast, the men of the program tended to be older (62 years old on average as compared to 49 years for women) and often took up urban agriculture as an activity once they had retired.

In terms of economic benefits of urban agriculture, there were stark differences between men and women (Table 11.3). While a higher proportion of women engaged in the sale of garden products (77%) compared to men (61%), men generally experienced greater economic returns on their garden sales. When women sold their garden products, they earned an average of $119 a month from sales. Men, in comparison, earned an average of $215 a month, nearly $100 more on average! Consequently, men were much more likely to be in the top quartile of earners: 41% earned over $200 in sales each month, compared to just 19% of women sellers. However, while men had greater absolute income from urban agriculture, women reported experiencing greater relative gains: 59% of women said their income had increased since they started a garden, compared to 42% of men. Importantly, 18% of women said their income had increased a lot, while just 5% of men gave the same response.

Beyond the differences in earnings, men and women also differed in the social benefits they experienced; women appeared to make more gains in developing relationships and engaging with others in their communities as a result of participating in the AGRUPAR program. When asked if they had developed

**Table 11.3** Gender summary

| | Women | | Men | | Total | |
|---|---|---|---|---|---|---|
| | % | *n* | % | *n* | % | *n* |
| | 84% | 161 | 16% | 31 | 100% | 192 |
| Sell garden products | 77% | 124 | 61% | 19 | 75% | 143 |
| Avg monthly earnings from sales (USD) | $119 | 124 | $215 | 19 | $131 | 143 |
| Top quartile of earners[a] | 19% | 22 | 41% | 7 | 22% | 29 |
| Experienced an increase in income since joining[a] | 59% | 72 | 42% | 8 | 57% | 80 |
| (Increased very much) | (18%) | (22) | (5%) | (1) | (16%) | (23) |
| Developed relationships with others in program | 81% | 125 | 55% | 17 | 77% | 142 |
| Developed relationships with others in community | 82% | 131 | 67% | 20 | 151% | 80 |
| More active in community | 82% | 130 | 55% | 16 | 78% | 146 |
| More confident speaking in public | 75% | 111 | 62% | 18 | 73% | 129 |
| More confident with others | 94% | 150 | 72% | 17 | 91% | 171 |
| (Much more confident) | 73% | 117 | 31% | 9 | 67% | 126 |
| More self-confident | 97% | 149 | 79% | 22 | 94% | 171 |
| (Much more confident) | 84% | 130 | 25% | 7 | 75% | 137 |

[a]Percentage calculated based on total number of producers who sell products, not total number of producers

relationships with others in the program, the vast majority of women (81%) said they had, while slightly more than half (55%) of men said the same. Similar responses were given regarding whether or not they had developed relationships with others in their neighborhood because of their participation in AGRUPAR, with 82% of women and 67% of men responding that they had. Women were also much more likely to say that they were more active in their communities, with 82% replying affirmatively compared to 55% of men.

In their personal lives, women also experienced more significant changes as a result of their participation in the program. Across all three measures of changes in personal confidence, women experienced substantially more gains than men. In the case of increased confidence speaking in public, men and women made comparable gains, with 75% of women and 62% of men saying they felt this had increased. In the two other measures, however, there was more difference between the responses of men and women. When asked if they had more confidence in their opinions when talking with family and friends, nearly all women (94%) said they experienced an increase, with 73% saying it had increased a lot. While a majority of men said this area of confidence had increased (72%), just 31% said it had increased a lot. Similarly, when asked if they had more confidence in themselves as a result of participation, nearly all women (97%) said yes, with 84% saying they had a lot more. Men experienced gains as well, with 79% responding that their confidence in themselves increased, but just 25% said it had increased a lot.

## 11.4 Discussion: How Producer Characteristics Influence Urban Agriculture

The findings from the AGRUPAR case study demonstrate how the personal characteristics of producers can influence how urban agriculture manifests and the benefits associated with it. Research showed that while participation in the AGRUPAR program had generally positive effects in the lives of participants across the board, the degree varied among different groups based on migration history, age, and gender. While this is a single case study in a specific context, findings underscore the importance of taking producer traits into consideration when studying urban agriculture.

In the first case, the migration status of producers appears to affect the economic benefits that are associated with urban agriculture. Producers who had migrated to the city, especially those who had arrived in the 15 years prior, were less likely to engage in the sale of their garden products. Even when they did sell their products, they tended to earn less than producers who were from Quito. One possible reason for this difference could be that migrants were less likely to sell at AGRUPAR's *bioferias*, which was where producers earned the most in sales. While it is unclear why migrants engaged in urban agriculture in ways that were distinct from native producers, a potential factor that may have contributed to differences could be that producers from Quito had more extensive and embedded social networks within their communities. Having grown up in the city, Quiteños were likely to be more integrated into their communities and have greater familiarity with the people, places, and resources that enabled them to sell their garden products.

Based on these findings, it is important to consider the migration status of urban agricultural producers, particularly when evaluating potential economic outcomes. For urban agricultural programs, such as AGRUPAR, it may be beneficial to consider how migrants are situated differently compared to native producers and identify what additional support they may need in order to more fully experience the benefits of urban agricultural production.

Age also emerged as an influential factor in how producers engaged in urban agriculture and the benefits experienced. Different age groups differed in their motivation for engaging in urban agriculture; younger producers were more likely to value the environmental aspects of urban agriculture, whereas older producers placed a higher value on the enjoyment derived from the activity of gardening itself. In terms of benefits associated with urban agriculture, younger producers realized fewer economic benefits; they were less likely to sell their garden products and tended to earn less than older producers. In contrast, younger and middle-aged producers appeared to have experienced greater social and personal benefits, as they became more active in their communities and made greater gains in personal confidence as a result of participation.

Recognizing that producers of different backgrounds (in this case age) have varying motivations and interests behind their drive to practice urban agriculture is an important factor to consider when engaging with producers. This is especially important for programs such as AGRUPAR; understanding what motivates participants to pursue urban agriculture, the benefits that they most value from it, and how these vary among different groups can help programs tailor the way they organize and deliver assistance to more effectively engage producers.

The third characteristic in this case study that had notable impacts on urban agriculture and the associated benefits was gender. While men and women both experienced benefits from engaging in urban agriculture, they benefited in distinct ways. Men experienced greater absolute economic benefits; they earned significantly more on average than women producers. However, it appears that women experienced greater relative economic benefits; although their income from sales was lower than men, they were more likely to say that their income had increased. Women also appeared to have experienced greater social and personal benefits. They were more likely to have developed relationships through participation in the program and had become more active in their communities as a result. They also experienced substantial, meaningful gains in personal confidence.

Considering how men and women are differentially situated to engage in and benefit from urban agriculture is essential for program administrators and evaluators alike. While the particular role that gender plays will vary from place to place, it is likely that in most contexts there will be notable differences between male and female producers. Understanding such differences is essential for accurately portraying the benefits associated with urban agriculture, as these will vary significantly by gender. Urban agricultural programs can acknowledge that men and women might engage in urban agriculture in distinct ways and develop an understanding of precisely what those differences are. This will enable programs to identify the unique needs of each group, in particular women, so that the program may provide additional, tailored support to help them realize greater benefits, such as the potential to bolster economic opportunities in this case.

## 11.5 Conclusion: The Many Faces of Urban Agriculture

In this chapter, the case study of the urban agriculture program, AGRUPAR, in Quito Ecuador provides an important lens for understanding urban agriculture more generally. While urban agriculture is practiced in some form by urban dwellers throughout the world, its ubiquity does not equate to uniformity. The form that urban agriculture takes, how it is practiced, who practices it, and the benefits derived from it are influenced by a multitude of factors. This case study focused on the influence of particular characteristics of urban producers, specifically the migration history of producers, their age, and, importantly, gender. The type of benefits urban agricultural producers of AGRUPAR experienced, and the degree to which they experienced them, varied among producers based on these individual characteristics.

In light of these findings, it is clear that the practice of urban agriculture is not a uniform phenomenon; because producers are situated differentially in the social context, the practice of urban production and the benefits associated with it do not accrue uniformly among all producers. Thus, when thinking about urban agriculture, it is important to reflect critically on how it interacts with other factors in a producer's life to lead to differential outcomes.

For those interested in promoting urban agriculture, these findings make clear the importance of considering differences among producers. People from varying backgrounds engage with urban agriculture in distinct ways and bring into this engagement different interests and different capacities. This insight is necessary in order to avoid a one-size-fits-all approach that fails to meet the needs of diverse producers. Understanding these differences among producers will enable programs and other actors to customize their efforts and maximize their effectiveness.

## References

Ackerman K, Conard M, Culligan P, Plunz R, Sutto M-P, Whittinghill L (2014) Sustainable food systems for future cities: the potential of urban agriculture. Econ Social Rev 45(2):189–206

Battersby J, Marshak M (2013) Growing communities: integrating the social and economic benefits of urban agriculture in Cape Town. Urban Forum 24(4):447–461. https://doi.org/10.1007/s12132-013-9193-1

Bradley K, Galt RE (2014) Practicing food justice at Dig Deep Farms & Produce, East Bay Area, California: self-determination as a guiding value and intersections with foodie logics. Local Environ 19(2):172–186

Brown KH, Jameton AL (2000) Public health implications of urban agriculture. J Public Health Policy 21(1):20–39. https://doi.org/10.2307/3343472

Bryld E (2003) Potentials, problems, and policy implications for urban agriculture in developing countries. Agric Hum Values 20(1):79–86

Cook J, Oviatt K, Main DS, Kaur H, Brett J (2014) Re-conceptualizing urban agriculture: an exploration of farming along the banks of the Yamuna River in Delhi, India. Agric Hum Values 32:265–279

Corrigan MP (2011) Growing what you eat: developing community gardens in Baltimore, Maryland. Appl Geogr 31(4):1232–1241

FAO (2019) FAO's role in Urban Agriculture. Food and Agriculture Organization of the United Nations website. http://www.fao.org/urban-agriculture/en/. Accessed 4 Jun 2019

Galluzzi G, Eyzaguirre P, Negri V (2010) Home gardens: neglected hotspots of agro-biodiversity and cultural diversity. Biodivers Conserv 19(13):3635–3654. https://doi.org/10.1007/s10531-010-9919-5

Hale J, Knapp C, Bardwell L, Buchenau M, Marshall J, Sancar F, Litt JS (2011) Connecting food environments and health through the relational nature of aesthetics: gaining insight through the community gardening experience. Soc Sci Med 72(11):1853–1863. https://doi.org/10.1016/j.socscimed.2011.03.044

Litt JS, Soobader M-J, Turbin MS, Hale JW, Buchenau M, Marshall JA (2011) The influence of social involvement, neighborhood aesthetics, and community garden participation on fruit and vegetable consumption. Am J Public Health 101(8):1466–1473

Olivier DW, Heinecken L (2017) Beyond food security: women's experiences of urban agriculture in Cape Town. Agric Hum Values 34(3):743–755. https://doi.org/10.1007/s10460-017-9773-0

Pudup MB (2008) It takes a garden: cultivating citizen-subjects in organized garden projects. Geoforum 39(3):1228–1240. https://doi.org/10.1016/j.geoforum.2007.06.012

Teig E, Amulya J, Bardwell L, Buchenau M, Marshall JA, Litt JS (2009) Collective efficacy in Denver, Colorado: strengthening neighborhoods and health through community gardens. Health Place 15(4):1115–1122 van Veenhuizen R (2006) Cities farming for the future: urban agriculture for green and productive cities. IDRC, Ottawa

Webber J, Hinds J, Camic PM (2015) The well-being of allotment gardeners: a mixed methodological study. Ecopsychology 7(1):20–28. https://doi.org/10.1089/eco.2014.0058

WHO (2016) Urban green spaces and health: a review of evidence. World Health Organization website. http://www.euro.who.int/en/health-topics/environment-and-health/urban-health/publications/2016/urban-green-spaces-and-health-a-review-of-evidence-2016

Zezza A, Tasciotti L (2010) Urban agriculture, poverty, and food security: empirical evidence from a sample of developing countries. Food Policy 35(4):265–273

**Kate Oviatt** has a PhD in Health and Behavioral Sciences, with a special emphasis on urban sustainability, from the University of Colorado Denver. Her education has focused on developing a multidisciplinary understanding of sustainable development, with an emphasis on urban agriculture, sustainable livelihoods, and international community development. In her research, she has explored urban agriculture in different contexts, from her local state of Colorado to India and Ecuador. Her work specifically focuses on how urban agriculture can be a vehicle for empowerment, especially among low-income women.

# Blurring the Boundaries: How an Emerging Group of Urban-Integrated Farmers in Singapore Are Changing the Profile of Farm Labour

12

Ching Sian Sia and Jessica Ann Diehl

**Abstract**

Food production is no longer seen as exclusively a rural function by urban planners; it has become a common trend in urban areas, with urban farms increasingly set up in spaces that are traditionally not used for agriculture. In Singapore, the formal urban agriculture industry began to emerge in early 2010s, with informal community urban gardening dating back to WWII with the victory garden movement an early precursor. The first urban farm in Singapore was founded in 2011, and urban farming has become an industry that is constantly growing and evolving. There is an on-going transition from traditional to high-tech approaches that is changing the way the farming industry requires labour—and the demographic profile of the urban farmer. To understand the unique qualities of the type of labour force hired by urban farms, this paper compares the labour profile of urban farms with the commercial farming industry in Singapore through a series of interviews with farmers. We conclude with a discussion on what it could mean for the future of labour force within the urban farm industry in Singapore, as well as potential broader implications.

**Keywords**

Urban farm · Urban farmer · Labour · Farm worker

C. S. Sia (✉) · J. A. Diehl
Department of Architecture, National University of Singapore, Singapore, Singapore
e-mail: ching.s@u.nus.edu



J. A. Diehl, H. Kaur (eds.), *New Forms of Urban Agriculture: An Urban Ecology Perspective*, https://doi.org/10.1007/978-981-16-3738-4_12

## 12.1 Introduction

Singapore is a food-crazed nation—with food stall cheap eats in "hawker centres" to fine dining and everything in between offering a range of local Peranakan to global western and Asian fusion dishes (Kong 2015). Despite being a country of foodies with an array of cuisines easily available to all Singaporeans, the majority of produce is actually imported from other countries. About 90% of food is imported, making Singapore heavily dependent on food imports (SFA 2020; Diehl et al. 2020). With less than 1% of farmland dedicated to agricultural purposes and only half of that zoned agricultural land dedicated to growing food (SFA 2020), Singapore appears to have a limited amount of space set aside for food production. With high-entry barriers to setting up farms in Singapore, a group of urban farming enthusiasts began sourcing for underutilized spaces such as vacant state land, rooftops, and indoor spaces as potential farming spaces in 2012, thereby decreasing the limitations due to lack of land available for urban farming (Elangovan 2019). Since 2012, at least 15, of what we in this paper call, urban-integrated farms have been set up, with 6 urban farm operators growing at multiple locations across Singapore. From soil-based farming on rooftops to vertical farming systems that utilize A-frames adopting hydroponic systems, as well as indoor-LED farming, many urban farms maximize the limited area available for food production either through a creative use of rooftop spaces or through a range of high-tech production methods. With the adoption of technology in production methods, urban farms are able to rely on a much smaller labour force (Ludher and Tan 2019). The transition from traditional to high-tech approaches has changed the way the farming industry requires labour. A number of urban farms in Singapore such as Edible Gardens City, Comcrop, and Vertivegies only require two to three employees to run the day-to-day farm operations. This is because many of the operations—from watering of plants to measuring the moisture of the soil—are computerized and automated, as compared to traditional farms that are often dependent on manual labour. This paper seeks to understand how farm labour is changing in Singapore by comparing urban-integrated farms with traditional commercial farms—specifically, differences in background and skills, educational level and wages, and motivation to work as farm labour. We define traditional commercial farms as farms that occupy zoned agricultural land (Fig. 12.1), while urban-integrated farms are defined as farms that occupy underutilized spaces such as rooftops and vacant state land (Fig. 12.2).

## 12.2 Overview of the Changing Farming Industry in Singapore

The traditional commercial farming industry is a sunset industry, with many young people shunning it due to the manual and intensive nature of work required (Kelsey 2015). This is a global phenomenon that is not unique to Singapore. Farming in Singapore has often been viewed as a job that is less glamorous compared to white-collar jobs (Mannan et al. 2017). With a population that is highly educated, many

**Fig. 12.1** Image of a traditional commercial farm. (Source: C. Sia)

**Fig. 12.2** Image of an urban-integrated farm. (Source: C. Sia)

**Fig. 12.3** A commercial farm that has adopted hydroponics system to grow. (Source: C. Sia)

Singaporeans would opt for a desk job rather than a job that requires long hours of manual labour under the relentlessly hot, tropical sun (Fig. 12.3).

To rectify this problem, the Singapore government has encouraged many farms to adopt high-tech systems to reduce reliance on manual labour (URA 2018). This is heavily encouraged through a $46 million USD Agriculture Productivity Fund introduced by the Singapore government for any commercial farms that would like co-funding for the high-tech farming systems the government would like to adopt. Commercial farms that have adopted a high-tech system (such as a hydroponics growing system) using the Agriculture Productivity Fund are more likely to have their leases renewed when they tender for new land. However, high-tech prioritization has created a new set of problems—with the cost of setting up farms to be of astronomical figures of around $1.1 million USD—creating huge financial barriers for young farmers who may want to enter the farming industry (Kok 2017; Michelin Guide 2019). On top of high-entry barriers to farming due to costly adoption of technology, commercial farm owners are also required to make an upfront payment of 10 years of land lease fees, a large investment making it difficult to start a farm.

Three quarters of land in Singapore is government owned, and the Singapore Land Authority (SLA) is responsible for development and regulation of land resources. Currently, agricultural land is being leased out in 20- to 30-year time periods by the Singapore Food Agency (SFA as of 2019; formerly the Agri-Food and Veterinary Authority (AVA) of Singapore), the government agency tasked with managing agricultural land leases. A combination of financial costs, uncertainties on land leases, and the physically demanding nature of the job has made it difficult for farms to sustain their livelihood in Singapore. And, on top of that, there are government mandated minimum productivity targets. To ensure there are enough farm workers to produce food crops in Singapore, many commercial farms have turned to foreign labour (refer to Chap. 14 for more details).

To gain insight on the farming industry related to the changing profile of the labour force in Singapore, semi-structured interviews were conducted with both traditional and high-tech commercial farms, specifically urban-integrated farms, between December 2018 and January 2020. Interviews were conducted with commercial farm owners or managers ($n = 12$) and urban-integrated farm owners ($n = 4$) and workers ($n = 8$) (analysed together). A list of 35 commercial farms were sourced, and more than 20 were contacted; all urban-integrated farms were invited to participate. Audio recorders were used to record the interviews and later transcribed. No interviews were carried out with commercial farm workers due to language barriers, but observations of the commercial farm workers were made that contribute to this paper. This research was approved by the National University of Singapore Institutional Review Board (NUS IRB).

## 12.3 Agricultural Background

To understand how background knowledge, experience, and existing agricultural skills contribute to joining the farming industry, we asked interviewees from the both the commercial farms and urban-integrated farms what their previous jobs were (if any), as well as any family involvement in farms that may have led to their employment in a farm.

### 12.3.1 Commercial Farms

One-third of the interviewed commercial farm owners ($n = 4$; 33%) had always worked in the farming industry—either at the present farm or at another farm operation. The majority of commercial farm owners and managers ($n = 10$; 83%) had prior experience in agriculture as most of them grew up in a farming family. They had either helped out during their school holidays when they were young, or had worked for a short period of time in their family farm or on another farm. Many pursued non-agricultural studies and careers, but eventually took over the family business again:

> *I was the son of the oldest farmer, which means that if you consider that, then I've been farming ever since I was born, but I really got involved in farming during my secondary school days. My first maiden bus was to the farm, and there I found my paradise and my wonderland—so many things to do, so much, so exciting. It's a very new horizon or new set of interests there when you know, so many things in the city or in the urban area, you cannot image to have this fun, this type of joy, talking about longkang* [drain] *fishing, catapult, hunting, and all these things. You can only do these things in the countryside, so I found my*

*joy there, my wonderland there. So holidays, that's my vacation area.* (Multi-generational commercial farm owner)

In comparison to the lead farmers who were often multi-generational Singaporean farmers, the majority of the hired farm workers working in commercial farms were immigrants from developing countries including Thailand, Bangladesh, and Myanmar. Many commercial farm workers were also farmers in their country of origin, either as commercial farmers or involved in subsistence farming as a way to supplement their income or to feed their families. As a result, many had experience in farming, with specific knowledge in traditional farming. The Singapore farm owners who currently still adopt traditional ways of farming have mentioned that their farm workers were able to give them some tips in terms of improving day-to-day operations within the farm. But, as many commercial farms in Singapore have adopted some form of high-tech system to improve crop yields and intensify crop production, the knowledge in traditional farming that many foreign farm workers have is not always transferable to the commercial farms in Singapore. Due to the vertical nature of the employment hierarchy within commercial farms, foreign-hired commercial farm workers generally do not get promoted to farm manager due to the transient nature of their employment; many of the migrant workers move to Singapore for a short-term period to gain employment, then move back to their home country after they have stayed for a period of at least 2 years. Some of the hired commercial farm workers do take on more of a "supervisor" role if they have stayed for a longer period and are able to communicate efficiently to other farm workers who are from the same country or hometown:

*Yes my workers back in Myanmar, they are all the country boys so at least they farm most of the time, but of course quail farming is something new to them. Yeah, we do 'old bird teaches the new bird' type of method, so that's how we train them. And we rotate our job scope, that means they must be well-versed in everything, so in case one is on leave, all of them can actually chip in and cover up the person on leave.* (Commercial farm owner)

### 12.3.2 Urban-Integrated Farms

When asked about previous jobs, urban-integrated farmers came from a variety of backgrounds ranging from those who worked in the food and beverage industry to those who were in the engineering industry. One-third ($n = 4$; 33%) were from the food and beverage industry, noting that being in the food and beverage industry had made them question where the ingredients of their dishes came from.

*I've been trained as an aerospace engineer, and then I went to culinary school—so I've never been officially trained at all as a farmer. Being in the culinary field, I was very interested in cooking, and slowly I realised I was interested in where food came from. So I wanted to explore how that journey would start from food being produced.* (Urban-integrated farmer)

Most of the respondents ($n = 5$; 42%) also mentioned that they wanted an entire change in industry, hence joining the farming industry. The majority of urban farmers ($n = 8$; 67%) who turned to urban farming as a career choice had no prior farming experience except volunteering short-term at a farm. Only one-third ($n = 4$; 33%) had volunteered on an urban farm before joining the industry, including one respondent who was previously in horticulture before switching to the high-tech urban agriculture industry.

> *I needed to look for a company to do an internship for my final year project, so I volunteered and did an internship here [at this farm] for my school project. So that was six months, and after that I renewed my contract; I continued my contract here. Actually my contract ended in May, I just continued working here. Yeah, it was my first job [in agriculture].* (Urban-integrated farmer)

## 12.4 Characteristics of Urban Farm Workers

To understand why farm workers to joining the farming industry, a series of questions on their educational level, wages, motivations as well as farm's involvement in the community were asked to gain a better understanding of what the pull factors are for joining the farming industry.

### 12.4.1 Education Level and Wages

Questions on educational background, educational level, and wages were asked to understand if educational background may have contributed to choosing farming as a career. We also wanted to understand how wages were compared to Singapore's median income.

#### 12.4.1.1 Commercial Farms

As mentioned, all farm workers at interviewed commercial farms were foreigners. High labour costs and the difficulties in attracting local Singaporeans to engage in manual and/or outdoor work have resulted in Singapore's commercial farm labour being almost entirely made up of foreign workers. Based on the information provided by commercial farm owners and managers, the starting pay for (foreign) farm workers ranged from $800 up to $1200 ($564 to $847 USD), with some drawing overtime pay or a bonus for work in additional to standard work responsibilities.

> *Compensation should be about $1000 SGD [$706 USD], it depends on the hours they work, it's an hourly rate. I don't remember the rate. It depends on performance, every farm has a supervisor, when their contract is about to expire and due to renewal, if they want to continue working and have a pay raise, they will mention it to the supervisor and the supervisor will approve/reject it depending on their performance.* (Commercial farm manager)

#### 12.4.1.2 Urban-Integrated Farms

Most urban-integrated farm owners ($n = 3$; 75%) and workers ($n = 5$; 62.5%) were degree holders and more than one-third of the workers were diploma holders ($n = 3$; 37.5%) in areas such as engineering, horticulture, social sciences, education, and business. They were all Singaporean citizens; none was foreign-born.

In Singapore, the median pay is approximately $4400 SGD ($3104 USD), while the median starting pay of graduate degree holders and graduate diploma holders are $3500 and $2350 SGD ($2469 and $1657 USD), respectively (Tan 2020). Despite more than half of urban-integrated farmers holding graduate degrees, starting pay is low with most pay ranging between $2000 and $2500 SGD ($1411 to $1764 USD) and less than a quarter commanding a pay above $3000 SGD ($2116 USD). Some receiving less than $2000 SGS ($1411 USD) as the urban-integrated farming industry, while growing, is still a niche business:

> *I get less than $2,000 [SGD] per month. For full-timers, the pay ranges from less than $2,000 to more than $3,000 and different people have different pay. The average working hours are 44 hours per week for a 5-day work week. The hours are similar to what you get in a corporate scene, from about 9am to 5pm.* (Urban-integrated farmer)

Three urban-integrated farmers also hired differently-abled individuals, as well as elderly, to be involved at different stages of the farming operation, from seeding to germination. On weekends, some farms also welcomed volunteers to help out on the urban farm. Some farms qualified for government subsidies for hiring differently-abled individuals as part of government schemes to encourage the employment of differently-abled individuals.

### 12.4.2 Motivation

Questions on why farm workers or managers chose to work in the farming industry were asked to find out what the motivations were due to the strenuous physical requirements to conduct the work.

#### 12.4.2.1 Commercial Farms

When asked why commercial farm owners had chosen to remain in the industry, many reported farming as a lifestyle choice ($n = 8$; 66.7%). The main reason to continue farming in Singapore was because their family had been involved in farming for many years ($n = 9$; 75%). However, most commercial farm owners also acknowledged the difficulties of sustaining farming in Singapore, and would rather not continue farming if they had the financial resources to exit the industry. On the other hand, similar to urban-integrated farmers, the motivations of commercial farmers were to produce enough to feed people and to be away from the "hustle and bustle" of the city. In the below excerpt from a commercial farm owner, he is motivated simply by being able to grow beautiful vegetables that taste good for people:

> *We are here, we are surviving because we are not motivated by how much I get [paid]. We enjoy people saying 'your vegetables taste good!' But, if kids are going to come here and say, 'I have a big paycheck, hao lui [hokkien for good many],' that's different altogether. I'm motivated by someone who comes and say, 'It's true, wahhh [Singaporean particle used for inflection] your vegetables are so nice!' Wahhh, I feel good you know?* (Commercial farm owner)

We did not directly ask what the motivations of the commercial farm workers were for farming in Singapore, but financial reasons could be a main factor. Many commercial farm workers are from developing countries and the wages in Singapore are substantially higher than in their home countries.

#### 12.4.2.2 Urban-Integrated Farms

Many urban-integrated farm owners ($n = 3$; 75%) cited improving food security and contributing to the food system as the main factor. On the other hand, urban farm workers were generally more motivated by a working environment that was drastically different from their previous line of work or to be part of the food system that served a social good or had an educational purpose. A number of urban farm workers cited the importance of educating the public about local produce being superior and just as good as imported produce:

> *We are so dislodged from the natural world. Our government did a very good job by planting up all the streetscape spaces, parks, whatever, just do as much as they can to spice up the greenery we have in our city. But ultimately we are still very not close together with nature, you see it but you don't really touch it or feel it or smell it. But by doing urban farming you have to take care of your plants and that's where you are actually in touch with nature rather than seeing nature.* (Urban-integrated farmer)

### 12.4.3 Community Involvement

Farms can be involved in the community to varying degrees. Community involvement enables the farmer to have a better understanding of the profile of the customer they are able to reach out to and to tap into part-time labour or volunteers.

#### 12.4.3.1 Commercial Farms

Commercial farms typically were unable to accommodate volunteers to come in due to strict food safety regulations by Singapore Food Agency (SFA). Hence, the level of community involvement at commercial farms was low. A small number of farms in Singapore did hold educational farm tours as a way to educate school-age children where food comes from. Agri-tourism through workshops and farm tours have been ways to help local commercial farms supplement their income in addition from sales of their produce.

#### 12.4.3.2 Urban-Integrated Farms

As mentioned, many urban-integrated farmers did not have prior farming experience. The benefit of this trend is that it has helped to lower the entry-level for volunteers who are keen on trying their hand at food production. A number of urban farms also hire individuals with special needs, as well as elderly, who would like to stay active. By involving people with different learning needs, as well as elderly, the community involvement of urban farms was quite high. Notably, farms involved with community activities were rooftop or land-based farms—not indoor-LED urban farms due to the stringent climate control requirements of indoor farms.

> *That one [farm] actually has a more direct impact on community. So when we were starting out the farm, actually our model is we want to create a circular economy within each neighbourhood, so for instance the vegetables we grow there, they are actually grown by residents that we hire from the nearby Asian Women Welfare Association organisation... we have a partnership there, so we hire the elderly from there on a part-time basis, they come over at certain timings when it's not too hot, where it's appropriate for them to do some work, and they help us with our farming process, they help us to maintain our farm. They come, then they cook for one another, they go about diligently doing their work. The oldest one is actually 80-years-old. So it's very heartening to see that, like we are able to provide this kind of opportunity for them to grow some vegetables instead of staying at home.* (Urban-integrated farm manager)

In a number of urban-integrated farms, that were not indoor, the community was welcomed to participate not only as a volunteer, but also to purchase and harvest fresh crops directly from the farm even as a consumer, giving people the opportunity to understand how crops were grown and to be part of the process:

> *If you would like to do your own harvesting, our self-harvesting session only begins at 5pm so do drop by after 5pm if you are keen to harvest your own pesticide-free lettuces!* (from an urban-integrated farm's Facebook page)

Based on the interviews with commercial farm owners and managers, as well as urban-integrated farm owners and workers, it seems that the profile of farm workers has shifted significantly from unskilled farm labour toiling in the commercial farms, to degree and diploma graduates working under the sun in urban-integrated farms. It is evident that urban-integrated farm graduates were not motivated primarily by wages as urban-integrated farm workers were compensated below average pay. But, they were strongly motivated by the concept of growing food locally and contributing to the local food system. The lack of experience did not hinder urban-integrated farm workers, as compared to commercial farm owners and managers who had grown up in farming families.

## 12.5 Food Crisis and Farm Labour

With the recent Covid-19 global pandemic, there is a need more than ever for local urban food production to be increased—to buffer disruptions caused by lockdowns and closure of borders between countries. While many people face the risk of unemployment, pay reduction, and potential loss of jobs and income in the future, the local farming industry offers a certain level of job stability. Singapore government policies are encouraging local farms to increase production in order for the country to with stand any disruptions, while it is business-as-usual case for many farms to continue working, and increase productivity to cope with several bouts of panic buying that happened in 2020 (Quek 2020). Similarly, the food price hike in 2007 to 2008 drove 44 million people worldwide into poverty, resulting in political and economic instability (World Bank 2008). By having a food system that is more self-reliant and more self-sufficient, a country is able to avert a similar food crisis as such, and able to provide its citizens a sense of security in terms of food supply.

With most research on urban-integrated farming related to yield and technology, this chapter investigated the socio-economic aspect of urban farming. The objectives were to look into the farming labour of both commercial farms as well as urban-integrated farms to gain insight into the people who are involved in farming, their motivations, and how they are involved in the community. By asking questions about their family background, existing skillset (if any), education level, wages, what drives them, as well as how the community could participate in the farms, it enabled us to draw some conclusions on who the traditional farming industry or the urban-integrated farming industry is likely to attract into its labour force.

We found that many traditional farm managers, as well as urban-integrated farm workers, were not only tertiary educated, but also driven by the passion to understand food, despite not studying in an agricultural-related field. Most farmers involved in the urban-integrated farming industry were also not from farming families that would have exposed them to food production activities since young. Rather, they typically became involved in urban farming initially via volunteering or through internships. Given that wages of those working in the urban-integrated farming industry are highly uncompetitive, passion for food and the desire to lead a more sustainable lifestyle seems to be the primary motivating factor to enter into the industry.

If more time was allowed, it would have been useful to hire an interpreter to interview the commercial farm workers largely made up of migrant workers from Bangladesh, India, Thailand, and Myanmar, and carry out similar interviews to understand their educational background, motivations, skillset, and any community involvement. A much bigger sample of respondents from both commercial farms and urban-integrated farms would also strengthen the results of this research.

## 12.6 Looking to the Future

Urban-integrated farming is a nascent industry in Singapore, but has the potential to provide employment for Singaporeans with or without prior experience or any expertise in farming. SFA has a 30 by 30 vision of increasing domestic food production to 30% and reducing reliance on food imports (SFA 2020). It requires increased output from both local commercial farms and urban-integrated farms, translating to a need for more people to get into the agriculture work force (Seow 2019). While many of these farms are required to adopt high-tech systems to increase productivity, it will likely still be necessary for commercial farms and urban farms to hire people. There is also a new Republic Polytechnic programme solely targeted at young adults studying urban agriculture, as well as government policies that are more favourable in terms of conditions for the establishment of urban farms as compared to commercial farms; there is potential for the urban farming industry to establish itself as a significant stakeholder in local food production, and increase demand for local food production. This in turn may increase the urban farming industry's starting pay, making it slightly more attractive to locals in Singapore to consider being urban farmer as a career, hence attract more local Singaporeans to get into the urban farming industry. This is important given Singapore's high cost of living, where a pay closer to the average starting pay of young adults may be more likely to retain urban farming talent committed to producing food for the local people.

The image of farming has also changed drastically with high-tech farms now declaring that "farms were not cool, but now they are" (Quek 2020). In this study, the results, based on interviews, demonstrate that the stereotype that the farm industry is a career for the less educated is being de-stigmatized; there is an increasing number of young and educated individuals entering the urban farm industry as farm workers, despite lower pay than the average graduate starting pay. They are motivated more by the intrinsic values of farming, as well as potential contribution to the local food system. However, if Singapore is looking to increase domestic food production to the target of "30 by 30," it needs to demonstrate a sustainable demand for local vegetables, that can in turn translate to sustainable profits for the urban farming industry and wages for the urban farmers. This will not only attract and retain even more individuals to join the urban farming industry, but also provide a case for the government to support the growth of the urban farming industry in Singapore.

## References

Diehl JA, Sweeney E, Wong B, Sia CS, Yao H, Prabhudesai M (2020) Feeding cities: Singapore's approach to land use planning for urban agriculture. Glob Food Sec 26:100377

Elangovan N (2019) Singapore needs to lift farm production, cut waste, as global protein shortage looms: study. https://www.todayonline.com/singapore/singapore-needs-boost-farm-production-reduce-waste-global-protein-shortage-looms-study. Accessed 6 Apr 2020

Kelsey J (2015) Reclaiming the future: New Zealand and the global economy. Bridget Williams Books, Wellington

Kok XH (2017) For sale by special tender: 12 plots for farming. The Straits Times. https://www.straitstimes.com/singapore/for-sale-by-special-tender-12-plots-for-farming-veg
Kong L (2015) From sushi in Singapore to laksa in London: globalising foodways and the production of economy and identity. In: Food, foodways and foodscapes: culture, community and consumption in post-colonial Singapore, pp 207–241
Ludher E, Tan PH (2019) Producing more with less. Urban Solutions Issue 14
Mannan S, Nordin SM, Rafik-Galea S, Rizal ARA (2017) The ironies of new innovation and the sunset industry: diffusion and adoption. J Rural Stud 55:316–322. https://doi.org/10.1016/j.jrurstud.2017.07.015
Michelin Guide (2019) Sowing the seeds of change. https://guide.michelin.com/sg/en/article/people/sowing-the-seeds-of-change. Accessed 10 Jun 2020
Quek E (2020) Grow local: greater urgency to produce 30% of Singapore's nutritional needs locally by 2030. https://www.straitstimes.com/lifestyle/food/grow-local. Accessed 7 Apr 2020
Seow BY (2019) Parliament: agri-food innovation park in Sungei Kadut to open from early 2021, says Koh Poh Koon. The Straits Times. https://www.straitstimes.com/politics/parliament-agri-food-innovation-park-in-sungei-kadut-to-open-from-early-2021-koh-poh-koon. Accessed 4 Mar 2019
SFA (2020) AVA's food security roadmap for Singapore. https://www.sfa.gov.sg/food-for-thought/article/detail/ava's-food-security-roadmap-for-singapore. Accessed 7 Apr 2020
Tan C (2020) IT grads get highest starting pay: survey. https://www.straitstimes.com/singapore/education/it-grads-get-highest-starting-pay-survey. Accessed 9 Apr 2020
URA (2018) Feeding cities of the future with agricultural technology. https://www.ura.gov.sg/Corporate/Resources/Ideas-and-Trends/Feeding-cities-of-the-future. Accessed 8 Apr 2020
World Bank (2008) Global food and fuel crisis will increase malnourished by 44 million. World Bank. https://www.worldbank.org/en/webarchives/archive?url=httpzzxxweb.worldbank.org/archive/website01052/WEB/0__CO-43.HTM. Accessed 13 Mar 2017

**Ching Sian Sia** is a PhD candidate in the Department of Architecture at the National University in Singapore (NUS). Her PhD research focuses on Urban Agriculture and Lefebvre's Rights to the City in Singapore. She holds a Master of Science in Integrated Sustainable Design (NUS), as well as a Postgraduate Diploma in Urban Design and Bachelor of Urban Planning and Development, both from the University of Melbourne in Australia. In her spare time, she engages in social activism particularly for gender equality and LGBTQ rights.

# Assessing Ecosystem Services and Job Opportunities in Peri-urban Agriculture Start-Up Projects

13

Ambrogio Zanzi, Federico Andreotti, Valentina Vaglia, Sumer Alali, Francesca Orlando, and Stefano Bocchi

**Abstract**

Significant socio-economic changes have occurred in the last decades: among all, increased migration from rural to urban areas. It appears clear that there is a need for resilient cities, capable of combining economic and environmental sustainability. Agroecological practices, as a reduction in agrochemicals input and extended use of living fences and tree rows, can improve environmental quality, assuring ecosystem services and urban food systems, and foster local productions and socio-economic tissue, improving the overall quality of life. That is the approach of the "Urban Innovative Action" OpenAgri project, aimed at the restoration of a 35-ha peri-urban area in Milan (Italy), thanks to the creation of a start-up incubator focused on food production and at the agroecological transformation of the area. This work focuses on the quantification and the evaluation of strategies for enhancing ecosystem services and investigating their link with job opportunities. Thanks to the Pareto Front algorithm and Principal Component Analysis, we were able to understand which start-up approach could provide both new job positions and better ecosystem services. In the research, OpenAgri emerges clearly as one of the first case studies which combine urban

A. Zanzi (✉) · V. Vaglia · S. Alali · S. Bocchi
Department of Environmental Science and Policy, Università degli Studi di Milano, Milan, Italy
e-mail: ambrogio.zanzi@unimi.it

F. Andreotti
Laboratory of Geo-Information Science and Remote Sensing, Wageningen University & Research, Wageningen, The Netherlands

CIRAD, UPR GREEN, Montpellier, France

GREEN, University of Montpellier, CIRAD, Montpellier, France

F. Orlando
Department of Molecular and Translational Medicine (DMMT), Università degli Studi di Brescia, Brescia, Italy

 
J. A. Diehl, H. Kaur (eds.), *New Forms of Urban Agriculture: An Urban Ecology Perspective*, https://doi.org/10.1007/978-981-16-3738-4_13

requalification with socio-economic issues, representing a scalable strategy in other areas, to solve the increasing need for sustainability. Our results highlight that a multidisciplinary approach is needed, both to stay on the market and to supply ecosystem services, combining productive, social and environmental initiatives, resulting in the more suitable solutions to enhance the value of urban and peri-urban ecosystems, while addressing the actual socio-economic themes and creating new jobs.

**Keywords**

Ecosystem services · Agroecology · Job opportunities · Sustainable peri-urban development

## 13.1 Ecosystem Services

In recent years, the concept of ecosystem services (ES) has gained primary attention in scientific communities, and, after the publication of the Millennium Ecosystem Assessment (MEA 2005), an increasing body of research is focusing on the quantification of ES in various environments (Bagstad et al. 2013; Costanza et al. 1997; Malinga et al. 2015).

We adopt the MEA definition of ES that, "Ecosystem services are the benefits people obtain from ecosystems. These include provisioning, regulating, and cultural services that directly affect people and the supporting services needed to maintain other services" (MEA 2005). This classification, updated recently, recognises at least 22 types of ES, divided into four categories: supporting, provisioning, regulating, and cultural services (Fig. 13.1) (Daily 1997; MEA 2005; TEEB 2010).

The classification of ES into unique categories is useful to understand how deep and vast are the benefits we receive every day from our natural environment. Indeed, different habitats and environments can provide multiple ES at the same time. For example, forests are crucial for carbon storage and sequestration, while agroecosystems, definable as a system characterised by both ecological and agricultural processes, are crucial for food production and supply. Therefore, it is clear that urban ecosystems are essential to improve the livability of our cities, assuring services with a direct impact on human health and well-being: e.g. air purification, climate regulation and noise reduction, as well as indirect and less tangible services, such as recreational and cultural activities dependent on the presence of nature in our urban agglomerate. Many researchers have now focused their attention on urban ES. The reason is logical apparent: since 2008, more than half of the world's population lives in cities. Moreover, by 2050, the percentage will grow up to 70%—compared to only 13% in 1900 (Salbitano et al. 2016)—due to increased migration from rural to urban areas, which is one of the significant socio-economic changes and the main challenge of our time (Kovats et al. 2014).

In order to reduce the disservices of the urban area (i.e. waterproofing, air quality problems) which affect both socio-economic conditions and natural capital, there is

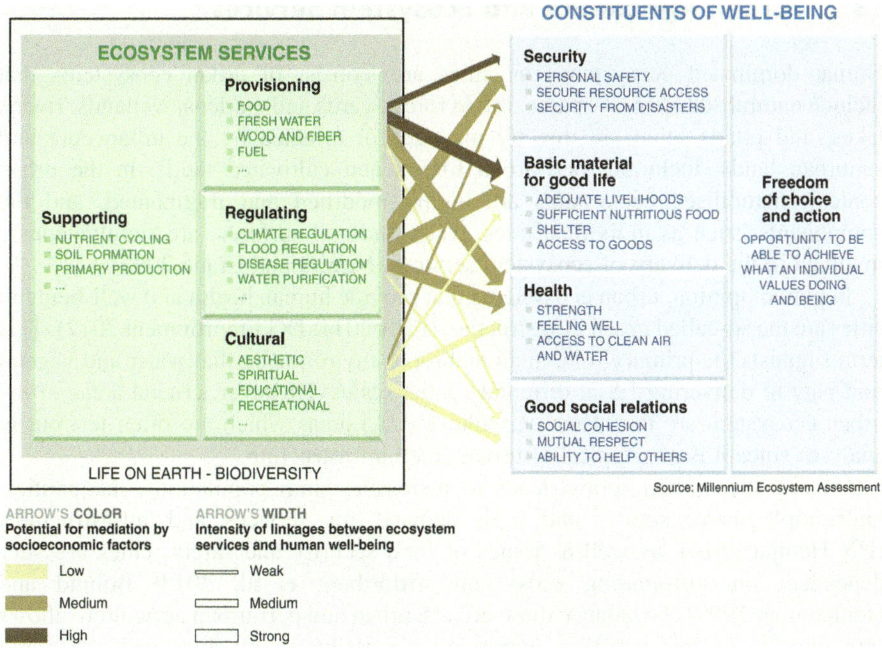


**Fig. 13.1** Ecosystem Services classification and link with Human Well-Being. The figure highlights the MEA classification of ES and several links with human life (Millennium Ecosystem Assessment 2005)

an urgent need for new strategies to improve the ES provided by our urban environments to reach the sustainable urban development goals as defined in Agenda 2030 (UN 2015). However, if compared to natural ecosystems, it is only recently that the scientific community has turned attention to the urban condition. Thus, the research in urban ES quantification is still in an initial phase (Gomez Baggethun and Baron 2013).

Starting from these assumptions, in this chapter, we analyse the link between the provision of ES and job opportunities in the urban and peri-urban context. Until today, few researchers have investigated this relation, which, in our opinion, is crucial in our cities, where the need for new jobs grows as fast as the demand for local food supply. This study focuses on the EU-funded project "OpenAgri" in Milan, in order to:

1. Explore the trade-offs and synergies among the different start-ups of the project
2. Estimate the potential contribution to the provision of ES by the start-ups under the OpenAgri within the Urban Innovative Action program
3. Quantify the ES provided by the ecological requalification of the project area.

## 13.2 Urban Agriculture and Ecosystem Services

Human-dominated ecosystems, as cities are, consist of urban ecosystems that include naturalised spaces—parks, urban forest, yards and gardens, wetlands, rivers, lakes, and ponds—that are directly managed or affected by the urban core and suburban lands, including peri-urban forests and cultivated fields. In the urban context, naturalised ecosystems are highly modified and fragmented, and the components, such as individual trees, water and soil surfaces, are simultaneously involved in the delivery of ecosystem services (Nowak and Crane 2002).

In public opinion, urban ecosystems that provide human health and well-being in cities are the so-called green infrastructure (EEA 2011; DG Environment 2012). This term suggests the primary role, in an anthropic environment, that water and vegetation play in delivering ES at different spatial scales. However, crucial areas of the urban ecosystem are the urban agriculture (UA) areas which are often left out in analyses since it is non-typical of urban green infrastructure.

Uncontrolled urban sprawl leads to increasing slum populations, inequalities, underemployment, sprawl and high demand for services and infrastructures (UN Habitat 2014), as well as issues of food security and safety; cities are very dependent on surrounding ecosystems (Mörtberg et al. 2013; Bolund and Hunhammar 1999). To balance these effects, urban and peri-urban agriculture allows providing ecosystem services, increasing the resilience of cities and enhancing human well-being (Alberti and Marzluff 2004). Urban and peri-urban agriculture provides cities with their local market of goods and services (Antrop 2000), having an impact on the mitigation of climate change, biodiversity loss, and land system change (Larondelle and Haase 2013).

This study explores the provision of ecosystem services provided by a case study of peri-urban agriculture in Milan, connecting with those studies that aim to assess the success of attempts to reduce the growing urban ecological footprint.

## 13.3 Urban Agriculture: New Opportunities

Urban agriculture plays a potential primary role in providing food supply to expanding cities while connecting urban populations to the rural landscape. UA considers area within the cities; instead, peri-urban agriculture is a form of agriculture at the fringes of growing cities, characterised by the transition zone between urban and rural areas (Piorr et al. 2011). Peri-urban agriculture refers to as "metropolitan agriculture" (Kittinger et al. 2016) or "urban fringe agriculture" (Adams et al. 2016). Data on the increasing role of UA are available all over the world (Orsini et al. 2013). Indeed, today, as estimated by the Food and Agriculture Organization (FAO) in the initiative "Food for Cities", UA is practised by 800 million people worldwide (FAO 2010), helping low-income urban residents save money on food purchases, often in still informal and disorganised ways.

In developing countries, UA is a strategy to address urban poverty improving health conditions and providing a more sustainable and stable economic growth at both family and community levels. In this context, the production derived by UA is complementary to the traditional agricultural production, since poor citizens can obtain perishable products such as vegetables, milk, and eggs (van Veenhuizen 2006). In the poorest context, urban farming is an activity mainly practised for subsistence reasons, improving food systems for city supply. In fact, with the rise of food demand in cities, small-scale farming is conducted for the commercial purpose creating new job opportunities for the enterprises related to food production, processing, and distribution (Cour 2001; Agbonlahor et al. 2007).

Smart and resilient city governance increasingly promotes a better range of activities for redesigning degraded peri-urban areas—a location where ecosystem restoration can provide more benefits than costs (De Groot et al. 2013). UA demonstrates the emerging ability of local start-ups and enterprises to develop new job skills that more efficiently connect the countryside with the city and better preserving diminishing patches of biodiversity (Kowarik 2011). Start-up incubators and labs can shape not only the economy of urban and peri-urban areas but can also provide several ecosystem services related to their activities, including providing local food supply (Gerster-Bentaya 2013).

### 13.3.1 Peri-urban Agriculture Case Study

The European Union, within the "Urban Innovative Action" program, has funded the "OpenAgri" project in the Milan urban area. The main scope of the project is to create an open innovation centre dedicated to the theme of peri-urban agriculture and the agri-food chain. Milan is known as the "financial capital" of Italy, rich in cultural and social activities, and leading many economic sectors in Italy. Anyhow, despite this economic energy, there are clear signs of inequity. For example, the youth unemployment rate is high (28.6%), but still, 10 percentage points below the national average; and the percentage of NEETs, defined as young population (aged 15–29) not Engaged in Education, Employment or Training, in the metropolitan area is at 17.6%, which is 2.0% below the national figure (UIA 2020).

The OpenAgri project study area considers the strategic peri-urban landscape, between the urbanised part of the city and the Parco Agricolo Sud, an agricultural and forested area connecting 61 municipalities, for a total 47,000 ha (Fig. 13.2). The improvement of the OpenAgri area will create an open innovation hub focused on peri-urban agriculture, including an ancient farm named Cascina Nosedo. Therefore, our case study deals with the agroecological requalification of an urban fringe area to serve as a living lab for social inclusion, jobs, and skills creation along the food supply chain while increasing the level of resilience and sustainability of the city (UIA 2020). In this context, agroecology is considered an approach and discipline that seeks to integrate science (e.g. agronomy, sociology, history), practices, and participation of the society (e.g. local knowledge, active indigenous participation) to guide research, policy, and action towards the sustainable transformation of the current agri-food system (Wezel and Soldat 2009; Gliessman 2015; Méndez et al. 2016).

The OpenAgri project attracts resources to address the challenges formulated in the Food Policy promoted by the City of Milan, following the goals of the Milan

**Fig. 13.2** Project area. OpenAgri is taking place in a 35 ha area, in the south part of Milan metropolitan area, which is in the central part of the Lombardy region

Urban Food Policy Pact (MUFPP). In 2015 the Municipality of Milan adopted the MUFPP to develop a sustainable food system to assure healthy and accessible food, while reducing food waste in order to make urban food systems more inclusive, resilient, safe, and diverse. Therefore, OpenAgri aimed to connect four different aspects: (1) sustainability, by fostering a local food production and providing new ecosystem services in the project area; (2) system innovation, by the creation of a start-up incubator and to the requalification of a lost peri-urban area; (3) creation of new job opportunities; and (4) multidisciplinary approach since it reconciles food production, peri-urban requalification, and economic upturn. This economic upturn is needed both in the agricultural sector and in the Milan suburban social context. Indeed, even if the sector only represents 2% of Italy's Gross Domestic Product (GDP), it directly occupies more than 20% of the workforce and contributes substantially to Italian exports (ISTAT 2017).

However, Italian agriculture has witnessed a contraction, both in production and in the workforce, by losing more than 100,000 people employed in the period 2013–2015 (ISTAT 2018). That situation is in the European trend, where in the last 10 years the agricultural workforce has marked a substantial decrease—with the loss of the 17.5% jobs (ISMEA 2018). The OpenAgri project in Milan can serve as an example of acceptable replicable practices to combine urban requalification, ecosystem services provided, together with the development of job opportunities and positive financial returns. This study opens a new field of research to better understand the relationship between economic performance and environmental aspects. The OpenAgri case study area consists of a network of agricultural fields, farms, and historical buildings, including the Maedieval Chiaravalle Abbey, linked together by the Vettabbia river. Thus, the area, with its long history, has witnessed many transformations, especially in the last century, influenced by urban sprawl and in a general abandonment and degradation (UIA OpenAgri report 2018). This scenario led to the quality decay for both the environment and lifestyle for the resident population, also causing the crisis of contraction of local farms and agricultural production. For these reasons, it is urgent the requalification of the area to pave the way for the revitalisation of peri-urban agriculture.

### 13.3.2 OpenAgri Ecosystem Services and Job Opportunities

We analysed the performance of the start-ups selected by Milan municipality during the selection of the OpenAgri projects (Table 13.1). In this section, we use a practical example to explain how we evaluated ecosystem services and job opportunities carried out by start-ups.

The evaluation of start-ups allowed us to address the following questions: (1) what are the ecosystem services (ES) and economic opportunities offered by start-ups in peri-urban areas? (2) Which start-up strategies can enhance the provision of multiple ecosystem services as well as economic incomes? In this chapter, we describe eight start-ups that were analysed to understand their potential contribution to ES provision and the relationship between ES and job opportunities.

We followed the Millennium Ecosystem Assessment (MEA) scheme (Fig. 13.1), considering the four ES categories. In our study, we have chosen to select and analyse several ES: in particular, six ES in order to investigate the link between their provision and job opportunities, and three ES supplied by the requalification of the area, thanks to the implant of new trees and shrubs (Table 13.2).

We evaluated one provisioning ES, primary production, measured as the planned monetary value produced by each start-up after 3 years of activity. The data were collected based on the business plans presented by each start-up during the selection process. We also evaluate one provisioning ES, secondary production, obtained from minor incomes of each start-up, i.e. cultural activities and training courses.

We started evaluating the link between regulating ES and job opportunities, by focusing on two out the five regulatory ES under overall analysis: crop pollination dependency and water use saved. Crop pollination dependency is an essential ES to agriculture as almost 65% of plant species need pollination by fauna (Klein et al. 2007). Concerning the essential animal-pollinated crops, over 40% depends on wild pollinators, highlighting how crop production depends mainly on pollinators. Pollinator dependency is an ES proxy already well used for financial terms (ISTAT 2018; Losey and Vaughan 2006; Gallai et al. 2009) and production level (ISTAT 2017; Aizen et al. 2009). We combined long-term data on global crop production and cultivated area provided by the FAO of the United Nations (FAOSTAT 2007) for assessing the pollinator dependence on crop types. As defined by multiple authors

**Table 13.1** List of start-ups working in the OpenAgri area and description of activities

| Start-up | Activities and target market |
|---|---|
| Start-up 1 | Spirulina algae production |
| Start-up 2 | Agri-technologies for crops and vegetable production |
| Start-up 3 | Wildflowers production and retail |
| Start-up 4 | Wheat cultivation for local bakers |
| Start-up 5 | Snail production |
| Start-up 6 | Seed production for local organic farmers |
| Start-up 7 | Wildflowers and edible plants production |
| Start-up 8 | Old cereal, hemp, and *Paulownia* sp. cultivation for the local market |

**Table 13.2** Indicators and measurements of different ES types by category

| ES category | ES type | Indicators/methods for measurement/sources |
|---|---|---|
| Provisioning | Primary production | The monetary value of the products |
| Provisioning | Secondary production | The monetary value of the training sessions, cultural activities |
| Regulating | Pollination dependency | Klein et al. (2007), Aizen et al. (2009) |
| Regulating | Water use saved (WUS) | Ostrom et al. (1999), Hess and Ostrom (2007) |
| Regulating | Air pollution removal | i-Tree software (Nowak and Crane 2000) |
| Regulating | Carbon storage and sequestration | i-Tree software (Nowak and Crane 2000) |
| Regulating | Oxygen production | i-Tree software (Nowak and Crane 2000) |
| Cultural | Job opportunities | Data obtained from the start-up's business plan |
| Cultural | Start-up cooperation | Data obtained from the start-up's business plan evaluating the sharing of working force and/or materials, using the same selling systems |

(Klein et al. 2007; Aizen et al. 2009), we defined five classes of pollinator dependence: (a) none (production does not increase with animal pollination; class 0), (b) little (0–10% production reduction; class 1), (c) modest (10–40% reduction; class 2), (d) high (40–90% reduction; class 3), and (e) essential (90% reduction without pollinators; class 4).

For water used saved, we considered water as a common resource shared by a group of people with constraints associated with its management. Common-pool resources, as indicated by E. Ostrom, refer to natural resources where one person's usage can subtract from another usage (Ostrom et al. 1999; Hess and Ostrom 2007). Highlighting the importance of water availability as an ES, we developed a "Water Use Saved" (WUS) indicator as a proxy. We assumed that water usage varied with crop water irrigation and that lower usage indicates the lowest irrigation data. We also assumed that by reducing water use, farmers not only saved money but also benefited from more efficient human/environmental resource management. The water requirement for each start-up was calculated, and then as suggested by E. Ostrom protocol, WUS was calculated as follows:

$$\mathrm{WUS}_i = (\mathrm{WU}_{\max} - \mathrm{WU}_i) \times 0.1 \quad (13.1)$$

where $\mathrm{WUS}_i$ is water use saved ($m^3$), $\mathrm{WU}_i$ is water use by a start-up project $i$, and $\mathrm{WU}_{\max}$ is the highest water use recorded among the start-ups participating in the study. The quantity of water used by each start-up was estimated using the business plans, where each start-up declared cultivated crops and extensions. The obtained value was further multiplied by 0.1 for computational convenience.

Finally, we evaluated two cultural services: job opportunities and start-up cooperation, which could enhance opportunities such as saving money by sharing materials cooperatively. Job opportunities data were obtained from each start-up's business plan. We defined four ranges of job security: (a) 1–3 workers employed; (b) 3–6; (c) 6–10, and (d) more than 10 people employed. The start-ups' cooperation within OpenAgri different projects, e.g. sharing the working force and/or materials or using the same selling systems, was evaluated from data declared in the business plans or which emerged in the first 2 years of the project.

## 13.4 Data Collection

Data were collected from November 2017 to March 2019—concurrent with the OpenAgri project. Data were collected using two sources. First, we had access to the business plans presented by each start-up during the public selection process. As official documents, this source provided economic data for each project. Second, we followed the selected start-ups by actively participating in debates and reunions with the representative of each of them. This informal participation was helpful to understand the actual and real state of action of each project, and to collect data about management organisation.

To formalise and track collected data, we prepared a short questionnaire that we sent to each start-up. The questionnaire had three sections: (1) the economic dimension, i.e. asking the aimed economic turn-over and the employed force labour; (2) the environmental dimension, useful to understand ES contribution, i.e. cultivated species, required amount of water, type of agronomic management (organic or conventional); and (3) the socio-relational dimension, to understand the start-ups' cooperation levels and the potential for conducting activities in addition to food production, i.e. recreational and cultural activities.

We then compared the data obtained from the business plans and the questionnaire. Assuming the questionnaire to be more accurate and up to date, we used this data for our study. In cases of incomplete responses, we assumed the data and information presented in the business plans.

To estimate the ES provided by the ecological requalification of the area, we built an internal database, with a realistic hypothesis of trees—species and numbers—that will be in the area. The database of the implant was the input for the I-Tree model simulation.

## 13.5 Pareto Front Algorithm and ES Performance

The selected indicators were useful to understand the relation between the ES provided and economical production, as well as to understand possible ES enhancing strategies. To answer the central questions of the study, we used the Pareto Frontiers algorithm to highlight ES trade-offs and to show which start-up was able to enhance ES and incomes simultaneously. The Pareto front algorithm is a tool widely used for selecting the best theoretical scenarios based on a large number of combinations of tested factors (Lafond et al. 2017). The Pareto front algorithm

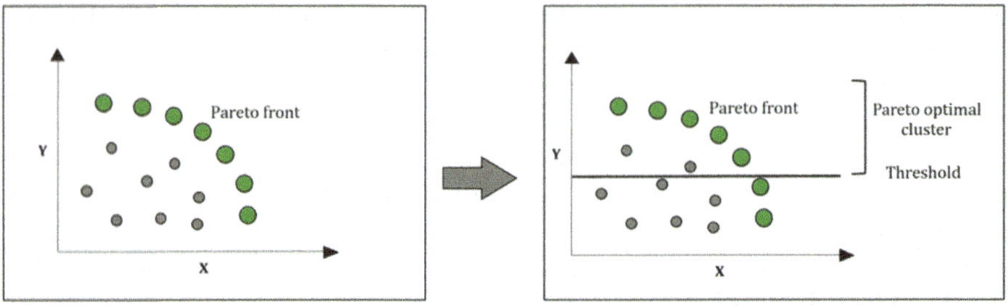


**Fig. 13.3** Conceptual framework using the Pareto Frontiers algorithm. Pareto analysis conceptual scheme illustrated for two objectives (*X* and *Y* indicators). The threshold splits the cluster into two: (1) below the threshold "Pareto front" and (2) above the threshold "Pareto optimal cluster". Each dot represents a start-up, and the green colour is the Pareto front and in grey is the non-Pareto Front (Andreotti et al. 2018)

subsets groups of scenarios which dominate others by maximising or minimising multiple factors. These theoretical optimal scenarios form a "front" on which one criterion cannot improve without deteriorating the others (Pardalos et al. 2008). In this study, we kept the conceptual idea of the Pareto front algorithm, applying it to real data in order to explore putative trade-offs and synergies within the set of eight start-ups (Table 13.1). By doing so, we aimed at identifying the best "performing" start-ups, i.e. those combining maximised levels of indicators. The indicators studied were assumed to be equally essential. Consequently, we did not set any weightage on the indicators studied when executing the Pareto analysis. The computed Pareto fronts were qualified as "Pareto optimal cluster" for the group of plots maximising the levels of indicators (Fig. 13.3).

A threshold was taken into consideration, corresponding to the last 3 years' average monetary value production of all the farms in Regione Lombardia (ISTAT 2018), which amounts to 108,823 €/year (approx. USD 120,745 $/year). That value is by far the highest in Italy, with marked differences from North Milan. To the South: in the same period, the farms in Molise have an average production of 11,904 €/year (approx. USD 13,208 $/year). Moreover, to better explore the results of the Pareto front algorithm, a Principal Component Analysis (PCA) was performed. The Pareto analysis and the PCA were carried out using R 2.13.0 with the packages Multiple Criteria Organization (MCO) and psych (Mersmann 2014; Revelle 2017).

The OpenAgri project also provided the opportunity for requalification of the ecological state of the area. Indeed, in addition to the creation of a food start-up incubator, the project's other goal was to create a net of living fences with tree rows and living hedges. The objectives of the net were: (1) better link the project area with the surrounding environment; (2) improve the provision of ecosystem services; and (3) naturally divide and define each field between the start-ups. Therefore, in our study, we quantify the ES provided by this agroecological net.

Following the project guidelines, we hypothesised a total area dedicated to the agroecological net of about 1.5 ha, planted with different plant species. Since the project was on-going at the time of writing, we created a database to be used for simulation of the ES resulting from the agroecological requalification. The database counts 3287 trees and shrubs, covering an area of 1.6 ha with a leaf area of 4.475 ha, composed of a mix of native and exotic tree species, to ensure high biodiversity and minimise the overall impact or destruction by species-specific insects or diseases. Dominant species in the database are typical of the existing landscape in the area: *Crataegus monogyna*, *Salix campestre*, *Morus alba*, and *Acer campestre*.

For the simulation of the ES provided by trees and shrubs, we assumed an initial height of plant between 0.5 and 2 m and a diameter between 3 and 5 cm. Using these parameters, we applied the I-Tree Eco (Nowak and Crane 2000) model to the database to estimate provided ES. The model uses the database, with the addition of local hourly air pollution and meteorological data, to quantify urban forest structure and provided ES (Nowak and Crane 2000). In this study, we focused our attention on the following regulating ES:

### 13.5.1 Air Pollution Removal

As air quality is highly important for human health, and many urban areas have a bad quality, it is clear that urban vegetation can play a crucial role in assuring a better air quality by removing pollutants. In this study, thanks to I-Tree Eco, pollution removal was calculated for ozone, sulphur dioxide, nitrogen dioxide, carbon monoxide, and PM2.5 (particulate matter $<2.5$ μm). I-Tree works estimating air pollution removal quantities calculating hourly tree-canopy resistances for ozone and sulphur and nitrogen dioxides, based on a hybrid of big-leaf and multi-layer canopy deposition models. Regarding PM 2.5, trees remove them when particulate matter lays on leaf surfaces (Nowak et al. 2013). Pollution removal by the agroecological net in the OpenAgri area was estimated using the built database and most recent pollution and weather data available, which are taken from Linate Airport weather station, close to the area.

### 13.5.2 Carbon Storage and Sequestration

Vegetation is able to sequester and store carbon in its tissue, thus lowering the level of carbon dioxide present in the atmosphere. In this study, the carbon storage evaluation derived from the biomass of each tree, calculated using equations from the literature in I-Tree Eco (Nowak and Crane 2000) and measured tree data, obtained from the available database. Carbon storage and carbon sequestration values were based on customised local carbon values. For this study, values were calculated based on a fixed value of USD 174 per metric ton, set as a current standard by I-Tree software.

### 13.5.3 Oxygen Production

As well as sequestering and storing carbon, trees and shrubs produce oxygen. The annual oxygen production of the vegetation is directly related to the amount of carbon sequestered by each tree, which is tied to the accumulation of tree biomass. The amount of oxygen produced derived from carbon sequestration based on atomic weights: net $O_2$ release (kg/year) = net C sequestration (kg/year) × 32/12 (i-Tree report).

## 13.6 Trade-Off Analysis

The trade-offs analysis indicated which start-up projects belonged (True) or did not (False) to the Pareto optimal cluster ("Belong to Pareto front" column) (Table 13.3) and if a start-up project was above the first production threshold ("Over the first production threshold" column) (Table 13.3). In general, there were no clear trade-offs or synergetic patterns between the ES indicators for the eight start-ups in the Pareto algorithm results or the principal components analysis (Fig. 13.4). On the other hand, we obtained clear clusters of start-up projects, which can belong, or not, to the optimal ones, meanwhile reaching the fixed threshold of economic turn-over.

### 13.6.1 Trade-Offs Analysis Results

Table 13.3 shows the findings of the eight start-ups analysed using both Pareto algorithm and the first production threshold. Only four start-ups produced more than the fixed threshold and were classified in the Pareto optimal cluster, meaning a simultaneous and positive performance in the provision of both ES and financial results. The Pareto front algorithms classified six start-ups in the Pareto optimal cluster. Using the established production threshold (ISTAT 2018), the optimisation deleted two of the six start-ups of Pareto top cluster. Regarding secondary production and pollinator dependence, the situation was not so clear as the start-ups could not have secondary production for diversification and/or crops that depend on pollination. For the Water Use Saved (WUS) indicator, differences were observed between the Pareto optimal cluster and no Pareto optimal clusters: for example, start-up #3 (Table 13.2) did not require water for its cultivation of wildflowers. Cultural services related to job opportunity guarantees from three to ten jobs for Pareto optimal clusters or not.

On the other hand, the first production threshold assesses the possibility—credibility—for the start-up to create job opportunities. Based on this analysis, start-up #8, even if it produced the highest number of job opportunities (ten job opportunities), it would be under the fixed threshold, therefore not qualifying for the optimal cluster. While start-up #1, which offered the lowest number of job opportunities (three job opportunities), was classified in the Pareto optimal cluster and above the first production threshold.

**Table 13.3** ES provision and threshold by OpenAgri start-up

| Start-up | First production | Secondary production | Water used saved (WUS) | Pollinator dependence | Job opportunities | Start-up cooperation | Belongs to Pareto front | Over the first production threshold |
|---|---|---|---|---|---|---|---|---|
| 1. Start-up 1 | 15.0 | 0 | 3.75 | 0 | 3 | 0 | True | True |
| 2. Start-up 2 | 14.5 | 7.75 | 0 | 3 | 6 | 0 | True | True |
| 3. Start-up 3 | 15.0 | 0 | 6.75 | 0 | 6 | 4 | True | True |
| 4. Start-up 4 | 4.8 | 0 | 0.55 | 3 | 6 | 0 | False | False |
| 5. Start-up 5 | 4.1 | 1.8 | 2.75 | 0 | 3 | 0 | False | False |
| 6. Start-up 6 | 0.5 | 0.51 | 0.25 | 2 | 6 | 8 | True | False |
| 7. Start-up 7 | 12.4 | 0 | 1.25 | 3 | 6 | 4 | True | True |
| 8. Start-up 8 | 3.6 | 0 | 3.12 | 3 | 10 | 6 | True | False |

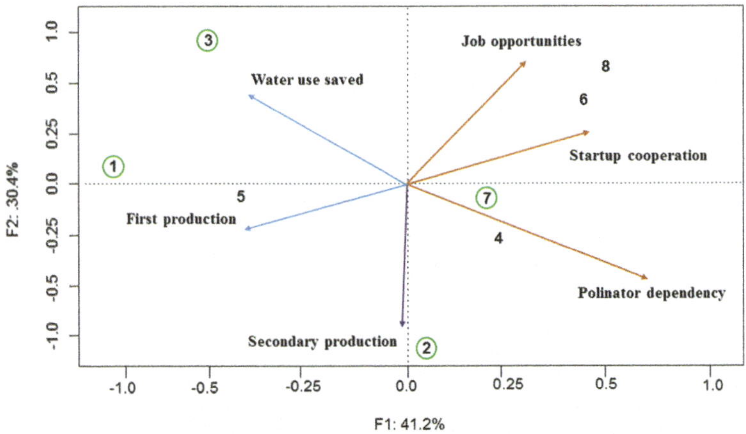


**Fig. 13.4** Principal Component Analysis of the studied ES and OpenAgri start-ups. Principal Component Analysis of the studied ES and OpenAgri start-ups (n° in Table 13.2). The green circles highlight the start-ups belonging to the Pareto optimal cluster above the primary production threshold. Each number corresponds to each singular start-up

We then combined the results from the Pareto optimal cluster with Principal Component Analysis. The first and the second axes of the Principal Component Analysis explained 41.2% and 30.4% of the total variance of the start-up studied, respectively (Fig. 13.4). The Principal Component Analysis discriminated accurately between the Pareto clusters on the first axis. On the one hand, Pareto top clusters were projected on the first axis, where the first production was one of the main contributors. However, on the other hand, the non-Pareto optimal front and low-yield intermediate clusters were mainly projected on the second axis. Any explanation or further conclusion to help readers interpret this?

## 13.6.2 Ecosystem Services Evaluation of the Semi-natural Hedgerow Network

### 13.6.2.1 Air Pollution Removal

I-Tree Eco has estimated that pollution removal was highest for ozone. It was estimated that trees remove 69.71 kg of air pollution (ozone ($O_3$)), carbon monoxide (CO), nitrogen dioxide ($NO_2$), particulate matter less than 2.5 μm (PM2.5), and sulphur dioxide ($SO_2$) per year with an economic value of USD 3906.

### 13.6.2.2 Carbon Storage and Sequestration

The gross sequestration of OpenAgri trees was estimated in 2.185 metric tons of carbon per year, with a value of USD 37,987. The i-Tree model estimates the fences in OpenAgri store 6.3 metric tons of carbon USD 1093. Of the species sampled, *Crataegus monogyna* stores and sequesters the most carbon: approximately 17.6% of the total carbon according to I-Tree Eco.

#### 13.6.2.3 Oxygen Production

Fences in OpenAgri are estimated to produce 5.825 metric tons of oxygen per year, thanks to photosynthesis. However, if we consider the overall production of oxygen and available reserve present in the atmosphere, the contribution of plants appears modest (Broecker 1970).

## 13.7 Key Findings

The study results confirm that, in a peri-urban context, start-ups can offer and provide different and multiple ES in addition to the production of food, including regulating services—crop pollination—as well as cultural and social services. This finding demonstrates the vital contribution potential of start-ups to urban agglomerates for job opportunities and better use of resources, while promoting new interactions and networking opportunities in the peri-urban framework of the area, in an overall context of urban requalification.

Multi-criteria optimisation methods such as the Pareto front algorithm were successfully applied and shown to be an effective method to support management decision-making processes (Lafond et al. 2017; Bugalho et al. 2016; Andreotti et al. 2018). It was, therefore, reasonable to further explore the potential of this methodology in various ecosystems, for example, in peri-urban areas, for which such an attempt could not be found in the literature.

While the OpenAgri project was still on-going at the time of this publication, from our results, it is clear that some of the start-ups analysed in this study (Start-ups 1, 2, 3, 7) represent best projects capable of combining the provision of multiple ES, as well as economic incomes. From our analysis, we conclude this is possible because of several aspects common to each of these four start-ups: first, a solid business plan, which tends to analyse and recognise new types of products with higher margins (i.e. spirulina production), as well as new trends (i.e. wildflowers), particularly requested by urban consumers. Second, each project promotes a more limited use of resources, and it takes into account diverse sources of income as secondary production or derived by cultural and social activities. These results are reflected by the Principal Component Analysis, which highlights the potential of start-up 7 (Fig. 13.4). Our findings highlight that a multidisciplinary approach for peri-urban start-ups is beneficial, to stay competitive in the market and to supply ecosystem services. Combining productive, social, and environmental initiatives can lead to more suitable solutions to enhance the value of peri-urban ecosystems, while answering to an increasing socio-economic issue, i.e. creating new jobs.

However, we note that for some ES—such as those regarding soil quality and the water cycle—not considered in the present study, useful results require a more extended period of data collection (5–10 years). Indeed, understanding and assessing the contribution of the project to improve the area (i.e. water cycle, air pollution removal, and several other regulating services) require that the project is complete and that all the start-ups are working and well-established. Moreover, dealing with an area with historical problems linked to environmental pollution, the results of

OpenAgri may need additional time to be visible and tangible especially for the results coming from the ecological requalification of the area. Based on the available project indication, we assumed the use of typical plant species with small-medium dimensions. The obtained results could vary if the number of plants increases and/or if the species change. It is also essential to understand that the ES provided by trees and shrubs can change during the time, also due to non-predictable factors (e.g. extreme weather events); therefore, continuous analysis of the situation should be put in place, to monitor change actively.

## 13.8 Conclusion

OpenAgri represents an optimal case study since it is a complex, yet familiar, peri-urban area, such as that found in many urban developed agglomerates, bringing environmental and socio-economic problems, as well as opportunities. Indeed, in the first 3 years of the project, start-ups faced significant global challenges in addressing the requalification of a lost area. Our results suggest that the requalification and the creation of start-up hubs can boost social inclusion and economic incomes, combining the provision of multiple ecosystem services and giving more value to urban agriculture, even in developed countries. However, start-ups and ecosystem services indicators should be analysed, considering a more extended time threshold in order to highlight the real economic and environmental sustainability of start-ups and the project as a whole. In this research, four out of eight start-up projects could produce more than the fixed threshold, meanwhile belong to the Pareto optimal cluster, meaning they can maximise ES provision and job opportunities. Our result depends on several factors, such as—the goodness of the business plan,—the rational use of the given resources,—the multidisciplinary approach, which allowed the start-ups to answer to the various and ever-changing requests of the local market.

Today the project is nearly concluded, and the start-ups are at the end of their training process. During the last months, the start-ups were involved in a pre-incubation process to develop their ideas and business plans. Thanks to this process, some start-ups changed their original project, adopting new strategies, in order to seize new opportunities and face unexpected problems, such as low availability of the needed amount of water to cultivate vegetables. We can thus affirm that each start-up has demonstrated a high level of resilience, being able to adapt itself to variable conditions. The requalification of the area is still on-going too, so in the next future also local citizens will benefit from the improvements, representing a scalable example in a similar place. In developed countries, at the margin of our growing cities, we can find a multitude of places that need a requalification, thanks to urban agriculture—providing in one solution several benefits. Therefore, urban agriculture in these complex contests should not be as subsistence food staple supplier, but as a single occasion to mark a socio-economic new start and to satisfy the latest urban needs in terms of sustainability, local food supply, and territorial cohesion. OpenAgri is one of the first endeavours linking together these related topics. It serves as a potential future case of study for further analysis and represents a replicable model in other urban contexts for the redevelopment of abandoned territories.

**Acknowledgements** This study was conducted as part of the EU Urban Innovative Action program "OpenAgri" project. We are thankful to project partners for their help and, in particular, we are grateful to each start-up that agreed to share data and information used in this study.

## References

Adams C, Rodrigues S, Calmon M, Kumar C (2016) Impacts of large-scale forest restoration on socio-economic status and local livelihoods: what we know and do not know. Biotropica 48(6): 731–744. https://doi.org/10.1111/btp.12385

Agbonlahor MU, Momoh S, Dipeolu AO (2007) Urban vegetable crop production and production efficiency. Int J Veg Sci 13(2):63–72. https://doi.org/10.1300/J512v13n02_06

Aizen MA, Garibaldi LA, Cunningham SA, Klein AM (2009) How much does agriculture depend on pollinators? Lessons from long-term trends in crop production. Ann Bot 103:1579–1588. https://doi.org/10.1093/aob/mcp076

Alberti M, Marzluff JM (2004) Ecological resilience in urban ecosystems: linking urban patterns to human and ecological functions. Urban Ecosyst 7(3):241–265

Andreotti F, Mao Z, Jagoret P, Speelman NE, Gary C, Saja S (2018) Exploring management strategies to enhance the provision of ecosystem services in complex smallholder agroforestry systems. Ecol Indic 94:257–265. https://doi.org/10.1016/j.ecolind.2018.06.048

Antrop M (2000) Background concepts for integrated landscape analysis. Agric Ecosyst Environ 77(1–2):17–28

Bagstad KJ, Semmens DJ, Waage S, Winthrop R (2013) A comparative assessment of decision-support tools for ecosystem services quantification and valuation. Ecosyst Serv 5(Sept):27–39. https://doi.org/10.1016/j.ecoser.2013.07.004

Bolund P, Hunhammar S (1999) Ecosystem services in urban areas. Ecol Econ 29:293–301. http://www.fao.org/uploads/media/Ecosystem_services_in_urban_areas.pdf. Accessed 31 Mar 2020

Broecker WS (1970) Man's oxygen reserve. Science 168(3939):1537–1538

Bugalho MN, Dias FS, Briñas B et al (2016) Using the high conservation value forest concept and Pareto optimization to identify areas maximizing biodiversity and ecosystem services in cork oak landscapes. Agrofor Syst 90:35–44. https://doi.org/10.1007/s10457-015-9814-x

Costanza R, d'Arge R, de Groot R et al (1997) The value of the world's ecosystem services and natural capital. Nature 387:253–260. https://doi.org/10.1038/387253a0

Cour JM (2001) The Sahel in West Africa: countries in transition to a full market economy. Glob Environ Chang 11(1):31–47. https://doi.org/10.1016/S0959-3780(00)00043-1

Daily GC (1997) Introduction: what are ecosystem services? In: Daily GC (ed) Nature's services: societal dependence on natural ecosystems. Island Press, Washington, DC, pp 1–10

De Groot RS, Blignaut J, Van Der Ploeg S, Aronson J, Elmqvist T, Joshua FJ (2013) Conserv Biol 27(6):1286–1129. https://doi.org/10.1111/cobi.1215

DG Environment (2012) Management plan 2012, European commission directorate-general environment. https://ec.europa.eu/dgs/environment/pdf/management_plan_2012.pdf. Accessed 31 Mar 2020

EEA, European Environmental Agency (2011) Annual report 2011 and environmental statement 2012. ISSN 1561-2120. https://doi.org/10.2800/52260

FAO (2010) Growing greener cities - how urban and peri-urban horticulture contributes to food and nutrition security. http://www.fao.org/ag/agp/greenercities/pdf/ggc-en.pdf. Accessed 31 Mar 2020

FAOSTAT (2007) http://faostat.fao.org/site/526/default.aspx. Accessed 31 Mar 2020

Gallai N, Salles J-M, Settele J, Vaissière BE (2009) Economic valuation of the vulnerability of world agriculture confronted with pollinator decline. Ecol Econ 68:810–821. https://doi.org/10.1016/j.ecolecon.2008.06.014

Gerster-Bentaya M (2013) Nutrition-sensitive urban agriculture. Food Secur 2013(5):723–737. https://doi.org/10.1007/s12571-013-0295-3

Gliessman SR (2015) Agroecology: the ecology of sustainable food systems, 3rd edn. CRC Press/ Taylor & Francis, Boca Raton, FL

Gomez Baggethun E, Baron D (2013) Classifying and valuing ecosystem services for urban planning. Ecol Econ 86:235–245. https://doi.org/10.1016/j.ecolecon.2012.08.019

Hess C, Ostrom E (2007) Introduction: an overview of the knowledge commons. In: Understanding knowledge as a commons. MIT Press, Cambridge, MA

ISMEA (2018) Istituto di servizi per il Mercato Agricolo Alimentare (ISMEA). Rapporto sulla competitività dell'agroalimentare italiano 2018. http://www.ismea.it/flex/cm/pages/ServeBLOB.php/L/IT/IDPagina/10303. Accessed 31 Mar 2020

ISTAT (2017) Istituto nazionale di ricerca L'ANDAMENTO DELL'ECONOMIA AGRICOLA Anno 2017. https://www.istat.it/it/files//2018/05/Andamento.economia.agricola.2017-1.pdf. Accessed 31 Mar 2020

ISTAT (2018) Istituto nazionale di ricerca RAPPORTO ANNUALE 2018 La situazione del Paese ISBN 978-88-458-1954-4. https://doi.org/10.1481/Istat.Rapportoannuale.2018. Accessed 31 Mar 2020

Kittinger J, Bambico T, Minton D, Miller A, Mejia M, Kalei N et al (2016) Restoring ecosystems, restoring community: socio-economic and cultural dimensions of a community-based coral reef restoration project. Reg Environ Chang 16:301–313. https://doi.org/10.1007/s10113-013-0572-x

Klein AM, Vaissie're BE, Cane JH, Steffan-Dewenter I, Cunningham SA, Kremen C et al (2007) Importance of pollinators in changing landscapes for world crops. Proc R Soc Biol Sci 274:303–313. https://doi.org/10.1098/rspb.2006.3721

Kovats RS, Valentini R, Bouwer LM, Georgopoulou E, Jacob D, Martin E et al (2014) Climate change 2014: impacts, adaptation, and vulnerability. Part B: Regional aspects. Contribution of Working Group II to the Fifth assessment report of the Intergovernmental Panel on Climate Change. Cambridge University Press, Cambridge, UK, pp 1267–1326. https://www.ipcc.ch/site/assets/uploads/2018/02/WGIIAR5-Chap23_FINAL.pdf. Accessed 31 Mar 2020

Kowarik I (2011) Novel urban ecosystems, biodiversity, and conservation. Environ Pollut 159: 1974–1983. https://doi.org/10.1016/j.envpol.2011.02.022

Lafond V, Cordonnier T, Mao Z, Courbaud B (2017) Trade-offs and synergies between ecosystem services in uneven-aged mountain forests: evidences using Pareto fronts. Eur J For Res 136: 997–1012. https://doi.org/10.1007/s10342-016-1022-3

Larondelle N, Haase D (2013) Urban ecosystem services assessment along a rural–urban gradient: a cross-analysis of European cities. Ecol Indic 29:179–190. https://doi.org/10.1016/j.ecolind.2012.12.022

Losey JE, Vaughan M (2006) The economic value of ecological services provided by insects. BioScience 56:311–323. https://doi.org/10.1641/0006-3568(2006)56[311:TEVOES]2.0.CO;2

Malinga R, Gordon LJ, Jewitt G, Lindborg R (2015) Mapping ecosystem services across scales and continents-a review. Ecosyst Serv 13:57–63. https://doi.org/10.1016/j.ecoser.2015.01.006

MEA, Millennium Ecosystem Assessment (2005) Ecosystems and human well-being: synthesis. Island Press, Washington, DC. https://www.millenniumassessment.org/documents/document.356.aspx.pdf. Accessed 31 Mar 2020

Méndez VE, Bacon CM, Cohen R, Gliessman SR (2016) Agroecology: a transdisciplinary, participatory and action-oriented approach. CRC Press, Boca Raton, FL. ISBN 978-1-4822-4176-1

Mersmann O (2014) MCO: multiple criteria optimization algorithms and related functions. 2014 R package version 1.0-15.1

Mörtberg U, Haas J, Zetterberg A et al (2013) Urban ecosystems and sustainable urban development—analysing and assessing interacting systems in the Stockholm region. Urban Ecosyst 16: 763–782. https://doi.org/10.1007/s11252-012-0270-3

Nowak DJ, Crane DE (2000) The Urban Forest Effects (UFORE) Model: quantifying urban forest structure and functions. In: Hansen M, Burk T (eds) Integrated tools for natural resources inventories in the 21st century. Proceedings of IUFRO conference. Gen. Tech. Rep. NC-212. U.S. Department of Agriculture, Forest Service, North Central Research Station, St. Paul, MN, pp 714–720. https://www.fs.usda.gov/treesearch/pubs/18420. Accessed 31 Mar 2020

Nowak DJ, Crane DE (2002) Carbon storage and sequestration by urban trees in the USA. Environ Pollut 116:381–389. https://doi.org/10.1016/S0269-7491(01)00214-7

Nowak DJ, Hirabayashi S, Bodine A, Hoehn R (2013) Modeled PM2.5 removal by trees in ten U.S. cities and associated health effects. Environ Pollut 178:395–402. https://doi.org/10.1016/j.envpol.2013.03.050

Orsini F, Kahane R, Nono-Womdim R et al (2013) Urban agriculture in the developing world: a review. Agron Sustain Dev 33:695–720. https://doi.org/10.1007/s13593-013-0143-z

Ostrom E, Burger J, Field CB, Norgaard RB, Policansky D (1999) Revisiting the commons: local lessons, global challenges. Science 284:278–282. https://doi.org/10.1126/science.284.5412.278

Pardalos PM, Chinchuluun A, Migdalas A, Pitsoulis L, Corporation E (2008) Pareto optimality, game theory and equilibria, optimization and its applications. Springer, New York, p 17. https://doi.org/10.1007/978-0-387-77247-9

Piorr A, Ravetz J, Tosics I (2011) Peri-urbanisation in Europe: towards European policies to sustain urban–rural futures. Synthesis report. University of Copenhagen, Forest and Landscape, Copenhagen. http://www.openspace.eca.ed.ac.uk/wp-content/uploads/2015/12/Peri_Urbanisation_in_Europe_printversion.pdf. Accessed 31 Mar 2020

Revelle W (2017) PSYCH: procedures for personality and psychological research, Version 1.7.8. North-Western University, Evanston, IL. https://CRAN.R-project.org/package=psych. Accessed 31 Mar 2020

Salbitano F, Borelli S, Conigliaro M, Chen Y (2016) Guidelines on urban and peri-urban forestry. FAO Forestry Paper No. 178. Food and Agriculture Organization of the United Nations. http://www.fao.org/3/a-i6210e.pdf. Accessed 31 Mar 2020

TEEB (2010) The economics of ecosystems and biodiversity TEEB. Ecological and Economic Foundations, Routledge, Abingdon, UK, p 410. http://www.teebweb.org/media/foundations.jpg. Access 31 Mar 2020

UIA (2020) OpenAgri - new skills for new jobs in peri-urban agriculture. https://www.uia-initiative.eu/en/uia-cities/milan. Accessed 31 Mar 2020

UN (2015) Transforming our world: the 2030 agenda for sustainable development, A/RES/70/1. United Nations, Geneva

UN Habitat (2014) Planning for climate change: a strategic, values-based approach for urban planners, cities and climate change initiative tool series. United Nations Human Settlements Programme. http://mirror.unhabitat.org/downloads/docs/PFCC-14-03-11.pdf. Accessed 31 Mar 2020

van Veenhuizen R (ed) (2006) Cities farming for the future, urban agriculture for green and productive cities. RUAF Foundation, IDRC and IIRR, Manila. ISBN 1-930261-14-4

Wezel A, Soldat V (2009) A quantitative and qualitative historical analysis of the scientific discipline of agroecology. Int J Agric Sustain 7:3–18. https://doi.org/10.3763/ijas.2009.0400

**Ambrogio Zanzi** is a PhD candidate in the Department of Environmental Science and Policy, State University of Milan, focusing his interest on agro-ecology, ecosystem service quantification, and landscape management. He holds a Bachelor's degree and a Master's degree in Agriculture and Plant protection science from the State University of Milan, followed with a specialization at the University of Massachusetts Amherst (USA). He was awarded the "FarmingBySat" third prize in 2016 by the European Environment Agency (EEA), in recognition of his Master's thesis about a new forecasting system to estimate quality and quantity of agriculture production using new technological tools.

**Federico Andreotti** is a PhD candidate at Wageningen University & Research (WUR) and the University of Montpellier. The double degree PhD project is in collaboration with the French agricultural research center CIRAD studying the process of co-constructing collective governance instruments to recognize and promote agricultural heritage systems in the global market.

**Valentina Vaglia** (Dr) is a PhD student at the Department of Environmental Science and Policy, University of Milan. Her research focuses on greenhouse gas emissions from the paddy field due to their management. She is also involved in wide-scale studies which deepen the integrated assessment of sustainability in agriculture and ecosystem services offered—her research and work as an Agronomist deal with innovative farming systems. She strongly believes in carrying on the best performances and low-impact strategies on environment and human health to enhance agriculture and is involved in research studies and activities of non-profit associations involved in agro-ecology.

**Sumer Alali PhD** is a fellow researcher at the Department of Environmental Science and Policy, University of Milan, Italy. He obtained his PhD in Agriculture, Environment, and Bioenergy from the University of Milan; he gained a Master's degree and a Bachelor's degree in Agriculture and Plant Protection Science from the University of Damascus, Syria. During his work, he went through topics related to Soil Biology, Entomology, and Biological Control. His latest work is focused on the assessment of agricultural practices on soil biodiversity through classic indices application and the implementation of molecular approaches for environmental studies.

**Francesca Orlando PhD** was researcher at the Department of Environmental Science and Policy, University of Milan, and then at the Department of Molecular and Translational Medicine, University of Brescia. She obtained her degree in Agricultural Science and PhD in Climatology and Soil Science from the University of Florence. During her work, she addressed different aspects involved in the integrated assessment of sustainability in agriculture. Her research deals with the assessment of the performances, impacts, and benefit of the agricultural practices and tools belonging to precision farming, organic farming, and conservative agriculture.

**Stefano Bocchi** (Prof) is a Full Professor of Agronomy and Cropping Systems at the Department of Environment Science and Policy, State University of Milan, Italy. He is Principal Investigator in several projects on cereals, fodder crops, agro-food systems analysis, and management both at national and international levels. Professor of graduate and undergraduate courses of agronomy, agro-ecology, agro-food system, landscape agronomy, and a member of the scientific board of PhD programs. He has been involved in several projects for international cooperation in Albania, Brasil, China, Ecuador, Egypt, Kenya, Lebanon, Peru, Philippines, Sierra Leon, and Tanzania. He is the delegate of the Dean for sustainability. He is author of more than 170 scientific papers and board member of various scientific societies. He was the scientific curator of the Biodiversity Park in EXPO 2015.

# Field Work: A Mixed-Methods Social Network Analysis of Urban Farmers and Hired Laborers in Four Cities

14

Jessica Ann Diehl

**Abstract**

As a managed living system, food production relies on resources of water, nutrients, and *labor*. Thus, if we discuss the ecological viability and sustainability of urban and peri-urban agriculture, we also need to acknowledge the farmer—the person who manages the plants and animals—as a fundamental resource. The focus of this chapter is twofold. First, to conceptualize labor as a resource and a social network as part of a larger goal of creating sustainable farming livelihoods. Second, to describe four case studies with diverse farmer-labor social networks in Delhi, India; Jakarta, Indonesia; Singapore; and Sydney, Australia. This research employed a mixed-methods research design to investigate commercial farmers' social relations with hired laborers. The case studies presented in this chapter were derived from two larger research projects to understand urban agriculture as a practice embedded in the urban system in selected case cities in Asia and Australia. Data were collected through farmer interviews, in-field observation, and photography. The four case cities demonstrate that even within one type of urban agriculture, namely commercial urban farms, labor requirements, and practices can be diverse. And, farmer-labor social networks profoundly impact the feasibility and sustainability of operating a farm. Given that urban farming is directly driven by farmers as a livelihood pursuit, and less directly an effort in sustainability or city-level food production, we must first identify the incentives and barriers that farmers face, as these factors will be primary drivers for behavior and decision-making among farmers.

J. A. Diehl (✉)
Department of Architecture, National University of Singapore, Singapore, Singapore
e-mail: akijac@nus.edu.sg



J. A. Diehl, H. Kaur (eds.), *New Forms of Urban Agriculture: An Urban Ecology Perspective*, https://doi.org/10.1007/978-981-16-3738-4_14

**Keywords**

Urban farming · Social networks · Qualitative research · Multiple-case study · Livelihoods

## 14.1 Introduction

New typologies of urban agriculture are emerging in cities worldwide in response to global threats to food security. In some trajectories, urban food production is trending toward reductivist, technologically advanced approaches, whereas in others, it is being integrated into city infrastructure in multi-functional ways. Importantly, while these two approaches seem disparate, both have the goal of producing food in the city. As a managed organic (i.e., living) system, food production relies on resources of water, nutrients, and labor. Anyone who has propagated a garden, or even tended a few houseplants, knows there are knowledge and skill requirements for cultivation, as well as time and labor commitments. What is more, there is a seasonality and a lifecycle to growing plants and raising animals; it is a dynamic system. Thus, if we discuss the ecological viability and sustainability of urban and peri-urban agriculture, we also need to acknowledge the farmer—the person who manages the plants and animals—as a fundamental resource.

Who is the farmer? Simply, the person who grows crops or raises animals for sale or self-consumption. But, *who* is the farmer? What skills, qualifications, licenses, land titles, formal contracts must the person possess? In the USA and across Europe and other advantaged countries where the local and organic food movement has gained significant traction, the farmer has diversified from that of a rural, lower educated, lower earning, older social group into that of an urban, higher educated, higher earning, younger social group. In this context, the urban farmer is often conceptualized as a socially conscious, environmentally sensitive, forward-thinking entrepreneur. But, when attention is turned to developing countries, our conceptualization of the urban farmer changes. Who is *that* farmer? Is she poor, is she educated? Is he literate, how many children does he have? Do they own a permanent home, do they have access to clean water? Do such ideas and images impact the attitudes of non-farming citizens in those places? And, how do such images impact how people of influence in those places *understand* who the urban farmer is—and how does that understanding impact related policies and land use regulations?

I grew up in a small New England town within walking distance of two dairy farms. I knew one family quite well. I would often go watch them bring in the cows for milking. Even with a mere 150 cows and automated milking machines, the farm employed a half dozen workers, mostly part-time and with high turnover. Fast forward to graduate school where I found myself entrenched in a research project to measure social networks of urban farmers in Delhi, India. Even with small vegetable plots averaging less than 2.5 acres, there was a complex system of hiring day laborers and other farmers for wages or in exchange for their own labor to complete daily tasks. It was evident to me through these experiences that the dynamic nature of farming requires a dynamic labor system.

The farmer managing the farm is rarely the only person tending the crops and animals. Who does the farmer hire? How does the labor required to operate a farm impact the sustainability of the farm? And, what are the social dynamics of farm-labor relations? The focus of this chapter is twofold. First, to conceptualize labor as a resource and a social network as part of a larger goal of creating sustainable farming livelihoods. Second, to describe four case studies with diverse farmer-labor social networks in Delhi, India; Jakarta, Indonesia; Singapore; and Sydney, Australia.

### 14.1.1 Diversity of Agriculture Participants

Urban agriculture can vary in orientation and scale from subsistence-oriented cultivation, to more recreational types of agriculture at the micro-scale, through small-scale, semi-commercial gardeners and livestock keepers, to medium- and large-scale commercial operations (van Veenhizen 2006; Mougeot 2005). The location of urban agriculture includes greenbelts, urban fringes, vacant or underutilized city lots, community gardens, fish farms, and greenhouses. These different types of urban agriculture require different lifecycle operations structure. In Designing Urban Agriculture, Philips (2013) summarizes four broad categories of urban agriculture: personal, public, non-profit, and for profit. Within each category, specific project types overlap with participants and objectives (Fig. 14.1). For example, a for-profit farm has an objective of for-profit production with participants who are employees. In contrast, a school garden has an objective of education with participants who are students. Thus, analogous to water and nutrient requirements that can vary dramatically depending on which crops are grown or animals raised, labor can likewise vary depending on the specific characteristics of the agricultural operation.

While urban agriculture manifests in diverse ways, this research focuses specifically on for-profit commercial urban farms because of their potential to make a meaningful contribution to the food system and food security at the city-scale. By definition, commercial farms are businesses that are intended to provide the farmer a livelihood. A livelihood is defined as "the capabilities, assets (including both material and social resources), and activities for a means of living. A livelihood is sustainable when it can cope with and recover from stresses and shocks, maintain or enhance its capabilities and assets, while not undermining the natural resource base" (Scoones 1998, p. 5). Given that urban farming is directly driven by farmers as a livelihood pursuit, and less directly an effort in sustainability or city-level food production, the day-to-day and operational incentives and barriers that farmers face will be primary drivers for behavior and decision-making.

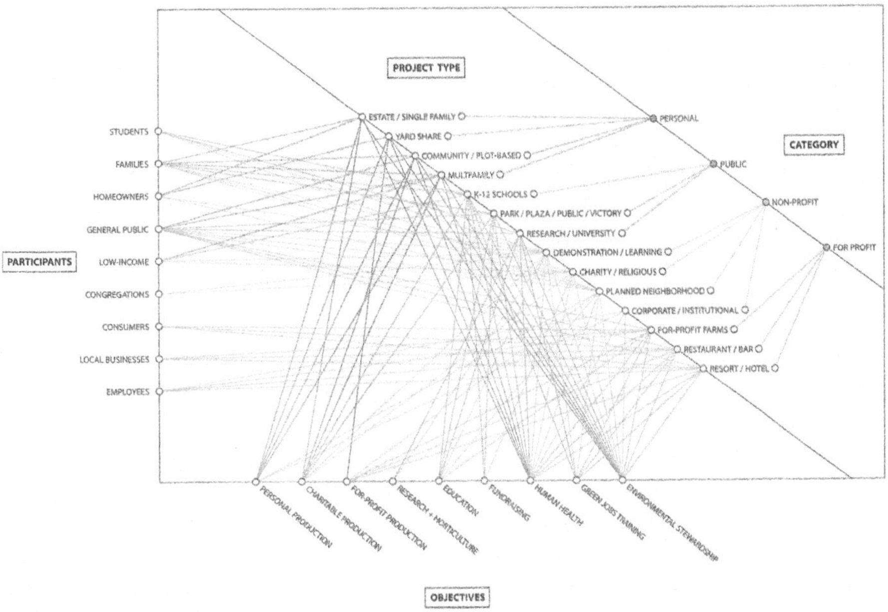


**Fig. 14.1** Urban agriculture categories linking participants by project type and objectives (Philips 2013)

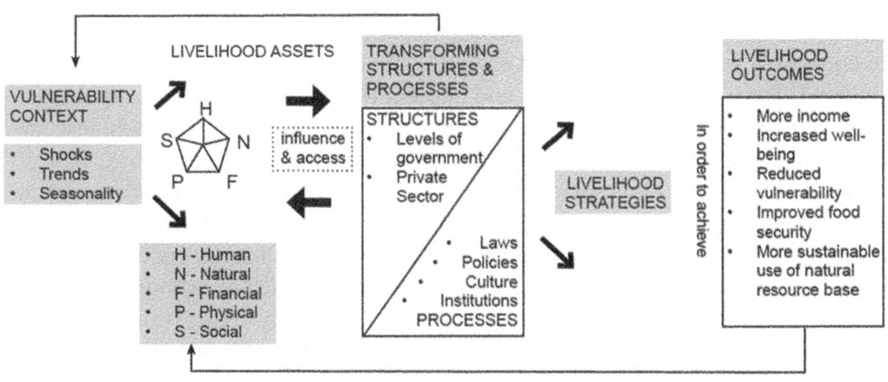


**Fig. 14.2** The Sustainable Livelihoods Framework by I. Scoones, adapted by D. Carney (Scoones 2009; Carney 2003)

### 14.1.2 The Sustainable Livelihoods Framework

This research applied the Sustainable Livelihoods Framework (SLF) as a way of conceptualizing farmers' livelihood strategies within the urban food system (Fig. 14.2). The SLF emerged from the vulnerability-resiliency literature in response to rural development, and was conceptualized as a framework for understanding the process of creating sustainable livelihoods (Scoones 2009). Notwithstanding the complexity of the SLF, this chapter focuses specifically on how *influence and access to livelihood assets* impact *livelihood strategies* in order to achieve *livelihood outcomes*. Translated within the context of this chapter, it is through the ability to access and influence human capital that the farmer is able to achieve for-profit production. Human capital is defined as knowledge and labor.

Farmers access and influence human capital as an intrinsic characteristic through their individual knowledge, skills, time, and effort. But, they also access and influence it through the people they hire to participate in daily operations, e.g., harvest crops, milk animals. Although not explicit in the SLF, social networks facilitate and constrain influence and access to livelihood assets.

### 14.1.3 Social Network Access to Human Capital

Social networks are a critical mechanism for mobilizing resources beyond the household (Blaikie et al. 1994; Knoke and Yang 2008). Social networks are comprised of agents and the relationships between them (Scott 1991). Agents are defined as individuals or collectives, and relationships are defined as the contact, connection, or tie between a pair of agents. A core assumption of social network theory is that direct or intensive contact exposes agents to better information, greater awareness, and higher susceptibility to influence by other agents, whereas indirect contact exposes agents to new ideas, and potential access to useful resources. Social networks act to channel information and resources to particular structural locations (i.e., agents), help create interests and shared identities, and promote shared norms and values.

The core theoretical problem in developing a methodology to measure social networks is to explain the occurrence of different relationships ("ties") between agents and account for variation in linkages. Often, social networks are measured quantitatively through a structure-function lens; metrics include identifying nodes and links (agents and relationships) and weighting them based on measures that include betweenness, bridging, centrality, closeness, and reach (Scott 1991). A quantitative approach provides information on the distribution and connectivity of relationships and knowledge exchange, but is limited in the ability to identify how or why the resource is mobilized. Social network analysis is, therefore, strengthened by a mixed-methods approach (Edwards 2010).

## 14.2 Methods

This research employed a mixed-methods research design to investigate commercial farmers' social relations with hired laborers. The case studies presented in this chapter were derived from two larger research projects to understand urban agriculture as an embedded practice in selected case cities in Asia and Australia. This chapter reports on findings from four cities: Delhi, India; Jakarta, Indonesia; Singapore; and Sydney, Australia. Data were collected through farmer interviews, in-field observation, and photography. Interviewees were consented[1] and the discussion was recorded with pen and paper for more vulnerable farmers in Delhi and Jakarta, and audio recorded with consent in Sydney and Singapore as well as documented through extensive note-taking on an iPad by a research assistant. Interviews in Delhi, Jakarta, and some interviews in Singapore required a translator.

Farmers represented discrete central social network agents and other farm workers (referred to interchangeably as workers or laborers) were conceptualized as a collective agent. In other words, I designated the farmer as an individual agent and workers/laborers as a group the farmer interacted with. Social networks were operationalized using two key social network theory assumptions according to Knoke and Yang (2008):

- Assumption #1: Structural relations exist at specific time-place locales. Structural relations are defined as the interaction between two agents. Variables included where, when, and frequency of farmer interactions with laborers.
- Assumption #2: The type of information exchanged (transaction) and degree of influence depends on whether the relation is direct or indirect. Variables included type of transaction, degree of influence, and direct/indirect relation.
  - The type of transaction was defined in terms of livelihood assets comprised of five capitals:
    Human capital, defined as knowledge and labor potential;
    Natural capital, defined as ecosystem assets;
    Financial capital, defined as money;
    Physical capital, defined as tangible "things";
    Social capital, defined as positive social connections (a type of social network, which can be comprised of positive or negative social connections).
  - The degree of influence was defined as the ways in which social structures facilitate or constrain access to resources (i.e., capitals) between agents. Degree of influence was measured as:
    Investing, defined as a transaction that provided resources to an agent with the expectation of a future return;
    Withdrawing, defined as gaining resources;
    Exchanging, defined as a transaction in which resources were both provided and gained;
    Blocking, defined as an agent preventing access to resources.
  - Direct relations were defined as bonding ties, a strong tie between immediate family members, neighbors, and friends, whereas indirect relations were defined as bridging ties, a connection with a person or people of different socioeconomic and/or cultural backgrounds (Lin 2001). Figure 14.3

[1]The Colorado Multiple Institutional Review Board (COMIRB) and corresponding Delhi research ethics authorities approved research in Delhi. And, the National University of Singapore (NUS) Institutional Review Board approved research in Singapore, Jakarta, and Sydney.

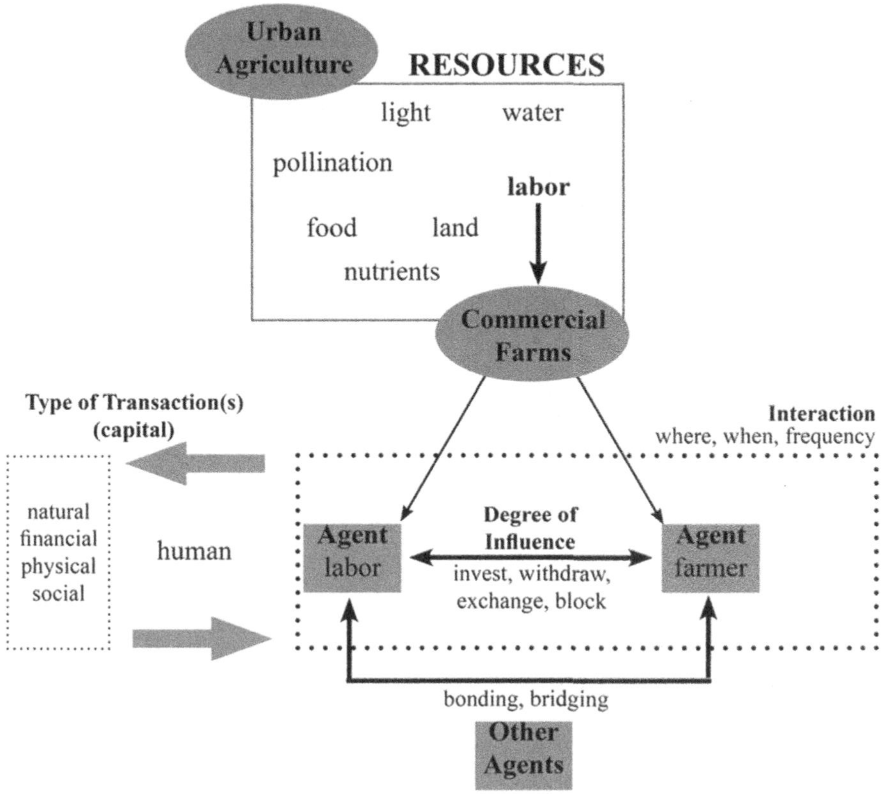


**Fig. 14.3** Conceptual diagram of farmer-labor social relations and access to resources (primarily human capital)

illustrates how social network analysis assumptions were used to operationalize how labor is accessed and influenced by commercial farmers in the form of human capital.

Semi-structured interviews were conducted at the farm with the lead farmer, farm manager, or head of household. In Sydney, interviews were also conducted at farmers' markets with the farmer. The interview guide was developed and further refined throughout the fieldwork to fit the local context (Diehl et al. 2019; Diehl 2020). Interview questions captured farm traits and access to resources through social networks. Social networks were measured using an ego-centric approach with the farmer as the central agent. The tie to agents (hired laborers as a collective worker/labor group) was measured as the location and frequency of interaction, the type of transaction, the degree of influence, whether the relationship was direct or indirect. The key interview question related to the farmer-labor social network was: *Do you hire any laborers/workers or people to help on the farm?* Farmers were asked to talk about the relationship with hired laborers/workers through the following prompts based on the two key social network assumptions outlined above:

- How do you find out who to hire and where do they come from? Where/how do you advertise?
- Do you teach them new skills or provide any resources?
- Do they tell you things or suggest improvements for your farm?
- Do they connect you with other people that can help improve your farm?
- How do you decide what to pay them? Do you ever negotiate? Would you share the wage/rate or range of pay?
- Do you talk about anything else besides work tasks?

Interviews were first analyzed qualitatively. Notes were digitized and audio recordings were transcribed, translated (if not in English), and then organized using the interview guide as a template. Interviews were imported into Atlas.ti for Mac (qualitative coding software) and coded. Codes and themes that could be categorized were summarized quantitatively. Results are reported descriptively in the following sections by case study city.

## 14.3 Case Study Delhi, India: Farmers Who Hire Labor but Also Work for Each Other

In Delhi, the selected case site was a single, large cultivated area on the Yamuna floodplain, which bisects the city. The site was located near a middle-income residential neighborhood and was within a few kilometers of the central business district of the city. It was approximately 2.5 $km^2$, supporting estimated 300 households. The larger research project interviewed 121 farming households selected using convenience and adjacent sampling methods, i.e., approaching the first person in the field, and moving on to the next-door neighbor. Interviews were conducted between February 2012 and June 2013.

The majority of farmers rented land ($n = 65$; 54%), 44 (36%) were employed in sharecropping,[2] and 7 (6%) had plots both on rent and tied to sharecropping ($n = 5$; 4% were missing data). Farmers reported six models of farm labor (Table 14.1). The most common model was for farmers to hire workers on a daily basis according to the needed tasks. It was also common for farmers to work on another farmer's plot for wages. Some farmers participated in a sharing economy of working on each other's plots as needed; they worked for each other ("labor for labor") rather than for

[2] Sharecropping is when a landowner allows a tenant to use the land in return for a share of the crops produced on the land. It is commonly practiced in India. A tenant/cultivator pays the landlord a specific proportion of the product as agreed on in the share contract. The tenant keeps only a fraction of the product, which is a disincentive to pay more for inputs. By contrast, a fixed rental contract gives the tenant incentive to maximize the produce as he gets to keep the entire output and makes only a fixed rental sum to the landlord (Sen 2016).

**Table 14.1** Summary of hired labor

| Hire | Freq. | Perc. |
|---|---|---|
| Other farmers only | 13 | 13 |
| Other farmers if available or outsiders | 21 | 21 |
| Hires outsiders | 10 | 10 |
| Residents-only (sharecropping) | 2 | 2 |
| Farmer works as temporary labor | 12 | 12 |
| Farmer works as permanent labor | 2 | 2 |
| Total # of interviews with responses[a] | 102 | |

[a]Interviews could have more than one response

pay. In the case of sharecropping, it was uncommon to hire workers—mainly because the landlord hired extra workers as needed. Lastly, some farmers had enough family members or a small enough plot that they did not require additional workers.

Some farmers only hired other farmers, whereas other farmers hired workers from outside the community (either in the case that other farmers were not available or as common practice)—usually finding outside workers waiting near the main road for a job opportunity for the day. Workers came from adjacent neighborhoods of Trilokpuri or Mayur Vihar or the peri-urban villages surrounding Delhi including Bhogal, Samaspur, Patparganj, Noida, Khondli, Kanyanpuri, Kundali, Ghadoli, and Nanglor. The majority of farmers were rural to urban migrants and a few would hire workers from their home villages, i.e., Badayun, Uttar Pradesh, and would host them for the growing season.

Most farmers living on nearby plots worked for each other, but if they were not available then knowing who to call or where to go for help was critical in protecting caring for the crop. How did farmers find workers? Some farmers asked around to find out who was available, whereas others would hire workers who roamed around and asked for work. Some workers were sent to farmers by relatives in their rural home village, whereas others were hired on the spot from a group of laborers waiting at the main road. Most farmers engaged laborers from inside the community; only one quarter of farmers hired workers from outside the community, whereas most farmers engaged laborers from inside the community. It is most common for farmers to hire and work for each other and only turn to "outsiders" when no one was available. Working "labor for labor" (i.e., farmers exchanging labor) as compared to working for pay instilled more trust and care in work completed; there was future relationship and obligation established.

Since laborers were lower in the hierarchy than farmers (who employed them), their bridging potential was limited—they lacked connections outside the community to other people with resources that could benefit farmers. However, bonding potential was high among farmers who worked for each other. Exchange of labor required trust that the hired person would improve the crops and investment in improving the crops of other farmers. Exchange of labor among farmers also saved money because they did not have to pay cash for labor. If farmers scheduled work on a rotating cycle, they could always be working either to till, plant, or harvest. They could also plan to plant different crops to reduce competition.

The relationship with hired workers had the potential for "investment" with the implied assumption of a future return. This was operationalized as "teaching skills or sharing knowledge." Farmers who "invested" in their workers expected more efficiency in time, better crop, etc. The farmers who hired other farmers had a degree of confidence that the worker already had requisite knowledge and skills. In contrast, for workers hired from outside the farm community, some farmers said that they told the workers what to do, or they had workers do the easiest tasks, "just clean the land, nothing technical." The majority of farmers that hired from outside, however, said that workers were typically from farming villages and already had the skills and knowledge required to farm. There was a sentiment that it was easy to hire workers because of the common agricultural background. But it was also clear that if the workers were hired from outside, they were usually asked to do simple tasks; whereas other farmers "know what to do so we give them more work." In some cases, farmers said that they gave and received advice from hired laborers. For example, they might discuss what should be planted or talk about other issues related to farming. At times, laborers told them what other farmers were doing and suggested what to grow, which seeds to buy, better practices, etc.

In terms of withdrawing resources in addition to human capital, farmers could acquire skills or knowledge from workers as a means of improving their own human capital and potential financial capital. Farmers fell into two general groups: those who said they did not need advice and those that said they did get advice. It was unclear in trying to disentangle who was giving and who was getting skills/knowledge because there was often an exchange rather than a one-directional transaction. This indicates that there was a similar level of power or even hierarchy in the farmer-labor social network. Families that were part of the exchange process were more embedded in the community social network than those who hired from outside. This could also indicate the power the family had within the community social network in terms of accessing other resources and information. Since many farmers worked for each other, further investigation could be conducted into how collaborative farming deepens trust among neighbors and acts as a bonding mechanism.

The primary capital that laborers provided was human capital. The main difference in compensation was whether the work was in cash or kind (i.e., labor for labor) (Table 14.2). There was not much negotiation of wages because the market dictated a relatively stable rate (ranging from 150 to 250 INR or $2.00 to $3.30 USD daily at the time of the research). One farmer reported that workers would come and negotiate the rates, but everyone paid 150 INR so she paid the same. Some farmers paid based on gendered work: women were paid 150 INR and men were paid 250 INR based on the kind of work assigned to them. Women were not hired to operate heavy tools; rather, they were hired for simpler work like cleaning the ground, weeding, etc. Other farmers paid based on workload: females and males received the same pay for labor—but usually men were assigned heavier work and were paid more for it. Generally, farmers preferred to avoid paying cash, rather working in return for the labor they receive.

**Table 14.2** Summary of the compensation for hired labor

| Hired labor compensation | Freq. | Perc. |
|---|---|---|
| Pays cash for labor | 64 | 59 |
| Works as laborer for cash | 31 | 29 |
| Exchanges labor for labor | 29 | 27 |
| Landowner pays laborers | 4 | 4 |
| Sharecropping—no additional labor | 2 | 2 |
| Sharecropping—residents hire/pay others | 2 | 2 |
| Sharecropping—share cost with landlord | 1 | 1 |
| Total # of interviews with responses[a] | 108 | |

[a]Interviews could have more than one response

In summary, farmers were at an advantage when they had the financial means to hire labor for pay. They were at an advantage when they had the connections to hire labor in kind when they did not have financial capability to pay cash. And, they had an advantage when they could invest time/labor in the system with the trust/agreement that they would receive equal labor in return in the future. In this way, the human capital they received through labors was linked to financial capital. But, there was also social capital within the farm community. Although farmers did not really have much to say about any other discussions with laborers beyond farming, since many farmers worked for each other they generally talked about crops and day-to-day issues. I did, notably, observe a lot of "hanging about" and chit chat about chores and family things.

## 14.4 Case Study Jakarta, Indonesia: Self-Sufficient Farms

In contrast to the continuous landscape pattern of Delhi's cultivated landscape, Jakarta is a peri-urban patchwork of discontinuous plots. The research sites were purposefully selected to meet the following criteria: public access (many agricultural plots are inaccessible from public streets), actively farmed, and farmers present on-site. Combining GIS (geographic information systems) mapping of satellite imagery and ground-sleuthing (on-the-ground visits) enabled selection of two clusters of agriculture fields within close proximity to one another. One cluster was located near the border of Kalideres and Cengkareng districts in northwest Jakarta, adjacent to *kampungs* (informal settlements or housing clusters) where some of the farmers lived. The second cluster was located in the west and northwest peri-urban fringe along the Chakung drain. A convenience sample of two to five households at each of ten sites within the two clusters was interviewed, 39 in total, between December 2017 and June 2018.

Jakarta urban farms were located within residential neighborhoods on privately owned or government land. The farmers had informal tenure status with insecure land title claim. The farms were semi-commercialized, small and to medium-scale, and market-oriented, producing both food crops and animal products. Farmers primarily grew low-risk and short cycle crops, which were more profitable. The farms were commonly managed by one farmer alone ($n = 20$; 51%), as a husband and wife ($n = 8$; 21%), or as a family ($n = 5$; 13%); few farmers hired workers when needed ($n = 1$; 3%), permanently ($n = 2$; 5%), or on sharecropping ($n = 3$; 8%).

Jakarta has experienced waves of rural to urban migration by those seeking better job opportunities, for example, during the 2008 global food crisis. Nearly three quarters of the farmers had migrated from across Java and Sumatra ($n = 28$; 72%), originating from more than a dozen different places across the two islands. Dates of migration ranged from as early as 1975 to as recently as 2013. Many migrant farmers had been in agriculture growing rice on rented land or in sharecropping.

Due to the small number of farmers that required and hired outside labor to help with field tasks, only an anecdotal report can be summarized. One farmer said that he hired two laborers per harvest who came from Java. He paid them 400–700 IDR ($0.03–0.05 USD) per *kangkong* (Chinese spinach) bundle. Another farmer said she did not have a fixed rate of pay—it depended on how much the laborers could harvest. One farmer said she occasionally asked other farmers to help her, but she provided only a meal in exchange for a day of labor.

Since the majority of farmers lived with their families in nearby houses or *kampong* dwellings, their family members were able to help in the fields as needed. Notably, many of the farmers' family members were employed in other city jobs. Therefore, farming was a supplementary rather than primary income—and, some farmers were retired and farming as a hobby. In the case of Jakarta, human capital was accessed primarily as an intrinsic resource rather than through labor networks—in other words, the farmers were farming themselves and not hiring laborers. At the micro-scale, the system of farming was a disconnected patchwork of operations; however, a consideration of the macro-scale provides a different picture.

The peri-urban fringe of a city is not fixed; it moves further away from the urban core as the city grows outward. While it is pushed away from the urban core, it is uniquely tied to the city and remains within the sphere of influence. As roads and infrastructure are constructed to better link the urban core and urban fringe for commuting, land values tend to increase (Lonard 2012). It is on the fringe where agriculture tends to intensify to take advantage of proximity to urban markets, reliable water supply, and labor. However, the consequence is that agriculture on the urban fringe faces rising land values, land speculation by developers, and conflict with new residential owners over farm practices, e.g., farm odors and pesticide spraying. While conflicts of agricultural land use on the peri-urban fringe are often highlighted, multi-functional benefits remain largely invisible. For example, the Jakarta farmers tended plots of land that otherwise would have been the responsibility of the city or an absent private owner to manage. In addition to producing food, they also managed vacant, underutilized, or pubic right of way land.

In the case of vacant land, a few of the interviewed farmers cultivated plots on large tracts of land owned by absent Indonesian Chinese families, who had not subdivided the land into residential plots like surrounding neighbors. Farmers were allowed to farm the land free of rent as long as it remained productive—giving the impression to the government that it was not neglected. In the case of underutilized land, a group of farmers tended plots located beneath a high voltage power line. The land belonged to the government utility agency, and the farmers were allowed to farm without paying rent. They provided an unacknowledged service of managing the land along the utility right of way. In the case of public right of way land, there was also a group of farm plots along a river bank. One farmer, a staff member of the Jakarta Environmental Agency (Dinas Lingkungan Hidup Jakarta), explained that the area was within his jurisdiction. He managed the farms while supervising the cleaning of the river by his team. The agency allowed and even encouraged the usage of vacant lands along the riverbanks as farm plots as the planting of the riverbanks helped to prevent erosion. He further explained that the farms needed to be well-maintained. And that anybody could farm there for free after acquiring permission from the agency.

While famers tapped into their human capital to produce food for nearby markets, as well as manage urban blue and green landscapes, their value to the urban system was tied to the dynamic movement of the peri-urban edge—one interviewee explicitly stated that the future of agricultural activities in Jakarta was insecure. The interviewee had various jobs besides farming, and lived in a non-permanent house adjacent to his field. He was well-known as the middleman for neighboring farmers, transporting their crops to a central market.

Another interviewee, who rented a house in a nearby *kampung*, said he was the head of *Kelompok Tani Subur*, a farming group who, he reported, won the best quality of crops in the 2016 annual Jakarta farming competition. He said he was a well-known farmer, supplying healthy and safe vegetables under the inspection of the Department of Agriculture. He said he aspired to preserve the agricultural land in Jakarta when the land he was farming on decreased significantly in size, declining from 50 ha to just around 2 ha. He said that his goal was to inform the government about the importance of agricultural land security, not only as an income generator but also as a flood prevention measure (as Jakarta is a flood prone city—so much so that the government decided in 2019 to move the capital to Borneo by 2024). He also wanted the government to include urban agriculture under the Jakarta annual provincial budget (APBD). He briefly contrasted how the food production from the Gang Hijau movement (community gardens located along alleyways; see Diehl and Oviatt 2019) did not provide income for the practitioners, while the selling of the crops from his farm did provide a livelihood for his family.

## 14.5 Case Study Singapore: Farmers Who Depend on Migrant Workers

Singapore is an island city-state classified as 100% urban. Land is state-owned rather than privatized like the other case cities. Large-scale urban farms are located in the northwest, occupying the less than 1% of land that is allocated for agriculture. Commercial farming is restricted to poultry, eggs, fish, and vegetables (AVA 2017). A random sample of 16 farms were selected from a publicly available list of 48 commercial farms. Interviews occurred between April 2019 and January 2020.

Singapore farmers depended on foreign workers for growing and taking care of farm operations. Most farms were managed by anywhere from 3 to 30 people, with half the staff tending crops (those of foreign descent) and half managing other operations (Singaporeans). Locally hired Singaporeans typically worked in-doors doing packaging, production, or administrative work. Foreign workers came from Bangladesh, Malaysia, Myanmar, Sri Lanka, India, Thailand, and China. One farmer described the hiring process, "We don't even bother starting with local [Singaporeans]. . .we just go straight to having foreigners because they sign the 2-year contract, they will be there every day, and they will be on time because they are staying on the farm. We need reliable help, so going through the staff network, we ask them to make recommendations, they may have their uncles, or in-laws who actually want to come and work. The committee that we have on our farm is a team of Bangladeshi that are related somehow or another to even some of the workers in other farms in the area." Although farmers could go through an agency, it was more common to seek recommendations from current foreign workers or ask other farmers for their foreign workers to recommend people from overseas—one farmer even hired married couples, saying they should not be separated. This ensured reliability; one farmer said that one of his workers had been employed on his farm for more than 30 years, having been hired by his father. Another farmer said that he used an agent; farmers would provide their criteria and the agent identified a match. He further explained that agents were formerly farmers who may have had their land taken or been displaced in some way and decided to work to connect farmers with foreign workers.

The northwest of Singapore, where commercial farms are located, is an anomaly in a city which boasts one of the best public transportation networks in the world. The northwest is poorly connected, making it difficult for Singaporeans, with low car ownership, to access for work. It is also far from the high concentration of hawker centers, markets, and restaurants that have put Singapore on the food map. The inaccessibility makes it difficult to attract Singaporeans to commute to work on a farm, which presents a challenge as the government sets quotas with preference for Singaporean workers. Many local Singaporeans now working on farms started as customers: "I came as a customer many years ago, as a university student, and then suddenly I've been here 10 years right. So a lot of our customers suddenly become staff, you know, and they connect us to other staff. . .I give you an example, on the weekends we have special activities, either outreach or markets, a lot of our customers suddenly become vendors—they become our collaborators and our partners." The younger generation is not interested in farming; the long hours, the hot tropical outdoor climate, and the weekend tasks do not fit the career expectations of the highly educated society. As a result, some farmers look for alternative worker options: one farm hired ex-offenders and had an autistic employee.

The primary benefit of hiring foreign workers was that they were farmers or had agricultural experience in their country of origin. However, in terms of "investment" in hired workers, farmers did discuss some of the skills and resources they provided to workers. First, as Singapore transitions to more high-tech agricultural systems, foreign workers have had to learn how to use the technology: "I think most of them had a farm back at home, but it's the traditional soil farm so we definitely have to teach. But, of course, it helps if they have farming background." Most farms use on-the-job-training, where a new worker will shadow a more experienced worker. One farmer further said, "We do an 'old bird teaches the new bird' type of method to train them—we rotate our job scope, that means they must be well-versed in everything, so in case one is on leave, all of them can actually chip in and cover for the person on leave." Safety was also a high priority—for both personal safety and following government food and beverage hygiene regulations. One farmer said that he gave time off in the case that a worker was interested in getting a driver's license—as drivers were needed for distribution and are expensive to hire locally. Another farmer said he occasionally helped with day-to-day stuff like helping a worker set up his own ibanking account, further saying, "It's not just about farming, it's also about relationships." A key challenge was in the communication because few foreign workers were proficient in English: "Every time you give them instructions, they don't understand, and even though they don't understand, they say 'yeah yeah' and then you find that they really don't understand so you even have to go every time to that spot and tell them and show them how to do."

Foreign workers were usually provided room and board as part of the work package. One farmer described his foreign workers: "They all know each other, and they are relatives. It's good to have a community here because first of all, they are either related by blood or kinship so they will take care of each other and they are not lonely here... you know you are helping a few families and it makes an impact on their lives and it's good to have a community rather than individuals, we feel, of course individuals also become a community but when they are happy here, they will work better also, they are more productive, they look out for each other, so we feel that our workers are happy."

Despite some of the challenges of hiring foreign workers, there were also benefits when hired workers were able to make suggestions or simply apply their own skills to improve farm operations. One farm explained, "Our foreign workers have deep knowledge on farming so they can grow a lot, the landscape you see is basically their doing. I think they have a lot of knowledge on the plants that we may not." Only one farmer said that the workers had to do what was planned without making any suggestions. But, for those farmers willing to let workers try new things, the benefits could be multiple—time savings, increased productivity, a more efficient method—all of which translated to increased income. One farmer put it simply: "Normally they ask, they have great idea, great innovation then I say, 'Okay you carry on.'" Because foreign workers are engaged in the day-to-day farming activities, they learn how to deal with particular pests, which crops to switch, when its time to rotate, when fertilizer is needed. On the other hand, farms moving toward automation do not require workers to have a farming background.

Foreign farm workers have low social status in Singapore. Living and working on a single farm limits their ability to participate in society. Their potential to provide bridging connections to their employers is in their ability to recommend other foreign workers. Recommendations carry weight because the capability of a new hire will reflect positively or negatively on the person who made the recommendation—and could impact contract renewal or salary. With strict employment restrictions, farmers were not able to share foreign workers if one farm needed extra labor when another did not. Within the farm, social status carried more weight—in some cases. When asked about any other interactions with workers, farmers fell into one of two categories: treating each other like friends by talking about families and sharing holidays or celebrations, or having little interaction besides work duties, often due to language barriers.

There is no minimum wage in Singapore, but farmers pay a levy plus wages for foreign workers. There was some hesitation when farmers were asked what they paid—one said, "I think we pay them well, we pay them fairly and we pay them what they accept and a little bit more so." In some cases, workers were provided room and board as part of their compensation, whereas in other cases, they were charged. Rates of pay ranged, with most farmers paying a monthly salary and few paying hourly wages. More than half the farmers said their starting monthly pay was $700–$800 SGD ($494–$564 USD). For workers with higher skills or long tenure, wages were reported to be as high at $1200–$1300 ($846–$917 USD) per month. A few farmers said they provided overtime pay, whereas others said they did not—that they paid fairly with the expectation that sometimes extra hours or fewer hours were required. In explaining the variation, one farmer said, "I feel that the employer always has the upper hand because there are so many of them. If the worker doesn't want to accept the rate, we can always go to somewhere else so that's the power reality. However, I do not feel that as a business employer we take advantage of them, I think that we are still very reasonable in what we remunerate them and on top of that also they do not have to pay us for their food and their lodging here, for certain companies sometimes they do, for foreign workers they charge the foreign workers for their lodging, but here we let them stay for free, and also I think we give them a fair wage and give them also fair working conditions. They earn $700 to $1000 [SGD] and then they stay here for free." Furthermore, one farmer explained that a lot of farm owners pay very high levies for foreign workers because it is difficult to get locals to do farm work: "[Locals] don't have the skills, they don't have the strength or stamina to be out in the field, even if they have the skills, they don't have the interest."

The main challenge to farming in Singapore, repeated again and again was the lack of human capital in the Singapore population—the local people either did not have the skills, wanted a salary greater than the farmer could afford, or did not want to work outside in the hot, humid climate.

## 14.6 Case Study Sydney, Australia: Farm to Farmer's Market Workers

Sydney is surrounded by peri-urban farmland that provides a buffer between the city and the Greater Blue Mountains World Heritage Area. There are an estimated 2000 vegetable growers with more than 80% from non-Anglo cultural and linguistic backgrounds (O'Neill and James 2014). Urban and peri-urban agriculture has provided streams of migrants with the opportunity to participate in farming (O'Neill and James 2014; James 2016). This history of agricultural pioneering evolved a strong link between peri-urban farmers and local markets (Merson et al. 2010). Local farmers' markets are highly active in Sydney, selling products from farms located in urban Sydney and the peri-urban and rural fringe. For this case, a roster of 25 commercial farms selling at Sydney farmers' markets, located within the urban or peri-urban fringe of the city, were invited via email, Instagram/Facebook, or in-person at a farmers' market to participate in interviews. In total, 13 farms/farm-related actors participated in interviews over a 2-week period in July 2019.

In total, nine (73%) farms were located within the urban boundary and four were located in the peri-urban/rural fringe (27%). Five urban farms were multi-generational operations run by migrants or first-generation citizens from diverse countries of origin. There were also two urban Chinese heritage farms, uniquely located within residential districts within the urban boundary. The farmers were Chinese migrants with strong ties to overseas family members as well as other farmers of the same origin. Similar to the Chinese Heritage farms, there were other recent migrant farms that were run by extended families including aunts and uncles, cousins, and close family friends, people looking for a stepping-stone for migrating to Australia in hopes of improving their job opportunities. Representing the opposite end of the spectrum, one urban farm was a start-up run by an Australian couple who left corporate jobs in the city in exchange for the slower pace of a family-run farm to raise their kids. The final urban farm was a social enterprise located in a public park and managed by a city government agency, operated at the time as a demonstration farm. The four peri-urban farms were large-scale operations farmed by multi-generational Australian families, growing a range of standard to specialized vegetables and fruits, including animal and dairy farms. All interviewed peri-urban farmers were university educated, with the youngest generation typically employed in professional or service sector jobs.

Most farms ($n = 9$) employed a range of seasonal and full-time workers although smaller operations ($n = 3$) relied solely on family labor. Having strong social ties with other farmers, neighbors, and/or the migrant community provided access to cheap labor. Workers were hired through word of mouth, paid advertisements, etc. For overseas workers, farmers relied both on recommendations from other farmers as well as their own workers, who could connect them directly to the overseas labor market through their friends and relatives there. It was common for the majority—or all—of the migrant workers on one farm to be from the same sending province.

Local workers typically lived nearby the farm, whereas many of the migrant workers were provided room and board as part of their compensation package. Migrant workers did not always have an agricultural background. Wages were based on skill, effort, and whether other resources were being provided (i.e., housing, food, education). One farmer explained that "the kind of people we employ are migrant workers, can't drive, can't speak English, you need to arrange transport and if they can arrange transport they can probably go somewhere else." For example, she said that there was a group from Cambodia who would send a bus to pick Cambodian workers up.

In the case of the four peri-urban farms, ranging from 15 to 320 acres, different labor strategies were required. Two farmers specifically talked about hiring local people, preferring employees with ties to the local community, which translated to long-term commitment. One farmer specifically mentioned hiring differently abled workers. He also said they had occasional bonfires for employees and families.

One farmer talked about how she decided on wages: "It isn't about money, I could do without the headache. With anyone coming in you might have problem getting them working at the speed you want. You have to consider how much you pay someone and how much they can do. Have to think about labor cost which is 80% of cost." In consolation for lower wages in the farm industry, there was a sentiment that despite the lower wages as compared to other work, a farm could provide workers a good environment—supportive of social bonds and family ties.

In terms of investing in workers, some farmers preferred to hire local people as a way to contribute to the local economy. Local workers had the potential to stay for a long period of time. One farmer said he had two staff members whom he had employed for over 15 years, and a third one reaching 8 years (out of ten employees). Although occasionally "backpackers" were hired, a few farmers stated they did not hire backpackers. Few farmers mentioned learning from their workers—contrarily, one farmer mentioned workers giving him tips for their own advantage (implying it was to his disadvantage). There was an undertone of negative interaction among a few interviewed farmers and their hired migrant labors. One farmer described practices at other migrant-run farms but said of her own operation: "[Some farmers I know] are very hard on the labor, sometimes I wish I could be like that. I am not like that. I won't be able to do it."

On the other hand, migrant workers often came from farming backgrounds and brought their own set of knowledge and skills. One farmer said he hired part-time workers through his (farming) friends because the workers had experience farming in Tasmania, Melbourne, although they had originated from China. One farmer stated: "We currently only have Chinese workers and mum knows Chinese people. A year ago we had five people who were all Thai. When Thai people were here I learnt a lot from them. They were very good in cutting and doing salads. We will get more workers soon. We will look for different people who can do greenhouses. Farm workers can be on tractor side or greenhouses etc."

One farmer with 30 employees noted that he hired differently abled workers who could be matched with their own skillset and physical capability but could also be given opportunities to learn new skills. As a fifth-generation farmer, his attitude toward his employees was to treat them like extended family. Another farmer said she had a worker that she hired through a system called restart. The worker had been incarcerated and the government provided a subsidy to employ such people.

Sydney's farmers, ranging from multi-generational pioneer families to recent migrants with diverse origins, employed a range of seasonal and full-time workers. Migrant farmers typically hired workers from overseas, often, but not exclusively, from their country of origin. On the other hand, multi-generational Australian farmers tended to hire local residents, seeking to invest in the local community.

Given the limited sample size for this research, it is difficult to generalize except to conclude that it was a diverse and complex farmer-labor system.

## 14.7 Conclusion

Commercial urban farms require labor to manage and sustain them as a productive, living system, but the farmer managing the operation is rarely the only person tending the crops and animals. This chapter described the people doing field work, specifically investigating farmer-worker social networks to shed light on the sustainability of urban farming as a livelihood pursuit. Labor was first conceptualized as a resource and a social network as part of a larger goal of accessing human capital to sustain the farm as a livelihood. By definition, commercial farms are businesses that are intended to provide the farmer a sustainable livelihood. But a livelihood is not simply about the earned income—a livelihood is about the capabilities, assets, and activities required to practice the work.

Four case studies were then described to illustrate diverse farmer-labor social networks in Delhi, India; Jakarta, Indonesia; Singapore; and Sydney, Australia. In summary, many Delhi farmers hired workers to help complete farm tasks. They preferred to hire neighboring farmers in exchange for their own labor when the other farmers needed help. In practice, there were multiple and diverse ways farmers engaged workers on their farms. The fluidity of identities of who was a farmer versus who was a laborer was evident and produced a social network that was more horizontal than hierarchical. Despite the marginal status of Delhi farmers, they had a complex system of social networks providing diverse access to human capital as needed. In addition to meeting farmers' needs, laborers without land had opportunities to seek out work by "roaming around" the cultivated area and suggesting tasks they could provide.

In contrast to Delhi, it was rare for Jakarta farmers to hire workers to help with farm tasks. Most plots were small enough for a single farmer to manage, with family members helping during key times such as for planting or harvesting. The Jakarta farmers tended plots of land that otherwise would have been the responsibility of the city or an absent private owner to manage. In addition to producing food, they provided unacknowledged multi-functional human capital in their management of vacant, underutilized, and public right of way land.

Moving to Singapore, farmers depended on foreign workers for growing and taking care of farm operations. Foreign workers came from surrounding developing countries and China. Although farmers could go through an agency, it was more common to seek recommendations from current foreign workers or ask other farmers for their foreign workers to recommend people from overseas. The isolation of urban farms from transportation and other urban amenities, high cost of hiring locals, and lack of local agricultural knowledge and skills were compounded barriers to hiring Singaporean workers. The primary benefit of hiring foreign workers was that they were farmers or had agricultural experience in their country of origin. However, language barriers limited interactions between farmers and workers. There was a distinct hierarchy in farmer-worker social networks.

Finally, Sydney had the most diverse farmer demographic among the four cities, ranging from multi-generational pioneer families to recent migrants with diverse origins. Farms employed a range of seasonal and full-time workers although smaller operations relied solely on family labor. Migrant farmers typically hired workers from overseas, often, but not exclusively, from their country of origin. In contrast, multi-generational Australian farmers tended to hire local residents, seeking to invest in the local community. Some workers were provided housing in addition to wages—true for both local and migrant workers. Among workers, some workers had farming experience, while others had no experience.

The four case cities demonstrate that even within one type of urban agriculture, namely commercial urban farms, labor requirements, and practices can be diverse. The farmer-labor social network profoundly impacts the feasibility and sustainability of operating a farm. Urban agriculture depends on urban resources including soil, water, pollinators, and labor. Different types of urban agriculture will vary in how the participants manage production tasks. Given that urban farming is directly driven by farmers as a livelihood pursuit, and less directly an effort in sustainability or city-level food production, we must first identify the incentives and barriers that farmers face, as these factors will be primary drivers for behavior and decision-making among farmers.

**Acknowledgments** Multiple grants funded this research. The research in Delhi was funded by The U.S. National Science Foundation Integrative Graduate Education and Research Traineeship (NSF-IGERT) Program (Award No. DGE-0654378) and the U.S. Dept. of State and the Government of India through the Fulbright-Nehru Fellowship funded the research in Delhi. The research in Singapore, Jakarta, and Sydney was funded by the Singapore Ministry of Education (MOE) Start-up Grant for New Faculty, funding # R-295-000-141-133.

## References

AVA (2017) Annual statistics. Agri-Food & Veterinary Authority of Singapore. https://www.ava.gov.sg/homepage

Blaikie P, Cannon T, Davis I, Wisner B, Cannon T, Davis I, Wisner B (1994) At risk: natural hazards, people's vulnerability and disasters. Routledge, New York

Carney D (2003) Sustainable livelihoods approaches: progress and possibilities for change. Department for International Development, London

Diehl JA (2020) Growing for Sydney: exploring the urban food system through farmers' social networks. Sustainability 12(8):3346

Diehl JA, Oviatt K (2019) Productive urban landscapes: emerging typologies of form and function in urban landscapes. In: Rinaldi BM, Tan PY (eds) High density cities. Birkhäuser Verlag, Basel

Diehl JA, Bose M, Main DS (2019) A social and spatial network approach to understanding beliefs and behaviors of farmers facing land development in Delhi, India. In: Ye X, Liu X (eds) Cities as spatial and social networks. Springer, Berlin, pp 73–99

Edwards G (2010) Mixed-method approaches to social network analysis. National Center for Research Methods, Manchester, England

James S (2016) Farming on the fringe: peri-urban agriculture, cultural diversity and sustainability in Sydney. Springer, Berlin

Knoke D, Yang S (2008) Social network analysis. Quantitative applications in the social sciences, 2nd edn. SAGE Publications, Los Angeles

Lin N (2001) Building a network theory of social capital. In: Lin N, Cook K, Burt RS (eds) Social capital: theory and research. Aldine De Gruyter, New York, pp 3–29

Lonard CS (2012) Urban fringe agriculture and its potential contribution to sustainable cities: a case study of Western Sydney

Merson J, Attwater R, Ampt P, Wildman H, Chapple R (2010) The challenges to urban agriculture in the Sydney basin and lower Blue Mountains region of Australia. Int J Agric Sustain 8 (1–2):72–85

Mougeot L (ed) (2005) AGROPOLIS. The social, political and environmental dimensions of urban agriculture. Earthscan, London

O'Neill P, James S (2014) Feeding Sydney: assessing the importance of the city's peri-urban farms. In: The security of water, food, energy and liveability of cities, Water science and technology library. Springer, Cham, pp 243–256. https://doi.org/10.1007/978-94-017-8878-6_18

Philips A (2013) Designing urban agriculture: a complete guide to the planning, design, construction, maintenance and management of edible landscapes. Wiley, Hoboken, NJ

Scoones I (1998) Sustainable rural livelihoods: a framework for analysis. IDS Working Paper 72. IDS, Brighton

Scoones I (2009) Livelihoods perspectives and rural development. J Peasant Stud 36(1):171–196. https://doi.org/10.1080/03066150902820503

Scott J (1991) Social network analysis: a handbook. Sage Publications, London

Sen D (2016) Sharecropping in theory and practice: a selective review. In: Understanding development. Springer, Berlin, pp 53–72

van Veenhizen R (2006) Cities farming for the future: urban agriculture for green and productive cities. RUAF Foundation, IDRC and International Institute for Rural Reconstruction, Leusden, The Netherlands

# Part IV

# Biodiversity

# Honey Bees, Wild Bees, and Beekeepers in Chicago's Community Gardens

15

Alexis D. Smith, Michael Roberts, Haley Lerand, Molly Doane, and Emily S. Minor

**Abstract**

Chicago, IL (USA) hosts a robust culture of both community gardens and urban apiculturists (keepers of honey beehives). Western honey bees (*Apis mellifera*) are not native to North America, and their impact on wild bees is not fully understood. Through interviews with beekeepers and biodiversity surveys in 24 of Chicago's community gardens (9 with beehives), we explored questions about urban apiculturists' perceptions and knowledge of wild bees, as well as the impact of urban apiculture on wild bees in community gardens. In the context of urban community gardens, our research suggests that although honey bees are an introduced species, beekeepers can play a positive role in wild bee conservation.

**Keywords**

Apiculture · Community gardens · Bees · Introduced species · Biodiversity

A. D. Smith (✉) · M. Roberts
University of Illinois at Chicago, Chicago, IL, USA
e-mail: asmit58@uic.edu

H. Lerand
Department of Anthropology, University of Illinois at Chicago, Chicago, IL, USA

M. Doane
Department of Anthropology, University of Illinois at Chicago, Chicago, IL, USA

The Field Museum, Chicago, IL, USA

E. S. Minor
Department of Biological Sciences, University of Illinois at Chicago, Chicago, IL, USA

Institute for Environmental Science and Policy, University of Illinois at Chicago, Chicago, IL, USA



J. A. Diehl, H. Kaur (eds.), *New Forms of Urban Agriculture: An Urban Ecology Perspective*, https://doi.org/10.1007/978-981-16-3738-4_15

## 15.1 Introduction

Bee populations are declining worldwide, which is a threat to biodiversity and food security (Potts et al. 2010). The primary causes of these declines are thought to be pesticides such as neonicotinoids (Goulson 2013), parasites such as the disease-transmitting mite *Varroa destructor* (Le Conte et al. 2010) and the microsporidian *Nosema* spp. (Klee et al. 2007; Meeus et al. 2011), and a lack of floral resources (Carvell et al. 2006). These stressors combined have an impact greater than any individual stressor would have alone (Goulson et al. 2015).

Although wild bees are impacted by these stressors, much of the focus on declining bee populations has been on honey bees (*Apis* spp.; Geldmann and González-Varo 2018). The western honey bee (*A. mellifera*) is native to Europe, Africa, and the Middle East (Han et al. 2012), but has been introduced around the world as a commodity. According to the most recent agriculture census, more than 2.8 million honey bee colonies on more than 60,000 farms were reported in the USA in 2017. The honey collected by these farms was valued at more than $320 million USD in total (USDA, National Agricultural Statistics Service 2019a).

Because of bees' small size and modest land requirements, apiculture (beekeeping) is better suited to cities than most other types of animal husbandry. In fact, amateur beekeeping has become a trend in major cities across the world such as Berlin (Germany, Lorenz and Stark 2015), Perth (Australia, Carmody 2017), London (England, Owen 2009), and Philadelphia (PA, USA, Sreenivasan 2016). Even where urban residents do not have backyards, they can keep bees in places such as rooftops and community gardens.

It is not well understood what impact beekeeping has on wild bee species, most of which are native. Some, but not all, studies have demonstrated competitive interactions between managed honey bees and wild bees over pollen and nectar resources (Paini 2004). Even in areas where there are a lot of flowers, wild bee fitness could be reduced if the wild bees have to spend more time foraging (Pyke et al. 1977) or if they are forced to forage on flowers with lower quality pollen (Huang 2012). Managed honey bees could also spread pathogens to wild bees. Although the varroa mite only infests honey bees (Le Conte et al. 2010), shared pathogens could be transmitted through contaminated pollen (Singh et al. 2010) or when bees make contact in shared spaces (Fürst et al. 2014). These interactions have the potential to have negative impacts on wild bee abundance or geographic ranges, but very few studies have demonstrated or even measured population-level impacts (Mallinger et al. 2017).

On the other hand, wild bees could benefit from resources provisioned by beekeepers or gardeners for the honey bees. This is especially likely in urban areas, where humans play a major role in shaping the environment. Out of a desire to provide for honey bees, beekeepers and gardeners may be motivated to plant more flowers, tolerate flowering weeds, or reduce use of pesticides and other potentially harmful chemicals (Maderson and Wynne-Jones 2016). Weeds such as dandelions (*Taraxacum officinale*) and white clover (*Trifolium repens*) are especially important to bees early in the spring before many other species bloom (Hicks et al. 2016). The beehive itself may also unintentionally create a "no-walk zone" around it, which could provide nesting habitat for ground-nesting bees.

For wild bees, cities can be important refuges that allow diverse bee communities to persist in spite of changes in surrounding natural and rural areas (Hall et al. 2017). While agricultural intensification (e.g., transitions to monocultures or innovations in systemic pesticides) makes rural areas increasingly inhospitable to bees (Hall et al. 2017), such changes are less common in urban agriculture. Urban agricultural spaces such as community gardens potentially offer important foraging and nesting resources for wild bees although their capacity to support bees may be limited by their size and the surrounding hardscape.

In this study, we focused on bees, beekeepers, and ecological resources in urban community gardens in Chicago, IL (USA). We echo the assertion that conservation issues require interdisciplinary, collaborative approaches (e.g., Suryanarayanan et al. 2018), especially in human-dominated spaces such as cities. Our study, a collaboration between ecologists and anthropologists, was carried out with this in mind. We wanted to know about the diversity of bees that Chicago's community gardens support, to understand the relationship between honey bee and wild bee abundance, and what factors predict these abundances. Finally, we wanted to know if urban beekeeping offered any potential benefits to wild bees. In 24 community gardens, 9 of which contained honey beehives, we recorded presence and identity of wild bees and honey bees and characterized floral resources and other aspects of the environment. In interviews with nine beekeepers who tend hives in community gardens, we asked about their perceptions of nature, wild bees, and honey bees. The results and discussion of both the social and ecological aspects of this study are combined to reflect the interconnectedness of the agents involved—wild bees, honey bees, beekeepers, and gardeners.

## 15.2 Methods

### 15.2.1 Study Sites

Chicago, IL (USA) is situated on the coast of Lake Michigan, on Potawatomi land in the Great Lakes Region (Greenberg 2002). With more than 2.7 million residents, Chicago is the third most populous city in the USA (US Census Bureau 2018). Chicago has hot summers and cold winters, offering a growing season that is short but potentially productive. On average, there are 175 days between the last spring frost and the first fall frost (Angel 2003), and different parts of the city fall in USDA hardiness zones 5b and 6a (USDA, Agricultural Research Service 2012). Chicago is home to many beehives, some of which are hosted in community gardens throughout the city. According to an interview subject, many hobbyist beekeepers do not formally report their hives, but in Cook county (the densely urban county where Chicago is located) 48 farms reported 1337 honey bee colonies in 2017 (USDA, National Agricultural Statistics Service 2019b).

Community gardens were found using the Chicago Urban Agriculture Mapping Project search engine (https://cuamp.org/) and Google Maps (https://www.google.com/maps). We contacted prospective study sites by sending an email to the garden managers, asking about the size of their garden, accessibility (i.e., whether the garden was fenced and locked), and whether they had beehives. If a garden's manager did not respond to our email, the garden was removed from the list of potential sites. Of the gardens for which managers responded, several garden managers were not able to grant us access or the gardens were very far from other gardens in the study. These gardens were excluded from the list of potential sites. We ultimately chose 24 gardens from a diversity of Chicago neighborhoods that each had at least 10 vegetable beds for food production, attempting to include as many gardens with honey beehives as possible. The final sample included 9 gardens with beehives and 15 without.

### 15.2.2 Bee Surveys

Bee surveys were conducted in July and August on days without rain, between 9 a.m. and 5 p.m., when most bee species are active. At each community garden, we selected ten random 1 $m^2$ sampling sites within vegetable beds and five 1 $m^2$ "pollinator garden" sampling sites. Pollinator gardens were defined as the areas in the community garden with the highest abundance of flowers and were not randomly selected. For 2 min, we observed each sampling site and recorded the identity and abundance of bees within the sample area. Bees were not collected, and thus many individuals could not be identified to the species level. We identified bees in the field to the genus level or lower when possible. If we encountered a bee and were unable to identify its genus but could rule out honey bees, we recorded it as "unidentified." Unidentified bees were included in our estimate of wild bee abundance. After observations were completed, we walked the garden for 5 min recording the identity (but not the abundance) of new bee taxa that we had not observed in our other samples. We conducted these surveys twice for a total of 30 sampled areas per garden; each sampled area was 1 $m^2$. Surveys at each site were separated by at least 30 days.

### 15.2.3 Garden Assessments

Vegetation surveys were conducted during the month of July 2017 because the majority of gardeners have their plantings established by then. At each garden, we randomly selected ten vegetable beds using dice as a random number generator and conducted a weed survey. Weeds included any plant not intentionally cultivated. At each vegetable bed, we estimated the percentage of a 1 $m^2$ quadrat covered by weeds and recorded the height of the tallest weed. We combined these measures into one measure of weed volume using a calculation based on the volume of a cone (weed area times 1/3 the height of the tallest weed). We calculated the mean "weed density" (in $L/m^2$) of the ten quadrats in each garden by dividing the weed volume by the area sampled.

We estimated floral resources in the gardens in two ways, resulting in the variables "floral richness" and "floral density." In July, during the weed survey described above, we counted the number of all plant species flowering in the common areas for the variable "floral richness." Common areas included all areas outside of the vegetable beds. During the bee surveys in July and August, we also counted the number of flowers in each 1 $m^2$ observation sampling site, and estimated mean "floral density" for each garden. Finally, we estimated woody vegetation density in each garden by counting all woody vegetation taller than 1 m in each garden and divided this count by the area of the garden.

Finally, we assessed the size of the gardens and the percentage of impervious surface within 1 km of the center of the gardens, which represents the foraging range of many bee taxa (Greenleaf et al. 2007). Garden size was calculated by tracing the boundary of the garden and creating polygons in Google Earth Pro. Percent impervious surface was calculated in ArcGIS 10.5 (ESRI 2017) using a classified satellite image of Cook County land cover (O'Neil-Dunne 2010). We reclassified building, road, and other paved surface categories in the Cook County land cover file as a single category, impervious surface. Then we used the ArcGIS Focal Statistics tool to calculate the percentage of impervious surface within 1 km of each garden's centroid.

### 15.2.4 Beekeeper Interviews

We conducted semi-structured interviews with nine beekeepers in the city of Chicago as one component of a larger, ongoing ethnography of Chicago's community gardeners carried out by researchers at The University of Illinois at Chicago and The Field Museum. Beekeepers were identified in the course of participant observation by anthropologists in the gardens and using "snowball sampling," when informants encountered in the gardens introduced researchers to other beekeepers in their networks. Four of the nine beekeepers were involved with the gardens where we were conducting our fieldwork. The other five beekeepers were recommended as interview subjects by beekeepers with whom we had spoken previously. We intentionally sought interviewees with a range of experience in beekeeping, from established beekeepers to those who had been beekeeping for less than 2 years.

The interviews were open-ended and conversational, but with a few guiding questions. We asked questions about how the beekeepers learned about beekeeping, their motivations for beekeeping, and how beekeeping had influenced their perceptions of nature in the city. Some questions (such as "What do you think about honey bees versus native bees?") were intentionally open-ended, and the beekeepers interpreted the question broadly. Interviews were recorded, transcribed, and then read thoroughly for any responses that provided insight to our questions. Conducting the interviews was approved by the Institutional Review Board (IRB) of The Field Museum.

### 15.2.5 Analysis

To determine which variables best explained wild bee and honey bee abundances, we ran single-variable generalized linear models for these two dependent variables with each of six explanatory variables. Explanatory variables included five environmental variables (Table 15.1) as well as the abundance of the opposing group (i.e., wild bees or honey bees). To make model coefficients easier to compare, we standardized each explanatory variable by subtracting the mean of the variable from each observation, and then dividing by the standard deviation. For comparison, we also ran a null model for each response variable. We then selected the best models for each response variable using Akaike's Information Criterion (Anderson 2008) with a correction for small datasets (AICc). The models best fit a negative binomial error structure. For these analyses we relied on R packages AICcmodavg (Mazerolle 2017) and MASS (Venables and Ripley 2002). We did not use bee generic richness as a dependent variable because we determined that two of our data points were overly influential and could lead to misleading results (i.e., the gardens had a narrow range of generic richness except for two gardens). To determine the relationship between honey bee and wild bee abundance in Chicago's community gardens, we plotted wild bee abundance versus honey bee abundance and compared linear and quadratic regressions (both with negative binomial error structures) to a null model. We again used AICc to select the best of these models.

We ran a series of tests to examine differences between gardens with and without beehives. First, to test whether the presence of beehives impacted whether a particular wild bee genus was observed in a garden, we compared proportions of gardens with beehives and gardens without beehives in which each genus was encountered using the prop.test function in R, which compares proportions using the chi-squared test (R Core Team 2018). Second, to test whether the presence of beehives affected abundance of wild bees or honey bees, we compared abundances of each group in gardens with beehives to the abundances in gardens without beehives using boxplots and the Wilcoxon test. Finally, we also used boxplots and the Wilcoxon test to compare two garden attributes related to gardener behavior that might be influenced by beehive presence, weed density, and floral richness, in gardens with and without beehives. Throughout the results and discussion below, we use the beekeeper interviews to provide insight into our questions, but especially into our question regarding whether beekeeping offers benefits to wild bees.

**Table 15.1** Candidate variables used in the generalized linear models to predict wild bee abundance and honey bee abundance

| Variable | Minimum | Median | Maximum |
|---|---|---|---|
| Floral density (flowers/m$^2$) | 8.54 | 33.45 | 79.00 |
| Weed density (L/m$^2$) | 0.81 | 6.23 | 41.08 |
| Woody vegetation density (plants/m$^2$) | 0.00 | 0.02 | 0.07 |
| Garden area (m$^2$) | 254 | 871 | 8939 |
| Impervious surface within 1 km (%) | 42.19 | 58.22 | 71.14 |

## 15.3 Combined Results and Discussion

### 15.3.1 How Abundant and Diverse Are Bee Communities in Chicago's Community Gardens?

In total, we encountered 2275 bees and 13 identified bee genera (including *Apis*) across all 24 gardens (Fig. 15.1). We were unable to identify 91 of the 2275 bees we encountered (0.04%). Honey bees were present at all 24 gardens, regardless of whether the gardens kept beehives. Many wild bee groups were nearly as widespread, such as bumble bees (*Bombus* spp.; 23 gardens), long-horned bees (*Melissodes* spp.; 23 gardens), sweat bees (*Lasioglossum* spp.; 22 gardens), and leafcutter bees (*Megachile* spp.; 22 gardens). The median generic richness across all gardens was 8 genera (range of 4–10 genera). The median honey bee abundance was 12.5 (range of 1–57 genera) and the median wild bee abundance was 70 (range of 29–151 genera); abundance is reported as the sum of the bees observed across $30 \times 1\ m^2$ sampled areas per garden.

Tonietto et al. (2011) collected bees in Chicago parks and green roofs, as well as prairies in the region. While the relative abundances of bees differ, the bee communities we observed in Chicago's community gardens were at least as diverse as those observed by Tonietto et al. in parks and green roofs at the level of genus. In some of the community gardens, we detected three genera that were not collected in parks or green roofs but were collected in prairies (*Augochlora*, *Coelioxys*, and *Colletes*). There were eight genera that they only collected in prairies that we did not encounter, and two that they collected in the city that we did not identify (*Halictus* and *Sphecodes*), although we encountered 91 bees that we were unable to identify due to the limitations of field identification. It is also possible that we misclassified

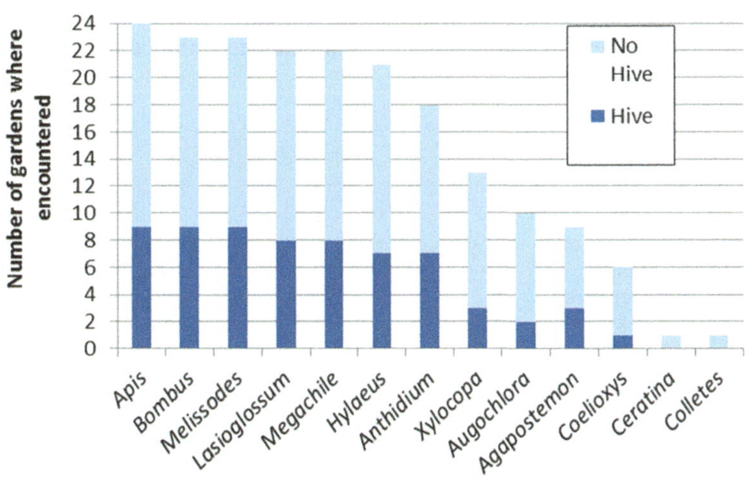


**Fig. 15.1** Bee genera encountered across all 24 gardens, ranked by the number of gardens in which they were encountered and coded by whether the gardens did or did not keep honey beehives

some *Halictus* bees as *Lasioglossum*, since these genera can be difficult to distinguish without a specimen in hand.

### 15.3.2 What Is the Relationship Between Honey Bee Abundance and Wild Bee Abundance, and What Factors Predict Wild and Honey Bee Abundances?

Because cities are largely artificial, the question of urban biodiversity conservation is often considered moot. One of the beekeepers we interviewed expressed, "I feel fine keeping bees in Chicago, because it's such an altered environment. There are so many [pollinator] resources and I don't feel like I'm displacing native bees. But I get really mad when people are like, 'Oh yeah I just got a house next to a massive nature preserve, and I want to like, put a bunch of beehives there.' Like, just let the native bees have it."

However, urban ecology research increasingly demonstrates that cities provide important habitat for wild bees. Many cities support an abundance and species richness of wild bees that are absent from surrounding rural areas (Hall et al. 2017). If honey bees depressed wild bee populations, then urban apiculture would be threatening valuable habitat for wild bees. Because a diversity of wild bees is important for crop pollination (Rogers et al. 2014; Mallinger and Gratton 2015; Lowenstein et al. 2015), this would also have negative implications for fruit production in urban agriculture.

We cannot answer this question definitively or globally, but in Chicago's community gardens there was no strong evidence that honey bees depressed wild bee communities or populations. The presence of beehives did not have a significant effect on whether a wild bee genus was encountered at a garden for any of the genera encountered (Fig. 15.1). Beehive presence did not affect abundance of either honey bees or wild bees; the differences were not significant according to the Wilcoxon test

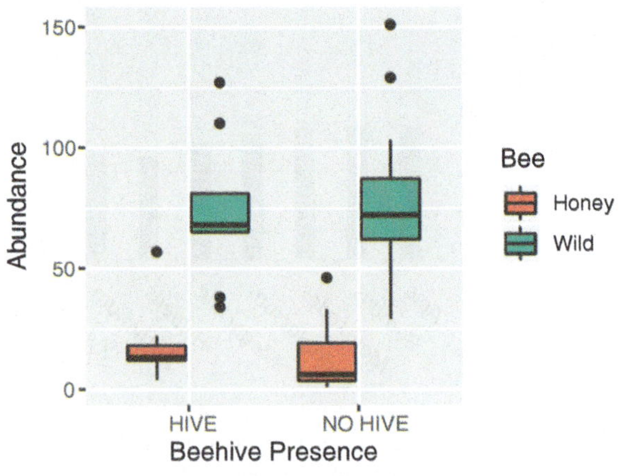


**Fig. 15.2** Comparisons of wild bee and honey bee abundances in community gardens with and without beehives. Beehive presence did not affect abundances of either wild bees or honey bees. Abundance is reported as the sum of the bees observed across 30 sampled areas per garden; each sampled area was 1 m$^2$

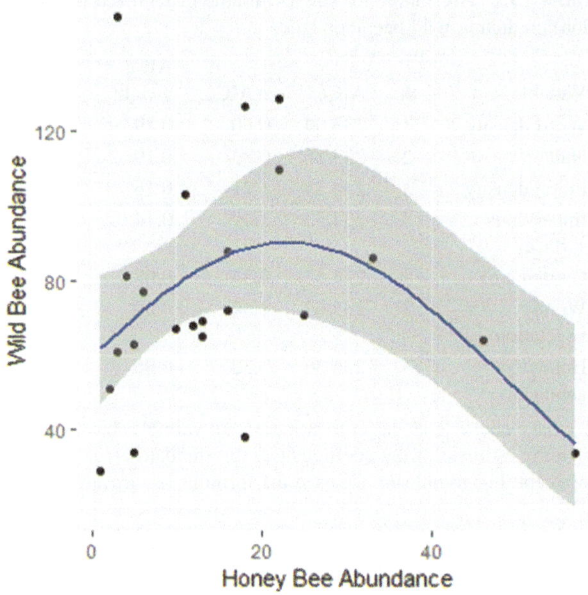


**Fig. 15.3** Fitted quadratic regression for wild bee abundance plotted against honey bee abundance (pseudo-$R^2 = 0.21$). Each data point represents one garden. The line represents a quadratic generalized linear model (negative binomial error structure), while the shaded area represents the 95% confidence interval. Abundance is reported as the sum of the bees observed across 30 sampled areas per garden; each sampled area was 1 $m^2$

(Fig. 15.2; $p = 0.23$ and 0.81, respectively). It is also possible that, due to honey bees' dispersal abilities and the prevalence of hives outside community gardens, honey bee abundance was elevated throughout the city and wild bee communities were depressed everywhere as a result; such background honey bee abundances would make the effect of beehives on wild bees difficult to detect.

There was also no consistently negative relationship between wild bee abundance and honey bee abundance. In most of the gardens, where there were more resources, there were more bees of all kinds. There were only two gardens where honey bee abundance was very high and wild bee abundance was low (and one garden where the opposite was true). The data fit a quadratic curve (Fig. 15.3; pseudo-$R^2 = 0.21$) better than a simple linear regression (both models with negative binomial distributions). This suggests the possibility that resource competition could become a problem at very high honey bee abundances although the small number of data points in the upper range provide only weak evidence of that possibility. We present this figure in accordance with the precautionary principle.

Wild bee abundance was positively related to weed density. None of the other single-variable models explaining wild bee abundance out-performed the null model (Table 15.2). Meanwhile, honey bee abundance was best explained by (and positively correlated with) density of woody vegetation (Table 15.3). This could suggest some degree of resource partitioning, where honey bees are attracted to the concentrations of flowers available on trees, vines, and shrubs such as hibiscus (*Hibiscus* spp.) and roses (*Rosa* spp.) while wild bees forage more often on the more scattered weedy species. Such resource partitioning could be a problem for wild bees if the nectar and pollen in weeds were of lower nutritional quality or if

**Table 15.2** AICc table for single-variable generalized linear models (negative binomial distribution) predicting wild bee abundance

| Variable | $K$ | AICc | ΔAICc | AICc weight | LL | Coefficient | Pseudo-$R^2$ |
|---|---|---|---|---|---|---|---|
| Weed density | 3 | 233.04 | 0.00 | 0.29 | −112.92 | 0.14 ± 0.08 | 0.13 |
| Null | 2 | 233.86 | 0.83 | 0.19 | −114.65 | 4.33 ± 0.08 | 0.00 |
| Floral density | 3 | 234.25 | 1.21 | 0.16 | −113.52 | 0.12 ± 0.08 | 0.09 |
| Impervious surface | 3 | 234.53 | 1.49 | 0.14 | −113.66 | 0.11 ± 0.08 | 0.08 |
| Garden area | 3 | 235.57 | 2.54 | 0.08 | −114.19 | −0.08 ± 0.08 | 0.04 |
| Woody vegetation | 3 | 235.62 | 2.59 | 0.08 | −114.21 | 0.08 ± 0.08 | 0.03 |
| Honey bee abundance | 3 | 236.36 | 3.32 | 0.06 | −114.58 | −0.03 ± 0.08 | 0.01 |

To make coefficients easier to compare, all variables have been standardized. For each variable we give the estimate of the coefficient ± the standard error. The coefficient for the null model is the intercept. For pseudo-$R^2$ we used the formula 1 − (residual deviance/null deviance)

**Table 15.3** AICc table for single-variable generalized linear models (negative binomial distribution) predicting honey bee abundance

| Variable | $K$ | AICc | ΔAICc | AICc weight | LL | Coefficient | Pseudo-$R^2$ |
|---|---|---|---|---|---|---|---|
| Woody vegetation | 3 | 180.45 | 0.00 | 0.58 | −86.62 | 0.38 ± 0.16 | 0.19 |
| Null | 2 | 183.18 | 2.73 | 0.15 | −89.31 | 2.72 ± 0.18 | 0.00 |
| Weed density | 3 | 184.65 | 4.20 | 0.07 | −88.73 | 0.21 ± 0.17 | 0.04 |
| Floral density | 3 | 184.89 | 4.45 | 0.06 | −88.85 | −0.23 ± 0.18 | 0.04 |
| Garden area | 3 | 185.04 | 4.59 | 0.06 | −88.92 | −0.15 ± 0.18 | 0.03 |
| Wild bee abundance | 3 | 185.69 | 5.24 | 0.04 | −89.25 | −0.06 ± 0.18 | 0.00 |
| Impervious surface | 3 | 185.81 | 5.36 | 0.04 | −89.30 | 0.00 ± 0.18 | 0.00 |

To make coefficients easier to compare, all variables have been standardized. For each variable we give the estimate of the coefficient ± the standard error. The coefficient for the null model is the intercept. For pseudo-$R^2$ we used the formula 1 − (residual deviance/null deviance)

foraging on weeds took more time or energy. If the bees are partitioning their resources in this way, there would likely also be fewer opportunities to transmit pathogens.

### 15.3.3 Does Urban Beekeeping Provide Any Benefits to Wild Bees?

In the gardens where people kept honey bees, there was a significantly higher number of plant species flowering than in gardens without honey beehives (Fig. 15.4). Gardens with hives also had a slightly higher median weed density,

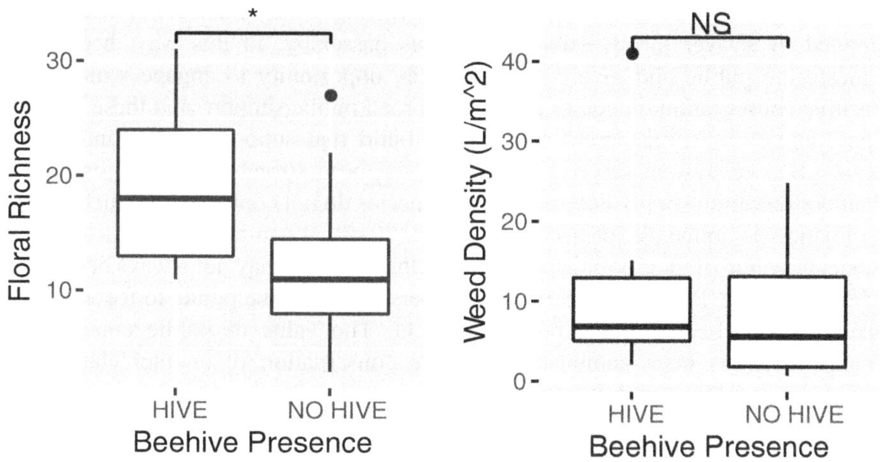


**Fig. 15.4** Boxplots comparing floral richness (left) and weed density (right) in gardens with and without honey beehives. The asterisk (*) in the image to the left indicates significance at $p < 0.05$. NS stands for "not significant"

but the difference was not statistically significant. The influence of beekeeping on human behavior and perception is better illuminated by the interviews. When asked whether beekeeping influenced the way he saw urban space, one interviewee responded, "Oh, definitely. Way more attuned. Like, dandelions are—I want to start a campaign to save the dandelions. Just so people see them as not a weed, but it's a food source for somebody for something. It's a beneficial plant, it's got its purpose, it's going to bloom for a month and it's going to go away. No amount of spraying is going to make it go away. So, it definitely made me more attuned with blooming plants and nature and agriculture in general."

Another interviewee, who managed a university greenhouse, said that before he started working there the greenhouse used pesticides for years (which can harm bees), but he eliminated pesticide use when he became manager. He also said he enjoyed growing plants that bees like, "There's something that's just fun about that. So, I pay attention to which plants are visited by bees, and I grow more of those plants, and I try to give more of those plants away. So, I know I'm a little bit of a Johnny Appleseed whenever I say, 'Take this plant, you're going to be amazed how many bees come to it.'"

Most of the beekeepers we interviewed participated as a hobby, rather than as paid work. Some expressed a love of beekeeping or an attachment to the bees themselves, viewing the bees as pets or even colleagues. One participant started volunteering with the beehives at a community garden just to have something to do, and it inspired him to build native bee habitats at both the community garden and his home garden.

Many of Chicago's beekeepers engage the public in bee education, both formally and informally. In fact, one of the interviews we conducted took place during an outreach event at a public conservatory. During other interviews, beekeepers stopped to answer the questions of curious passersby. In this way, beekeepers educate the public and give city-dwellers an opportunity to engage with nature. Positive conservation outcomes depend on broad public support, and these kinds of opportunities for engagement can help to build that support (Miller and Hobbs 2002). Bee ecosystem services alone may not be a strong enough argument for their conservation simply because rare bee species do not contribute as much to crop pollination as abundant species (Kleijn et al. 2015). When bees are reduced to the monetary value of their pollination services, the numbers may not always be in favor of conservation. Perhaps our urban beekeepers can teach the public to feel the same attachment to bees that the beekeepers feel. The value of public concern and engagement has been demonstrated in the conservation of another charismatic pollinator, the Monarch butterfly (*Danaus plexippus*). In short, nature needs cities, and urban residents are important stakeholders in biodiversity conservation (Derby Lewis et al. 2019).

Some of the education the beekeepers provide also benefits wild bees in more direct ways. At the aforementioned event at the conservatory, beekeepers were demonstrating how they detect and reduce varroa mites on the honey bees. When a member of the public asked them what people who are not beekeepers can do to help honey bees, one beekeeper replied, "People who are not beekeepers, what can be done is the more. . .plants you plant the better nutrition the bees get, the better their immune system is, the more they can resist the mites. At the end of the day, it's all about the immune system and nutrition. Just like us." This response is not only astute, but it gives actionable advice that benefits all bees. Like us, bees require certain essential amino acids which they get from their food. A diverse diet of high-quality pollen helps bees meet those needs and improve their resistance to parasites and other stressors (Huang 2012; Di Pasquale et al. 2013) and increase their longevity (Schmidt et al. 1987).

## 15.4 Conclusion

One beekeeper we interviewed expressed that she had experienced hostility from people who are concerned with native species conservation. She gave an example of an email she had received: "She was like, my milkweed are covered in honey bees and what are you going to do about this?" Our research suggests that such antagonism between "wild bee people" and "honey bee people" might be unwarranted. We cannot speak authoritatively on all habitat types where honey bees and wild bees interact, but in Chicago's community gardens there is evidence that these two groups coexist, possibly as a result of floral resource partitioning. At moderate honey bee abundances, wild bees and honey bees were positively correlated with each other. In general, we encountered more bees of all kinds where there were more floral resources, and we found evidence that beekeeping positively influences floral resource availability.

In spite of the potential for competitive interactions between wild bees and honey bees, both parties (beekeepers and conservationists) are stakeholders in bee health and floral resource availability. Combining our efforts, through public engagement and plantings in community gardens and elsewhere, is likely to achieve better conservation outcomes than if we treat other stakeholders as adversaries. Both parties can advocate for and provide resources for both managed honey bees and wild native bees in cities. Community gardens are important urban spaces for bees, both in providing habitat and inspiring advocates; beekeepers have influence in both of these areas.

## References

Anderson DR (2008) Model based inference in the life sciences: a primer on evidence. Springer Science and Business Media, New York

Angel J (2003) Illinois growing season. Illinois State Climatologist Office, Illinois State Water Survey, U of I. https://www.isws.illinois.edu/statecli/Frost/growing_season.htm. Accessed 12 Jun 2019

Carmody J (2017) Spike in backyard beekeeping in Perth as people embrace organic foods. ABC News. https://www.abc.net.au/news/2017-10-27/backyard-beekeeping-is-growing-in-popularity-perth/9090466. Accessed 28 Jun 2019

Carvell C, Roy DB, Smart SM et al (2006) Declines in forage availability for bumblebees at a national scale. Biol Conserv 132:481–489. https://doi.org/10.1016/j.biocon.2006.05.008

Derby Lewis A, Bouman MJ, Winter AM et al (2019) Does nature need cities? Pollinators reveal a role for cities in wildlife conservation. Front Ecol Evol 7:220. https://doi.org/10.3389/fevo.2019.00220

Di Pasquale G, Salignon M, Le Conte Y et al (2013) Influence of pollen nutrition on honey bee health: do pollen quality and diversity matter? PLoS One 8:e72016. https://doi.org/10.1371/journal.pone.0072016

ESRI (2017) ArcGIS Desktop: 10.5. Environmental Systems Research Institute, Redlands, CA

Fürst MA, McMahon DP, Osborne JL et al (2014) Disease associations between honeybees and bumblebees as a threat to wild pollinators. Nature 506:364–366. https://doi.org/10.1038/nature12977

Geldmann J, González-Varo JP (2018) Conserving honey bees does not help wildlife. Science 359: 392–393. https://doi.org/10.1126/science.aar2269

Goulson D (2013) Review: An overview of the environmental risks posed by neonicotinoid insecticides. J Appl Ecol 50:977–987. https://doi.org/10.1111/1365-2664.12111

Goulson D, Nicholls E, Botias C, Rotheray EL (2015) Bee declines driven by combined stress from parasites, pesticides, and lack of flowers. Science 347:1255957. https://doi.org/10.1126/science.1255957

Greenberg J (2002) A natural history of the Chicago region. University of Chicago Press, Chicago, IL

Greenleaf SS, Williams NM, Winfree R, Kremen C (2007) Bee foraging ranges and their relationship to body size. Oecologia 153:589–596. https://doi.org/10.1007/s00442-007-0752-9

Hall DM, Camilo GR, Tonietto RK et al (2017) The city as a refuge for insect pollinators: insect pollinators. Conserv Biol 31:24–29. https://doi.org/10.1111/cobi.12840

Han F, Wallberg A, Webster MT (2012) From where did the Western honeybee (*Apis mellifera*) originate? Ecol Evol 2:1949–1957. https://doi.org/10.1002/ece3.312

Hicks DM, Ouvrard P, Baldock KCR et al (2016) Food for Pollinators: quantifying the nectar and pollen resources of urban flower meadows. PLoS One 11:e0158117. https://doi.org/10.1371/journal.pone.0158117

Huang Z (2012) Pollen nutrition affects honey bee stress resistance. Terr Arthropod Rev 5:175–189. https://doi.org/10.1163/187498312X639568

Klee J, Besana AM, Genersch E et al (2007) Widespread dispersal of the microsporidian Nosema ceranae, an emergent pathogen of the western honey bee, Apis mellifera. J Invertebr Pathol 96: 1–10. https://doi.org/10.1016/j.jip.2007.02.014

Kleijn D, Winfree R, Bartomeus I et al (2015) Delivery of crop pollination services is an insufficient argument for wild pollinator conservation. Nat Commun 6:7414. https://doi.org/10.1038/ncomms8414

Le Conte Y, Ellis M, Ritter W (2010) Varroa mites and honey bee health: can Varroa explain part of the colony losses? Apidologie 41:353–363. https://doi.org/10.1051/apido/2010017

Lorenz S, Stark K (2015) Saving the honeybees in Berlin? A case study of the urban beekeeping boom. Environ Sociol 1:116–126

Lowenstein DM, Matteson KC, Minor ES (2015) Diversity of wild bees supports pollination services in an urbanized landscape. Oecologia 179:811–821. https://doi.org/10.1007/s00442-015-3389-0

Maderson S, Wynne-Jones S (2016) Beekeepers' knowledges and participation in pollinator conservation policy. J Rural Stud 45:88–98. https://doi.org/10.1016/j.jrurstud.2016.02.015

Mallinger RE, Gratton C (2015) Species richness of wild bees, but not the use of managed honeybees, increases fruit set of a pollinator-dependent crop. J Appl Ecol 52:323–330. https://doi.org/10.1111/1365-2664.12377

Mallinger RE, Gaines-Day HR, Gratton C (2017) Do managed bees have negative effects on wild bees?: a systematic review of the literature. PLoS One 12:e0189268. https://doi.org/10.1371/journal.pone.0189268

Mazerolle M (2017) AICcmodavg: model selection and multimodel inference based on (Q)AIC(c)

Meeus I, Brown MJF, De Graaf DC, Smagghe G (2011) Effects of invasive parasites on bumble bee declines: parasites and bumble bee declines. Conserv Biol 25:662–671. https://doi.org/10.1111/j.1523-1739.2011.01707.x

Miller JR, Hobbs RJ (2002) Conservation where people live and work. Conserv Biol 16:330–337. https://doi.org/10.1046/j.1523-1739.2002.00420.x

O'Neil-Dunne J (2010) Cook County land cover

Owen (2009) The bee business: an amateur apiary revolution. The Independent. http://www.independent.co.uk/environment/nature/the-bee-business-an-amateur-apiary-revolution-1752307.html. Accessed 28 Jun 2019

Paini DR (2004) Impact of the introduced honey bee (Apis mellifera) (Hymenoptera: Apidae) on native bees: a review. Austral Ecol 29:399–407. https://doi.org/10.1111/j.1442-9993.2004.01376.x

Potts SG, Biesmeijer JC, Kremen C et al (2010) Global pollinator declines: trends, impacts and drivers. Trends Ecol Evol 25:345–353. https://doi.org/10.1016/j.tree.2010.01.007

Pyke GH, Pulliam HR, Charnov EL (1977) Optimal foraging: a selective review of theory and tests. Q Rev Biol 52:137–154. https://doi.org/10.1086/409852

R Core Team (2018) R: a language and environment for statistical computing. R Foundation for Statistical Computing, Vienna, Austria

Rogers SR, Tarpy DR, Burrack HJ (2014) Bee species diversity enhances productivity and stability in a perennial crop. PLoS One 9:e97307. https://doi.org/10.1371/journal.pone.0097307

Schmidt JO, Thoenes SC, Levin MD (1987) Survival of honey bees, Apis mellifera (Hymenoptera: Apidae), fed various pollen sources. Ann Entomol Soc Am 80:176–183

Singh R, Levitt AL, Rajotte EG et al (2010) RNA viruses in Hymenopteran pollinators: evidence of inter-taxa virus transmission via pollen and potential impact on non-apis Hymenopteran species. PLoS One 5:e14357. https://doi.org/10.1371/journal.pone.0014357

Sreenivasan (2016) Why urban beekeeping is a rising trend in major cities. PBS NewsHour. https://www.pbs.org/newshour/show/urban-beekeeping-rising-trend-major-cities. Accessed 28 Jun 2019

Suryanarayanan S, Kleinman DL, Gratton C et al (2018) Collaboration matters: honey bee health as a transdisciplinary model for understanding real-world complexity. BioScience. https://doi.org/10.1093/biosci/biy118

Tonietto R, Fant J, Ascher J et al (2011) A comparison of bee communities of Chicago green roofs, parks and prairies. Landsc Urban Plan 103:102–108. https://doi.org/10.1016/j.landurbplan.2011.07.004

US Census Bureau (2018) American FactFinder - Results. https://factfinder.census.gov/faces/tableservices/jsf/pages/productview.xhtml?src=bkmk

USDA, Agricultural Research Service (2012) USDA Plant Hardiness Zone Map. https://planthardiness.ars.usda.gov/PHZMWeb/Default.aspx. Accessed 12 Jun 2019

USDA, National Agricultural Statistics Service (2019a) 2017 Census of Agriculture-United States Summary and State Data. https://www.nass.usda.gov/Publications/AgCensus/2017/Full_Report/Volume_1,_Chapter_1_US/usv1.pdf. Accessed 12 Apr 2019

USDA, National Agricultural Statistics Service (2019b) Colonies of honey bees - inventory and honey sales: 2017 and 2012. In: 2017 Census of agriculture - county data. pp 501–502. https://www.nass.usda.gov/Publications/AgCensus/2017/Full_Report/Volume_1,_Chapter_2_County_Level/Illinois/st17_2_0021_0022.pdf. Accessed 12 Apr 2019

Venables WN, Ripley BD (2002) Modern applied statistics with S, 4th edn. Springer, New York

**Alexis D. Smith** recently received their PhD in Biological Sciences from the University of Illinois at Chicago. Their research is in urban ecology with a focus on understanding how urban habitats can best support biodiversity as well as the human dimensions of urban habitats.

**Michael Roberts** is a graduate student in the Department of Biological Sciences at the University of Illinois at Chicago. His main areas of interest are what traits make an insect species well-suited to an anthropogenic environment and how species already in these environments continue to adapt.

**Haley Lerand** graduated with her Bachelor of Arts degree from the University of Illinois at Chicago, where she worked with the Department of Anthropology researching the ways in which community gardening contributes to the well-being of individuals and communities. She is the Project Manager at Global Garden Refugee Training Farm, a non-profit organization that helps refugees from agrarian backgrounds establish themselves as farmers in the USA.

**Molly Doane** is Associate Professor of Anthropology at UIC. She is an Environmental Anthropologist who studies community conservation, community food systems, and organic gardening in Mexico and the USA. She has written about community forestry in Oaxaca, Mexico, the fair-trade coffee movement in Mexico and the USA, organic farming in Wisconsin, and community gardening in Chicago.

**Emily S. Minor** is an Associate Professor at the University of Illinois at Chicago, with a joint appointment in the Department of Biological Sciences and the Institute for Environmental Science and Policy. She is an ecologist who studies ecological communities and human–environment interactions in cities.

# The *Rurban* Elephant: Behavioural Ecology of Asian Elephants in Response to Large-Scale Land Use Change in a Human-Dominated Landscape in Peri-Urban Southern India

16

Nishant Srinivasaiah, Srinivas Vaidyanathan, Raman Sukumar, and Anindya Sinha

**Abstract**

An impetus for growth and development in India, with the second largest human population in the world, has resulted in rapid changes in land use across the country, especially over the last two decades. While the land area under agriculture has only slightly increased, there have been significant changes in the shift from single-cropping to double- or multiple cropping every year and an overall increase in built-up areas. We assess the impacts of such transformations on the lives of India's largest land mammal, the Asian elephant, at a time when about 400 people and 150 elephants succumb annually to human–elephant conflict

N. Srinivasaiah
Frontier Elephant Programme, Foundation for Ecological Research, Advocacy and Learning, Pondicherry, India

Frontier Elephant Programme and Animal Behaviour and Cognition Programme, National Institute of Advanced Studies, Bangalore, India

SOAS University of London, London, UK

S. Vaidyanathan
Frontier Elephant Programme, Foundation for Ecological Research, Advocacy and Learning, Pondicherry, India

R. Sukumar
Asian Nature Conservation Foundation, Bangalore, India

Centre for Ecological Sciences, Indian Institute of Science, Bangalore, India

A. Sinha (✉)
Frontier Elephant Programme and Animal Behaviour and Cognition Programme, National Institute of Advanced Studies, Bangalore, India

Indian Institute for Scientific Education and Research Kolkata, Mohanpur, India

Centre for Wildlife Studies, Bangalore, India
e-mail: asinha@nias.res.in



J. A. Diehl, H. Kaur (eds.), *New Forms of Urban Agriculture: An Urban Ecology Perspective*, https://doi.org/10.1007/978-981-16-3738-4_16

across the country. Although elephants generally prefer forested habitats, the increased availability of nutritious crops and forest cover in the form of agroforestry plantations outside protected forests has led them to move extensively across peri-urban areas and successfully adapt to this novel anthropogenic ecological regime. We discuss these unique, inexorable processes of synurbisation, adaptation in a nonhuman species to human-induced change, with a particular focus on how the successful exploitation of *rurban* agricultural resources has allowed for the appearance of spatially and temporally flexible behavioural innovations that, in turn, impact the life-history strategies of a threatened elephant population in a peri-urban region of southern India.

**Keywords**

Movement ecology · Behavioural flexibility · Synurbisation · Peri-urban agriculture · Human–elephant conflict management

## 16.1 Changing Landscapes, Shifting Behaviours?

We live in a world that is increasingly urbanising, with the ratio of urban to rural human population decreasing from 1:6.7 in the 1900s to less than 1 today (Bettencourt and West 2010). It is also well documented that the human footprint on the environment, especially its negative impact on global biodiversity is on the rise (Venter et al. 2016). Urban sprawl itself has been predicted to increase from 0.3% of the total land area in 2000 to about 1.1% by 2030, with a concomitant reduction in forested areas (Angel et al. 2007; Santini et al. 2019). Recent studies, however, suggest that the earth is greening (Chen et al. 2019). Although this is a positive development, 82% of the increase in green cover in India has been attributed to agriculture and only 4.4% to forest cover, mostly monoculture plantations (Chen et al. 2019). In addition, linear infrastructure development, of roads and railways in and around forest areas, between 2000 and 2015, has resulted in further fragmentation of wildlife habitats of large ranging mammals, such as the Asian elephant, with more than 90% of the forest patches in India being reduced to less than 1 $km^2$ in area (Nayak et al. 2020).

Nonhuman species are known to exhibit behavioural adaptations in response to the environmental constraints they face in their daily lives (Stephens et al. 2007). Studies globally have shown, however, that, in a human-centric world, animals are forced to either adapt to these human-induced changes or perish, a process known as synurbisation (Santini et al. 2019). Animals that can adapt and, at times, even thrive in increasingly urbanised areas are able to vary their time-activity budgets (Sih et al. 2011; Tuomainen and Candolin 2011; Wong and Candolin 2015), search out and feed in nutrition-rich patches such as cropfields and, increasingly in response to anthropogenic pressures, increase their movement rates in human-dominated areas or completely move out of high human-density sites to settle in alternate areas (see Gaynor et al. 2018 for a review). Such novel habitats may also be an ecological trap (Simon and Fortin 2020). This is especially true for large mammals with widespread home ranges that extend across human-dominated areas, such as the Asian elephant,

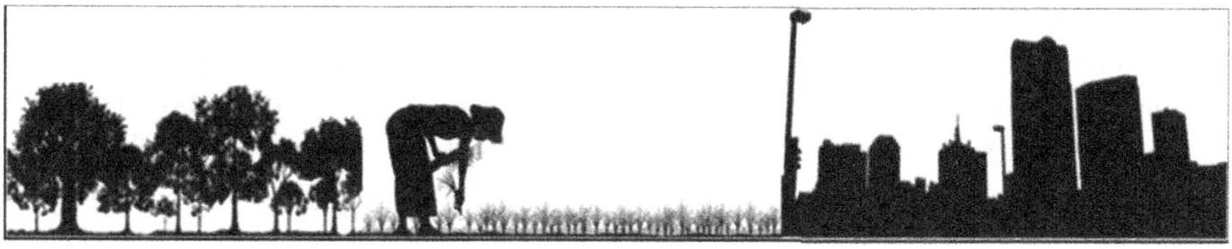

**Fig. 16.1** Forested habitats of elephants are rapidly transforming into agricultural and urbanised areas

which has been known to feed on nutritious crops for more than three decades now (Sukumar 1989; Graham et al. 2009; Srinivasaiah et al. 2019).

The Asian elephant *Elephas maximus* has home ranges varying from 250 to 1000 km$^2$, depending on the forest vegetation type and season (Sukumar 2003, 2006; Choudhury et al. 2008). The ranging behaviour of this species in such landscapes across southern Asia is also likely to be influenced by ecological as well as anthropogenic factors (Johnsingh and Williams 1999; Srinivasaiah et al. 2012; Kshettry et al. 2020). Large-scale extraction of resources from natural habitats has resulted in habitat degradation and loss across Asian elephant ranges (Venter et al. 2016), with the consequent depletion of resources for the elephants themselves (Sukumar 1989; Calabrese et al. 2017). Habitat fragmentation has, in fact, been recognised as a major reason for the endangerment of this species, with individual elephants becoming restricted to patchy refuges linked by corridors that are under high conservation threat (Leimgruber et al. 2011; Sukumar 2003). Additionally, human activities such as land use change and poaching have resulted in increased human–elephant conflict, with associated mortalities for both species (Sukumar 2003; Hauenstein et al. 2019).

With human activities expanding into natural forested habitats (Venter et al. 2016), both elephants and humans increasingly find themselves competing over resources (Pimm et al. 1995; Balmford et al. 2001; Sukumar 2003). It is noteworthy that although elephants and humans have possibly shared habitats and resources over centuries now, more recent and extensive changes in land use, mostly the advent of agriculture (Chen et al. 2019), may have significantly impacted the behaviour of elephants (Buij et al. 2007; Graham et al. 2009; Srinivasaiah et al. 2019). In extreme cases, they have even ceased to occur in very high human-density landscapes (Parker and Graham 1989a, b; Eltringham 1990; Barnes et al. 1991; Fay and Agnagna 1991; Happold 1995; Hoare and Du Toit 1999; Buij et al. 2007; Kshettry et al. 2020). An understanding of the impact of such large-scale anthropogenic changes, occasionally even in peri-urban areas (Fig. 16.1), on the behavioural ecology of big mammals, such as elephants, is, however, essential for landscape-level conservation planning, management of forest areas and for human welfare across the Indian subcontinent.

In this chapter, we assess the impact of changing land use and management practices on the distribution of elephants in a human-dominated landscape over a period of 16 years. Our study area spanned the states of Karnataka and Tamil Nadu, specifically the districts of Bangalore, Tumkur, Ramanagara and Krishnagiri. Here, we first discuss the impact of different resource-availability regimes on the residence time of individual elephants in patches and the influence of biological variables such as gender, age and group type at the population level. We then discuss how the daily activity patterns of male elephants can change, as they transition from a low human-density habitat, such as a protected forest to a high human-use production landscape. By detailing the modifications in ranging and behavioural decisions made by this population of Asian elephants in a fragmented, largely agricultural landscape of southern India in response to varying levels of resource availability and human activity, we intend to showcase the behavioural adaptability of elephants that can help them persist in the *rurban*.

## 16.2 The Study Landscape and Its Elephants

Our study landscape, spanning over 10,000 km$^2$, comprised several Protected Areas (PAs) including the Bannerghatta National Park and Cauvery North Wildlife Sanctuary, and their surrounding human-dominated areas in the districts of Bangalore, Ramanagaram, Tumkur and Krishnagiri in the states of Karnataka and Tamil Nadu with an average density of ~350 people/km$^2$ (Census 2011; Srinivasaiah et al. 2012, 2019). The terrain in this landscape is highly undulating, with a mean altitude of 865 m above mean sea level and an average annual rainfall of 937 mm. The vegetation within the forest ranges from predominantly deciduous to scrub woodland with riparian patches along the streams (Srinivasaiah et al. 2012). Geographically, the PAs are contiguous with larger patches of forests located southeast and southwest, forming part of the Nilgiris—Eastern Ghats Elephant Reserve. More importantly, what concerns us here is that the PAs are surrounded by well-irrigated croplands and human settlements that dot the landscape (Anand et al. 2009).

The high-density human population in and around these PAs largely comprise subsistence farmers, livestock-grazers, and manual labourers engaged in sand-mining and granite-quarrying. The communities depend on the forests for their non-timber forest produce, firewood and livestock-grazing. Amongst the farming communities, the majority are marginal farmers practicing subsistence agriculture alongside farmers who grow commercial crops and have plantations. The Pas have distinct dry and wet seasons with a mean monthly rainfall of 5 mm in January and 170 mm in October, receiving both the southwest and northeast monsoon rains in most years (Srinivasaiah et al. 2019, in press).

The study landscape was classified into different land use types, based on geospatial data obtained from the National Remote Sensing Agency of the Government of India (downloaded from http://bhuvan3.nrsc.gov.in/cgi-bin/LULC250K.exe). The original 19 Land Use and Land Cover (LULC) categories (NRSA 2006) were merged to derive eight LULC categories: Deciduous Forest, Degraded Forest, Plantation (including orchards), Crop (seasonal and multicrop), Current Fallow, Wasteland, Waterbody, and Built-Up Area (Fig. 16.2; Srinivasaiah et al. 2019). Within a smaller area of ~350 km$^2$, known to be an area of intense use by elephants within the larger landscape, a detailed assessment of resource availability and threats

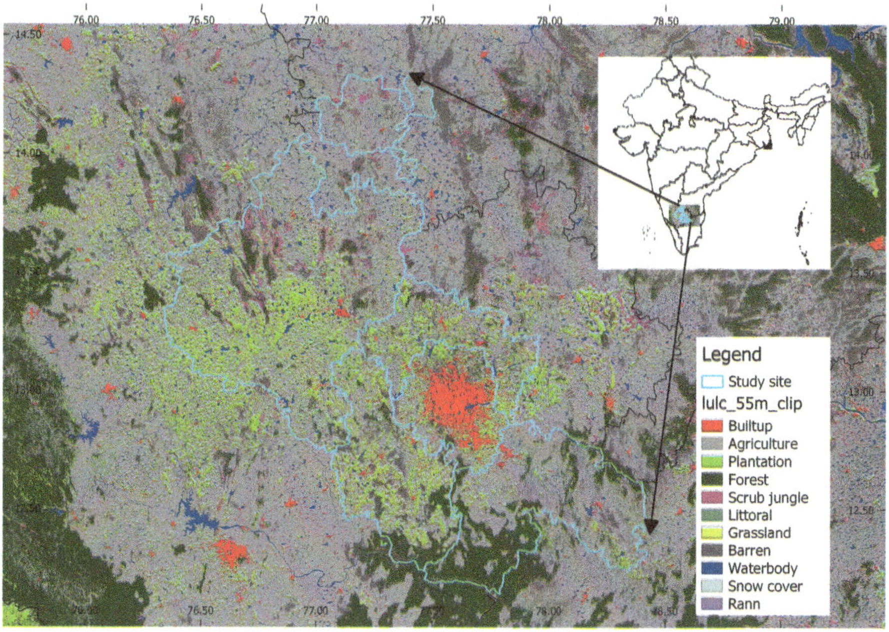


**Fig. 16.2** Map of the study landscape in southern India showing LULC types at 55-m resolution and the important urban centres of the region

was undertaken. The maximum average values of Normalised Difference Vegetation Index (NDVI) and Leaf Area Index (LAI) were extracted with MODIS data products, using Quantum GIS (QGIS 2009) and Geographic Resources Analysis Support System (GRASS 2007) and their values, obtained during the observation period, used as surrogate measures of forage (Druce et al. 2008; Okello and D'Amour 2008; Young et al. 2009) and shade availability, respectively. The values of NDVI, LAI, Number of Waterbodies per 4 km$^2$ Area and a Human Disturbance Index (encounter rate of human- or human-associated disturbances × proportion of sampling segments in which these disturbances were found) were thus, respectively, used as surrogate measures of the four study variables: forage, shade, water and human presence. Quantile classification was used to classify each of the four variables into low, medium and high strata, with their cut-off values being distributed such that the entire study area was classified into equal numbers of low, medium and high strata for each variable.

The Asian elephant has been estimated to number 35,000–50,000 individuals, spread across a range of 13 Asian countries (Blake and Hedges 2004). India has approximately 50% of the total population of the species (20,000–25,000), with southern India supporting around 10,000 elephants in the wild (Project Elephant 2017). Traditionally, significant levels of hunting and poaching for tusks, and, increasingly, habitat loss, fragmentation and degradation of habitats have led to a drastic decline in elephant populations, with the species currently endangered

(Choudhury et al. 2008). The estimated density of elephants in our study landscape was 1.0 elephant $km^2$ (Project Elephant 2017).

The study population consisted of 78 individuals, distributed in 13 herds, 4 all-male groups and 12 solitary males, roving over an area of ~600 $km^2$. The population-level analyses that we conducted were based on demographic data collected from all the individuals across all the dry and wet seasons since 2009. In addition, 25 adult (>20 years of age) and subadult (10–20 years of age) males, representing a subset of these 78 individuals, were observed more closely, between 2009 and 2015, in an area of ~80 $km^2$, comprising a PA and a human-use agricultural area (HA). The land use characteristics of the HA and PA have been depicted in Table 16.1.

The observed elephants were classified into three group types, namely solitary or single male elephants; all-male group (AMG), a coalition of male elephants alone; and herd, consisting of one or more family units. Individual elephants were grouped into four age classes: adult (>15 years), subadult (5–15 years), juvenile (1–5 years) and calf (<1 year). Data on diurnal activity of elephants was collected using instantaneous scan sampling, in which all visible individual male elephants, in herds, all-male groups or as solitary, were scanned once every 15 min in order to record six mutually exclusive behavioural states, namely Feeding, Moving, Standing, Watering, Socialising and Other Behaviours (Altmann 1974). Behavioural observations of elephants at night in the HA, however, were conducted using extensive ad libitum sampling. Moreover, we conducted a total of 3628 instantaneous scans of demographic structure and behavioural states of all individuals in our study population, once every 15 min during the observation periods, to assess the intensity of habitat use, total range area and distribution patterns. We used the *G*-test of independence to compare the behavioural activities of elephants across different age-gender categories, habitat variable strata and seasons of the year, while the Wilcoxon rank sum test was employed to compare behavioural profiles between different habitats in the study landscape.

## 16.3 Life in an Urbanising Landscape

Elephant populations mostly occur in high densities within well-established PAs across the Indian subcontinent (Jathanna et al. 2015); these constituted about 51.8% and a meagre 1.9% of the PA and HA in our study landscape, respectively. The rural areas, with cropfields, forest plantations and grazing pastures, covered over 60% of the land area. Often adjacent to populous human habitations, these are known to provide adequate space and resources to accommodate a wide array of wildlife that live alongside burgeoning human populations (Western et al. 2009). It is important to note that cropfields accounted for approximately 18.33% of our study area, a significant proportion of this being peri-urban in nature, adjacent to several important urban centres such as Bangalore and Hosur of southern Karnataka and western Tamil Nadu, respectively. Moreover, degraded forests, scrublands, rocky areas and culturable wastelands, among others, amount to more than 15% of the total

**Table 16.1** LULC characteristics of the Human-use Agricultural Area (HA) and the Protected Forest Areas (PAs) in the study landscape

| | Mean percentage | | Minimum percentage | | Maximum percentage | | Standard deviation | | Standard error | |
|---|---|---|---|---|---|---|---|---|---|---|
| Cell LULC characteristic | HA | PA | HA | PA | HA | PA | HA | PA | HA | PA |
| Built-up area | 3.01 | 7.12 | 0.97 | 0.00 | 5.02 | 22.74 | 2.05 | 10.52 | 1.03 | 5.26 |
| Crop | 21.15 | 15.50 | 17.12 | 3.59 | 25.25 | 33.69 | 04.55 | 13.05 | 2.27 | 6.52 |
| Current fallow | 0.83 | 1.48 | 0.22 | 0.07 | 1.32 | 4.67 | 0.56 | 2.16 | 0.28 | 1.08 |
| Deciduous forest | 1.90 | 51.80 | 1.34 | 34.88 | 3.00 | 67.52 | 0.75 | 18.02 | 0.38 | 9.01 |
| Degraded forest | 0.10 | 7.50 | 0.02 | 2.95 | 0.22 | 12.28 | 0.10 | 4.39 | 0.05 | 2.20 |
| Plantation | 56.60 | 3.27 | 47.96 | 1.39 | 62.55 | 6.06 | 6.39 | 2.11 | 3.19 | 1.05 |
| Wasteland | 0.27 | 13.00 | 0.00 | 9.51 | 0.61 | 14.89 | 0.31 | 2.37 | 0.16 | 1.18 |
| Waterbody | 16.15 | 0.37 | 9.90 | 0.00 | 22.47 | 0.75 | 5.13 | 0.43 | 2.57 | 0.21 |

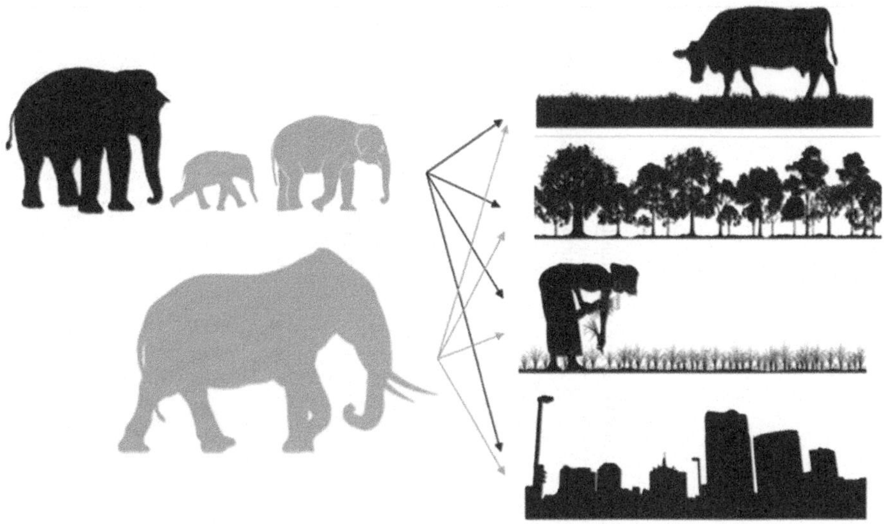

**Fig. 16.3** Herds and adult male elephants preferably reside in forested habitats but are increasingly moving to inhabit other land use types, such as cropfields, grazing pastures and peri-urban or even urban areas

geographical area of the country (State of Indian Agriculture 2016). These, often human-dominated, areas harbour elephants as well, even though elephant populations may be in significantly lower densities (Srinivasaiah et al. 2019). Thus, elephants can choose from a diverse ecological and anthropogenic setting to spend their time in (Fig. 16.3).

## 16.4 Decision-Making in a Dynamic Natural-Anthropogenic Landscape Matrix

As rainfall is seasonal in our study area, occurring only from late May to early October, a change in forage quantity and quality is expected across the year. Such a change in forage quality was indeed reflected in the increase in NDVI across the dry and wet seasons during our study ($G$-test of independence, $G = 44.215$, df $= 2$, $p = 0.000$). Conversely, although the local streams flow mostly during the wet season within the park, there was no change in water availability detected across the two seasons. This could be attributed to the numerous waterholes constructed by the state Forest Department. Human activity in the farmlands neighbouring the forests usually increases during the monsoon or the wet season and in the post-monsoon period. Such activities also continue into the post-harvest season subsequent to the cultivation of subsistence crops. Plantation farming, however, was prevalent throughout the year and there was no substantial variability in human presence or preoccupations observed in the study area across seasons (Surface water: $G = 0.368$, df $= 2$, $p = 0.804$; Human activity: $G = 0$, df $= 2$, $p = 0.054$).

Asian elephants, given their slow metabolic rates and large body size, usually need to forage over long periods of time, often up to 18 h a day, feeding on up to 200 kg of food daily (McKay 1973; Sukumar 2003; Srinivasaiah 2019). Our analysis revealed that the ranging behaviour of the study elephants was indeed primarily driven by the availability of forage-rich sites in the landscape, followed by the occurrence of human activities and availability of water. The tracking of forage-rich sites was more prominent during the dry season than in the wet season, with water being the least influential parameter governing elephant movement patterns or habitat use in the landscape across the year. The degree of shade availability was strongly positively correlated to forage availability and hence, not analysed further.

Elephants spent their time (100%) exclusively in the high-forage areas during the wet season, but only 64% of their time in the dry season, a significant difference ($G = 1883.091$, $\text{df} = 2$, $p = 0.000$). In response to human activity levels, elephants used the low-human-activity areas significantly more during the wet than in the dry season ($G = 267.398$, $\text{df} = 2$, $p = 0.000$). Their usage of the high-human-activity areas, however, remained the same across seasons. In response to the distribution of water, as could be expected, the study elephants used high-water-availability areas significantly more during the dry season than in the monsoon ($G = 1217.448$, $\text{df} = 2$, $p = 0.000$).

In support of our initial findings, we observed the study elephants to preferentially seek out high-forage areas during the dry season, in which they spent 64% of total observed time; forage was thus of prime importance to the elephants in the region. In the wet season, however, they spent significantly more time in low-human-activity areas, which accounted for 35% of the study area. This indicated that if forage was available in areas without human presence, such areas would be chosen by the elephants over human-use areas; overlap in resource use with humans was thus avoided. These results provide evidence of a trade-off between resource acquisition and risk avoidance displayed by the elephants in this region.

## 16.5 Decision-Making in Novel Human-Dominated Habitats

The elephants in the study landscape typically exhibited important behavioural adaptations at the level of the population, manifest principally in age-, gender- or group size-based differences in seasonal habitat use, but also at the individual level through the display of idiosyncratic behaviours (Srinivasaiah et al. 2012).

A comparison of the decisions made by the two genders, for example, revealed that female elephants, represented in our study by herds or mixed-sex groups, adopted a more risk-averse strategy (Sukumar and Gadgil 1988), with their occurrence and duration of stay limited to habitats with the best available natural resources but least human activity. The males, in contrast, ($G = 490.874$, $\text{df} = 2$, $p = 0.000$), foraged across a spectrum of land use and anthropogenic activities, their decisions possibly dependent on the particular stage of their life history (Sukumar and Gadgil 1988; Chiyo et al. 2011; Srinivasaiah et al. 2012). Sexually and socially immature males, for example, were probably influenced by their mothers to range in

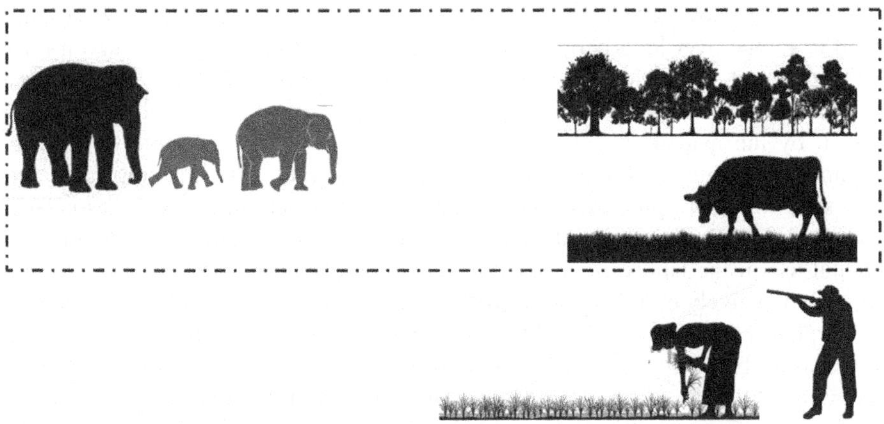

**Fig. 16.4** Female elephants usually adopt a risk-averse strategy by ranging in areas with requisite forage availability but least human activity

low-human-activity areas, such as the forests, while the sexually mature but socially immature adolescent males and the sexually and socially mature adult males moved between the resource-rich cropfields and the forests, the latter especially when in *musth* to mate with the females.

The general strategy of female elephants is to be risk-averse while seeking out requisite levels of forage availability (Fig. 16.4). This phenomenon changes in the dry season when forage levels are typically low. The study herds then tended to occur in medium-forage–medium-human-disturbance areas ($G = 357.459$, df $= 2$, $p = 0.000$). Most high-forage areas during the dry season were cropfields and plantations with high-human activity, as compared to the medium zones with least variability, both in resource availability and threats. Females typically used medium-forage–medium risk areas as a strategy to ensure access to now-scarce resources although they faced the unavoidable risk of encountering humans—a threat to themselves as well as their dependent calves and juveniles (see also Baskaran 1998). This decision was in stark contrast to that shown by the males, who invariably sought out high-forage areas across seasons, regardless of the threat levels involved (Fig. 16.5, $G = 0.92$, df $= 2$, $p = 0.608$).

## 16.6 Elephants of the Rurban

Although the study elephants, particularly females in herds, generally avoided high human-use areas in the landscape for most of the year, they often ranged across cropfields and fragmented forest habitats during the cropping season. More than half of the annual location data points of females were thus confined to these areas for 3 months in a year, between November and January. This is also the time when crops such as *ragi* or finger millet are in their vegetative stage, ready for harvesting. Males,

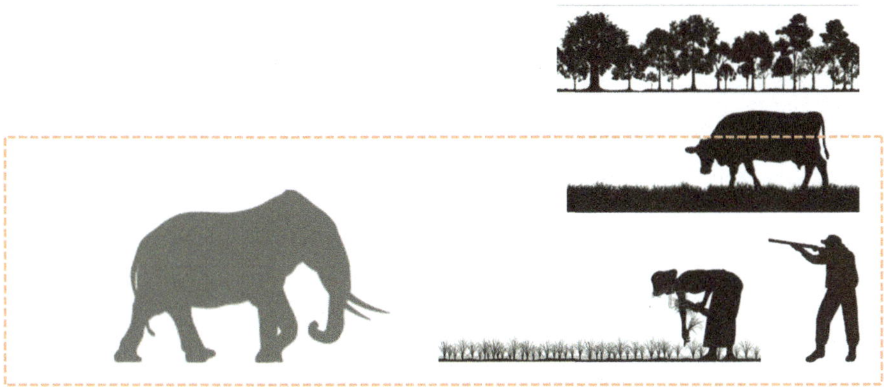

**Fig. 16.5** Male elephants often adopt a high-risk strategy by ranging in areas, typically cropfields and human habitations, with high-forage availability, irrespective of human activity levels

**Table 16.2** Minimum convex polygon (MCP) and kernel density estimates (KDE) of the home range size of the study elephant herds between 2000 and 2016

| Period | Years | MCP estimate (km$^2$) | KDE estimate (km$^2$) |
|---|---|---|---|
| 1 | 2000–2001 | 1034.23 | 3126.25 |
| 2 | 2002–2003 | 506.53 | 3006.56 |
| 3 | 2004–2005 | 1919.61 | 5279.70 |
| 4 | 2006–2007 | 1706.73 | 2212.81 |
| 5 | 2008–2009 | 1093.36 | 1403.41 |
| 6 | 2010–2011 | 3620.96 | 693.48 |
| 7 | 2012–2013 | 1650.66 | 2742.54 |
| 8 | 2015–2016 | 3649.70 | 5543.49 |

however, typically ranged outside forested areas and across cropfields throughout the year (Srinivasaiah et al. 2019).

Asian elephants are known to vary their range depending on the habitat characteristics and resource availability. They have also been observed to disperse from their natal ranges to move to a different region in response to climatic events such as El Niño, characterised by reduced rainfall and drought-like conditions, or La Niña, with its cyclones and increased rainfall, especially in the Indian context (Sukumar 2003). We estimated the range area of our study elephant herds by tracing the movement patterns of 47 adult females, resident in these herds and constituting about 60% of the 78 individual elephants being monitored, during the period from 2000 to 2016. They ranged over a mean (±SE) area of 1897.72 (±411.12) km$^2$, as obtained through the minimum convex polygon (MCP) method or 3001.03 (±601.15) km$^2$, as per the kernel density estimation (KDE) technique, during these years (Table 16.2, Fig. 16.6); the maximum range size recorded was in 2015–2016.

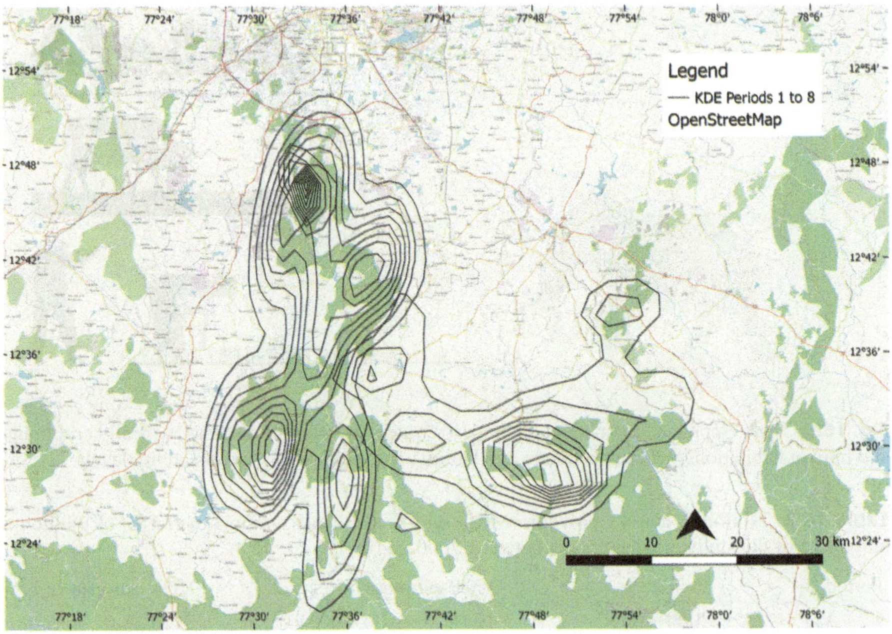


**Fig. 16.6** Kernel density estimates (KDE) of the home range size of the study elephant herds over a time period of 16 years, between 2000 and 2016, with each Period consisting of 2 consecutive years (Periods 1–8, Table 16.2)

The variation in the intensity of use of the landscape across land use types by the study elephant herds, both inside the protected forest areas and in the human-dominated landscape surrounding the forests, from 2000 to 2016, is shown in Fig. 16.7. There was a significant eastward expansion in herd movement primarily outside the forested areas, with an increase in ranging in the highly fragmented forest patches, during three periods—2004–2005, 2010–2011 and 2015–2016—that appeared to mark critical points in the ranging decisions of female elephants in this population (Sukumar 2003).

The female-led herds typically spent October to February in the western regions of the study landscape, in and around the Bannerghatta National Park, until 2003–2004, after which they began to move eastwards and explored the areas adjoining the Cauvery North Wildlife Sanctuary. By the end of 2016, a complete switch in the annual temporal usage of the landscape had set in, with the study population spending October to March in the eastern regions of the study landscape, having increased their range almost three times since 2003–2004. One of the most important factors promoting this movement in the short term appeared to be the "drives"—the long-distance driving away of elephants from the Bannerghatta National Park to the forests down south, in order to reduce human–elephant conflict in the western parts of Bannerghatta, close to the city of Bangalore, the IT capital of India (Manjunath NB, pers. obs.).

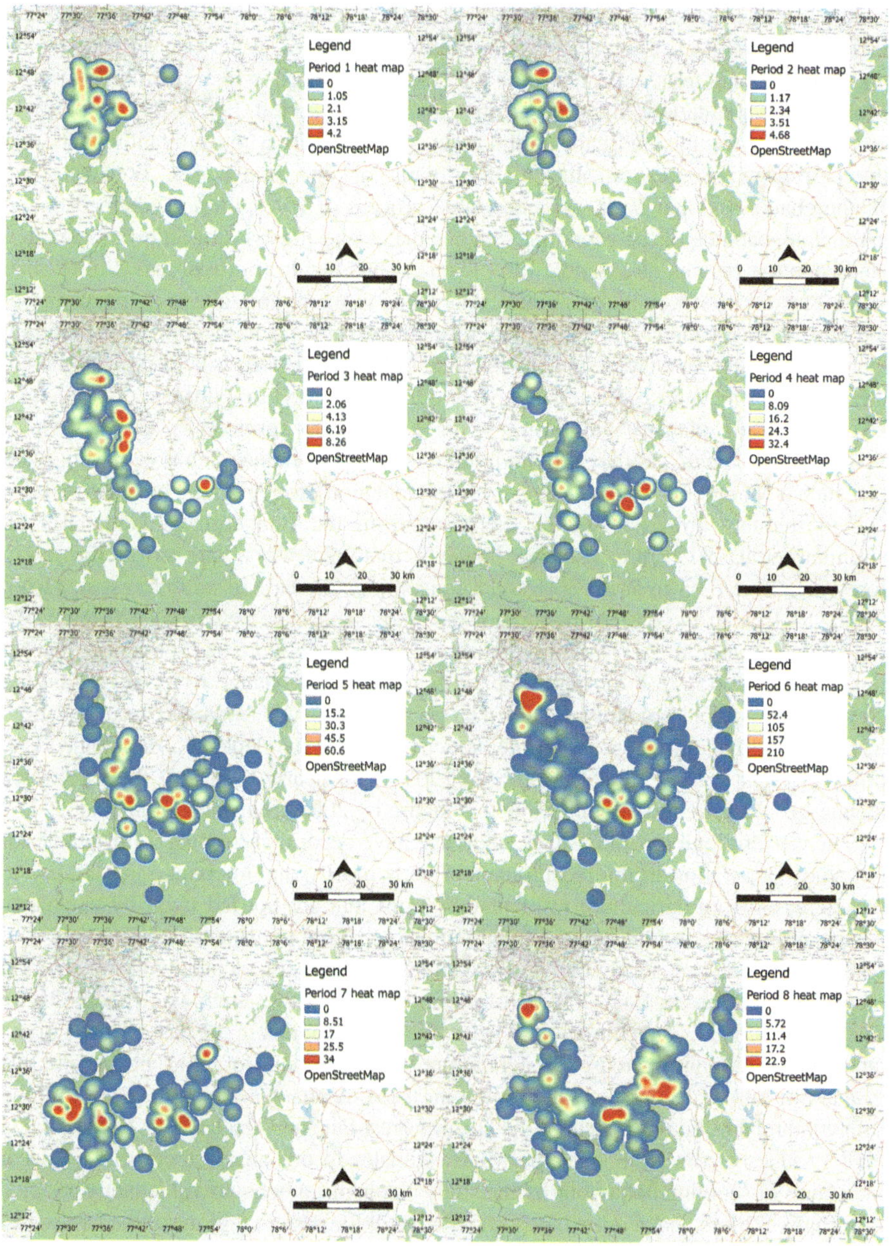


**Fig. 16.7** Heat map showing the distribution and intensity of use of the landscape by the study elephant herds during Periods 1–8, from 2000 to 2016

While these drives may have helped elephants explore new habitats, a significant factor that influenced this decision to establish themselves in the eastern parts of the region in the long term appeared to be the land use changes in their distribution range. Monoculture plantations increased from 2.8% in the period 1973–1992 to 5.3% in 1992–2007 across the landscape (Adhikari et al. 2015), with a concomitant reduction in the crop-growing areas in the western parts of our study area, bordering the Bannerghatta National Park. An inexorable process of *rurbanisation*, defined as the gradual appearance of the economic characteristics and lifestyles of an urban area in a landscape that retains its essential rural features, had begun (Sorokin and Zimmerman 1929; Parsons 1949). A boom in real estate value in the region, due to its proximity to the rapidly urbanising city of Bangalore, resulted not only in a number of farmers selling their land but also an increase in out-migration of famers to work as landless labourers in the neighbouring towns. These lands were either converted to industrial or residential layouts or were planted with fast-growing commercial agroforestry crops, such as *Acacia* or *Eucalyptus*, as they did not require much maintenance. It has been suggested that such a change from subsistence crops to plantation cropping may have partly been due to frequent crop damage by elephants from the Bannerghatta National Park (Adhikari et al. 2015). It is also true, however, that there was a long-term reduction in forest area from 53% to about 26% and a concomitant increase in scrub forest area from 0.55% to 15% in the eastern regions of the study landscape between 1920 and 2015 (Ramachandran et al. 2017). Although the extent of agriculture had remained more or less constant at 45–46% during this period, a significant change from single rain-fed cropping to multiple groundwater-aided cropping, augmented by canal-based irrigation, had possibly made this region a more attractive habitat for elephants. Simultaneously, however, their natural habitats were being lost, with nearly 0.08% of the forest area having been diverted to agriculture in this region by 2005 (Ramachandran et al. 2017).

The western regions of the study landscape have become increasingly unsuitable for our study population in the recent years. This is mostly due to change in land use in these areas—from agriculture to plantation crops that are not fed upon by elephants, persistent increase in built-up areas and human densities, construction of animal-proof fences and trenches along the protected areas to prevent elephants from entering cropfields, and the regular adoption of more active measures such as elephant drives. The study elephants have thus found better foraging grounds in the eastern parts of the landscape. With an increase in agriculture to meet their foraging needs and the availability of scrub forest to take refuge in, these elephants seem to have also found safety in and around the forests of the Hosur Forest Division since 2008. The relatively higher values of the KDE over the MCP estimates in our analyses constitute a clear indication of scouting behaviour by the study elephants, a behavioural strategy adopted to perhaps explore and acquire better knowledge of the eastern regions of the landscape.

## 16.7 The Rurban Male Elephant

In polygynous species, the physical and physiological condition of males often determine their positions in the dominance hierarchy and influence the outcome of intrasexual competition (Poole 1987; Rasmussen et al. 2007; Chiyo et al. 2011).

Male elephants need to build body mass, come into the energetically demanding state of *musth*, and seize opportunities to mate—reproductive demands that require them to continuously track both forage and females (Poole 1982).

Solitary males in *musth* in our study population tended to occur along with oestrous females in their herds. The relatively younger males were also primarily observed within PAs and other forests as they associated with the herds. It is noteworthy that a number of adolescent and older males, nevertheless, continued to frequent agricultural, human-use habitats outside the forested areas. Male Asian and African elephants are known to inhabit agricultural cropfields primarily to forage on the available crops (Sukumar 1989; Chiyo et al. 2012). As dominance in male elephant society is usually determined by *musth* and body condition (Chelliah and Sukumar 2013), individuals could use a high-risk high-gain strategy by feeding on nutritious crops to improve their body condition (Pokharel et al. 2017). In support of these earlier suppositions, we indeed observed a number of males in the human-use areas to be in excellent body condition.

Our studies on rurban male elephants have led to the important discovery of a significant number of all-male groups, primarily distributed at relatively high frequencies in the intensely human-dominated agricultural regions of the study landscape, particularly around the peri-urban areas of Tumkur, Ramanagaram and Hosur (Srinivasaiah et al. 2019). The remarkable adaptability of these all-male groups of elephants is best exemplified by a group of 11 male elephants that ranged in a large expanse (~10,000 km$^2$) across a rurban landscape, about 100 km to the west of the Bannerghatta National Park (Srinivasaiah et al. 2016).

A comparative analysis of the diurnal time-activity budgets of the 11 male elephants in the HA and that of 18 adult males that ranged exclusively in the PA indicated that the former spent a mean (±SE) of 2.54 (±0.28)% and 6.95 (±0.69)% of their time Feeding and Moving in contrast to 40.93 (±3.08)% and 39.55 (±3.10)% displayed by the latter, respectively. These differences are statistically significant (Wilcoxon rank sum test, $W = 0$, $n = 29$, $p = 0.000$ for both behaviours). The males in the HA, however, spent a relatively high proportion of their time budget in Standing (42.42 ± 0.55%), Watering (24.87 ± 0.68%) and Socialising (18.56 ± 0.52%). These behavioural states (together with Other Behaviours) were cumulatively exhibited by the males of the PA only to a meagre extent of 19.52% of their total time, these proportions also being significantly different (Standing and Watering: $W = 198$, $n = 29$, $p = 0.000$; Socialising: $W = 187$, $n = 29$, $p = 0.000$; Other Behaviours: $W = 196$, $n = 29$, $p = 0.031$).

It is, however, striking that the proportion of observed time spent Feeding (53.83 ± 1.07%) and Moving (40.43 ± 1.13%) by the 11 males in the HA increased significantly during the night. The diurnal time-activity budget of individual male elephants in the HA departed significantly from that exhibited by individuals in the adjacent PA. In the HA, the elephants were highly spatially restricted by human threats, at least during the daytime. These males thus often took refuge in village waterbodies and agroforestry plantations during the daylight hours (Srinivasaiah 2019). Feeding and moving then became largely nocturnal, when human activities reduced, and the elephants could feed on the unguarded crops. The proximity of

**Fig. 16.8** Changing behavioural strategies of the study elephants across the day in human-dominated habitats

these males to one another in spatiotemporally restricted areas, such as waterbodies, surrounded by human habitations, may have also facilitated increased social interactions between these individuals, not typically seen in the males within the PA (Fig. 16.8).

In the resource-rich HA, we speculate that the spatiotemporal restrictions due to human presence and the perceived risks from them may have also compelled the study male elephants to coordinate their behavioural activities and move together as a single unit while they foraged on high-risk cultivated crops. This is supported by our observation that male associations displayed significantly low inter-individual variance in the proportion of time spent in the important behavioural states in the HA (Srinivasaiah, unpubl. obs.). In contrast, males in the PA, with little or no human presence, were free of such risks, and hence showed higher variance in their time-activity budgets. The importance of perceived risk from humans in determining the behavioural activities of the study elephants was also supported by our observation that at night, when human presence was minimal in the HA, the study elephants exhibited relatively greater variability in the proportion of time spent in different behavioural activities, notably feeding and moving (Srinivasaiah, unpubl. obs.).

## 16.8 Conclusions: The Synurbised Asian Elephant

Elephants are known to range across large areas in search of resources, such as food and water (Baskaran and Desai 1996). These behavioural activities were under threat in the human-dominated habitats often occupied by our study elephant population. Living in these high-risk areas, however, had their own benefits, mainly in the form of nutritious, human-origin food resources. In order to successfully persist in these areas, therefore, male Asian elephants needed to be strategic and adapt to the prevailing anthropogenic ecological regimes through processes of synurbisation. Solitarily pursued trial-and-error methods of habitat exploration could prove to be costly, as the landscape, such as the one inhabited by our study elephants, was

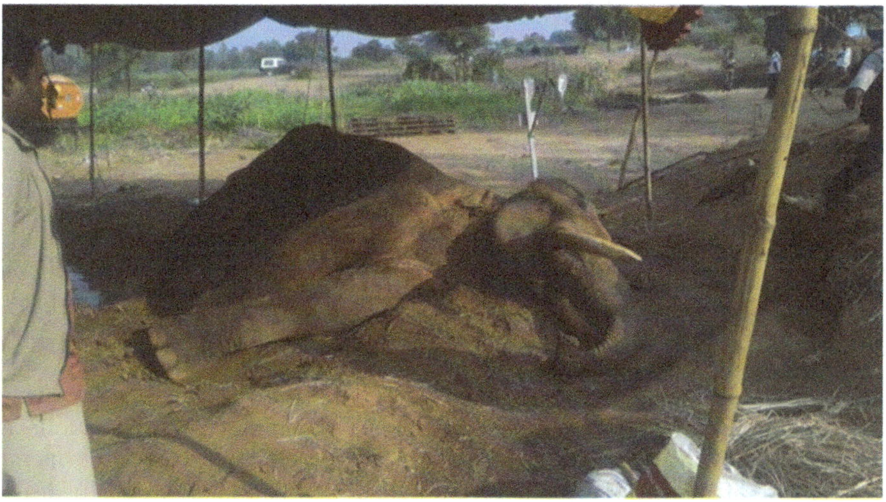

**Fig. 16.9** A young adult male elephant, which died from a fractured leg after falling into a storm water drain in the Bangalore Urban district of Karnataka state

crowded with linear intrusions, including highways, rail lines, innumerable electric wires and trenches that could prove fatal to individual elephants on the move (Fig. 16.9; Nayak et al. 2020).

The increasing agroforestry practices in the villages across the study landscape facilitated the elephants to approach human habitations, as they often used these patches as resting sites during their movements (Krishnan et al. 2019; Srinivasaiah 2019). Moreover, the availability of highly nutritious crops throughout the year led to the rapid escalation of the frequency of visits and duration of stay by elephants in human-dominated areas, with concomitant damage to different crops often ranging from 20% to 50% (Agrawal et al. 2016). In several instances, young adult and adolescent male elephants, who we refer to as the millennial males (Srinivasaiah 2019), have even become resident in these areas. The crop-loss claims, consequent to elephant depredation and registered under the state Forest Department on the eastern side of Bannerghatta, has increased from 2 to 1500 between 2000 and 2010, a staggering 750-fold increase in a decade. Such a drastic increase in the levels of mostly antagonistic interactions between people and elephants is unfortunate, as the conflict is consumptive in nature and seriously impacts rural livelihoods, threatening the food security of the closely adjoining urban areas in the process (Agrawal et al. 2016). Moreover, encounters between the two species often become critical, with approximately 400 people and 150 elephants succumbing to human–elephant conflict across the country annually.

The novel, but stable, all-male groups of elephants, with relatively large numbers of young adult and adolescent individuals, which have begun to emerge recently, appear to constitute a new form of social organisation in the species, possibly in response to highly fragmented habitats with poor inter-patch connectivity

(Srinivasaiah et al. 2019). Individuals in these all-male associations seem to coordinate their behavioural activities and strategies in such a way that it promotes more efficient crop-foraging behaviour in the high-risk, high-resource areas. While associating in all-male groups, we believe, could be an adaptive, synurbic social strategy (Srinivasaiah et al. 2019), elephants in regions of intense human use, including rurban or peri-urban areas, may, in addition, even need to modify their proximate behaviours, such as change their circadian rhythms, thus becoming more behaviourally active at night. Finally, it is entirely possible that such behavioural coordination could, in the future, lead to the establishment of stable cultural traditions in elephants that typically inhabit these rapidly evolving anthropogenic landscapes.

We have clearly seen that our study elephants prefer forage-rich sites, while displaying significant flexibility in the selection of these sites across the study landscape, be they natural forests or cropfields. Such remarkable behavioural flexibility, manifest also in the dramatic modification of short-term behavioural patterns or long-term life-history strategies, and developed within the lifetime of individuals, makes elephants one of the most resilient mammalian species on earth. Indeed, it is this very adaptable nature of elephants that may help them survive climate change and other perils of the Anthropocene (Kanagaraj et al. 2019). This is also reminiscent of strategies, such as finding new refugia, which ancestors of the modern-day elephant may have explored in order to survive mega-extinction events of the past (Davis et al. 2013). The movement of elephants to explore novel ecological regimes, due to destruction or fragmentation of their current home ranges, as observed in our study population, is not devoid of human–elephant conflict, especially as more forest habitats are lost or diverted to non-forest land use.

It is imperative that a holistic elephant management plan takes into account the negative human influences on natural environments that challenge or otherwise threaten the lives of individual elephants. The conservation of most elephant populations across the Indian subcontinent, endangered as they are by the obtrusive presence of humans and their various anthropogenic activities, must also incorporate, more ideally, non-human-centric land use planning and landscape design if at all we desire to survive into the future alongside this remarkable species with which we have shared the planet since times immemorial.

**Acknowledgements** The authors would like to thank the officials and administrative staff of the Forest Departments of Karnataka and Tamil Nadu states in India for their constant support and involvement in the study and for the necessary research permissions. NS expresses his gratitude to the field staff of the respective Forest Departments, farmers living alongside the elephants in the study area and the elephants themselves for their patience during the many hours of interactions. He would also like to thank Manjunath NB and his team from Nature and Wildlife Conservation Committee, Bannerghatta for collating part of the crop compensation claim records from the Forest Department and Anisha Jayadevan for her immense help with data analysis. RS was a JC Bose National Fellow during the tenure of this study. The authors express their sincere gratitude to the administrative staff, colleagues and friends at the National Institute of Advanced Studies, Indian Institute of Science, Asian Nature Conservation Foundation and the Foundation for Ecological Research, Advocacy and Learning for their constant support and encouragement during the long years of this study.

## References


Adhikari S, Southworth J, Nagendra H (2015) Understanding forest loss and recovery: a spatiotemporal analysis of land change in and around Bannerghatta National Park, India. J Land Use Sci 10:402–424

Agrawal PK, Verghese A, Radhakrishna S, Subaharan K (2016) Human animal conflict in agro-pastoral context: issues and policies. Indian Council of Agricultural Research, New Delhi

Altmann J (1974) Observational study of behavior: sampling methods. Behaviour 49:227–267

Anand VD, Varma S, Gopalakrishna SP (2009) Conservation of Asian elephant through mitigation of human-elephant conflict in Bannerghatta National Park, southern India. Technical report II: 2009. A Rocha India, Bangalore

Angel S, Parent J, Civco D (2007) Urban sprawl metrics: an analysis of global urban expansion using GIS. Int J Appl Earth Obs Geoinf 5:29–39

Balmford A, Moore JL, Brooks T, Burgess N, Hansen LA, Williams P, Rahbek C (2001) Conservation conflicts across Africa. Science 291:2616–2619

Barnes RFW, Barnes KL, Alers M, Blom A (1991) Man determines the distribution of elephants in the rain forests of northeastern Gabon. Afr J Ecol 29:54–63

Baskaran N (1998) Ranging and foraging behaviour of Asian elephants in Mudumalai Wildlife Sanctuary. Doctoral Thesis, Bharathidasan University, Tiruchirappalli, India

Baskaran N, Desai AA (1996) Ranging behaviour of the Asian elephant (*Elephas maximus*) in the Nilgiri Biosphere Reserve, south India. Gajah 15:41–49

Bettencourt L, West G (2010) A unified theory of urban living. Nature 467:912–913

Blake S, Hedges S (2004) Sinking the flagship: the case of forest elephants in Asia and Africa. Conserv Biol 18:1191–1202

Buij R, Mcshea W, Campbell P, Lee ME, Dallmeier F, Guimondou S, Mackaga L, Guisseougou N, Mboumba S, Hines JE, Nichols JD (2007) Patch-occupancy models indicate human activity as major determinant of forest elephant *Loxodonta cyclotis* seasonal distribution in an industrial corridor in Gabon. Biol Conserv 135:189–201

Calabrese A, Calabrese JM, Songer M, Wegmann M, Hedges S, Rose R, Leimgruber P (2017) Conservation status of Asian elephants: the influence of habitat and governance. Biodivers Conserv 26:2067–2081

Census (2011) District census handbook Karnataka and Tamilnadu. Ministry of Home Affairs, Government of India; Office of the Registrar General and Census Commissioner, New Delhi

Chelliah K, Sukumar R (2013) The role of tusks, musth and body size in male-male competition among Asian elephants, *Elephas maximus*. Anim Behav 86:1207–1214

Chen C, Park T, Wang X, Piao S, Xu B, Chaturvedi RK, Fuchs R, Brovkin V, Ciais P, Fensholt R, Tommervik H, Bala G, Zhu Z, Nemani RR, Myneni RB (2019) China and India lead in greening of the world through land-use management. Nat Sustain 2:122–129

Chiyo PI, Archie EA, Hollister-Smith JA, Lee PC, Poole JH, Moss CJ, Alberts SC (2011) Association patterns of African elephants in all-male groups: the role of age and genetic relatedness. Anim Behav 81:1093–1099

Chiyo PI, Moss CJ, Alberts SC (2012) The influence of life history milestones and association networks on crop-raiding behavior in male African elephants. PLoS One 7:e31382

Choudhury A, Lahiri Choudhury DK, Desai A, Duckworth JW, Easa PS, Johnsingh AJT, Fernando P, Hedges S, Gunawardena M, Kurt F, Karanth U, Lister A, Menon V, Riddle H, Rübel A, Wikramanayake E (2008) (IUCN SSC Asian Elephant Specialist Group) Elephas maximus. In: IUCN red list of threatened species 2008:e.T7140A12828813. https://doi.org/10.2305/IUCN.UK.2008.RLTS.T7140A12828813.en

Davis J, Pavlova A, Thompson R, Sunnucks P (2013) Evolutionary refugia and ecological refuges: key concepts for conserving Australian arid zone freshwater biodiversity under climate change. Glob Chang Biol 19:1970–1984

Druce H, Pretorius K, Slotow R (2008) The response of an elephant population to conservation area expansion: Phinda Private Game Reserve, South Africa. Biol Conserv 141:3127–3138

Eltringham SK (1990) Wildlife carrying capacities in relation to human settlement. Koedoe 33:87–97

Fay JM, Agnagna M (1991) A population survey of forest elephants (*Loxodonta africana cyclotis*) in Northern Congo. Afr J Ecol 29:177–187

Gaynor KM, Hojnowski CE, Carter NH, Brashares JS (2018) The influence of human disturbance on wildlife nocturnality. Science 360:1232–1235

Graham MD, Hamilton ID, Adams WM, Lee PC (2009) The movement of African elephants in a human-dominated land-use mosaic. Anim Conserv 12:445–455

GRASS (2007) GRASS 6.2 users' manual. GRASS Development Team, ITC-irst, Trento, Italy. https://grass.osgeo.org/development/

Happold DCD (1995) The interactions between humans and mammals in Africa in relation to conservation: a review. Biodivers Conserv 4:395–414

Hauenstein S, Fattebert J, Gruebler M, Naef-Daenzer B, Pe'er G, Hartig F (2019) Calibrating an individual-based movement model to predict functional connectivity for little owls. Ecol Appl 29:e01873

Hoare R, Du Toit J (1999) Coexistence between people and elephants in African savannas. Conserv Biol 13:633–639

Jathanna D, Karanth KU, Kumar NS, Karanth KK, Goswami VR (2015) Patterns and determinants of habitat occupancy by the Asian elephant in the western Ghats of Karnataka, India. PLoS One 10:e0133233

Johnsingh AJT, Williams AC (1999) Elephant corridors in India: lessons for other elephant range countries. Oryx 33:210–214

Kanagaraj R, Araújo M, Barman R, Davidar P, De R, Dinesh DDK, Gopi GV, Johnsingh AJT, Kakati K, Kramer-Schadt S, Lamichhane BR, Lyngdoh S, Madhusudan MD, Najar MUI, Parida J, Pradhan NMB, Puyravaud J-P, Raghunath R, Rahim PPA, Selvan KM, Subedi N, Trabucco A, Udayraj S, Wiegand T, Williams AC, Goyal SP (2019) Predicting range shifts of Asian elephants under global change. Divers Distrib 25:822–838

Krishnan V, Kumar MA, Raghunathan G, Vijayakrishnan S (2019) Distribution and habitat use by Asian elephants (*Elephas maximus*) in a coffee-dominated landscape of southern India. Trop Conserv Sci 12:1940082918822599

Kshettry A, Vaidyanathan S, Sukumar R, Athreya V (2020) Looking beyond protected areas: identifying conservation compatible landscapes in agro-forest mosaics in north-eastern India. Glob Ecol Conserv 22:e00905

Leimgruber P, Min Oo Z, Aung M, Kelly DS, Wemmer C, Senior B, Songer M (2011) Current status of the Asian elephant in Myanmar. Gajah 35:76–86

McKay GM (1973) Behavior and ecology of the Asian elephant in southeastern Ceylon, Smithsonian contributions to zoology, vol 125. Smithsonian Institution Press, Washington, DC

National Remote Sensing Agency (2006) Manual of national land use land cover mapping using multi-temporal satellite data. Department of Space, Government of India, Hyderabad

Nayak RR, Karanth KK, Dutta T, Defries R, Karanth KU, Vaidyanathan S (2020) Bits and pieces: forest fragmentation by linear intrusions in India. Land Use Policy 104619. https://doi.org/10.1016/j.landusepol.2020.104619

Okello MM, D'Amour DE (2008) Agricultural expansion within Kimana electric fences and implications for natural resource conservation around Amboseli National Park, Kenya. J Arid Environ 72:2179–2192

Parker ISC, Graham AD (1989a) Elephant decline: downward trends in African elephant distribution and numbers (Part I). Int J Environ Stud 34:287–305

Parker ISC, Graham AD (1989b) Elephant decline: downward trends in African elephant distribution and numbers (Part II). Int J Environ Stud 35:13–26

Parsons T (1949) Essays in sociological theory: pure and applied. Free Press, New York

Pimm S, Russell G, Gittleman J, Brooks T (1995) The future of biodiversity. Science 269:347–350

Pokharel SS, Seshagiri PB, Sukumar R (2017) Assessment of season-dependent body condition scores in relation to faecal glucocorticoid metabolites in free-ranging Asian elephants. Conserv Physiol 5:191–206

Poole JH (1982) Musth and male-male competition in the African elephant. Doctoral Thesis, University of Cambridge, Cambridge, UK

Poole JH (1987) Rutting behaviour in African elephants: the phenomenon of musth. Behaviour 102: 283–316

Project Elephant (2017) Estimation of wild elephant population in the year 2007 and 2012. Ministry of Environment, Forest and Climate Change, Government of India, New Delhi. http://www.moef.nic.in/division/introduction-4

QGIS (2009) Quantum GIS 1.4.0: geographic information system user guide. Open Source Geospatial Foundation Project. Quantum GIS Development Team. http://www.qgis.org/en/site/

Ramachandran R, Roy P, Vishnubhotla C, Sanjay J, Joshi P (2017) Long-term land use and land cover changes (1920–2015) in Eastern Ghats, India: pattern of dynamics and challenges in plant species conservation. Ecol Indic 85:21–36

Rasmussen HB, Okello JB, Wittemyer G, Siegismund HR, Arctander P, Vollrath F, Douglas-Hamilton I (2007) Age- and tactic-related paternity success in male African elephants. Behav Ecol 19:9–15

Santini L, González-Suárez M, Russo D, Gonzalez-Voyer A, von Hardenberg A, Ancillotto L (2019) One strategy does not fit all: determinants of urban adaptation in mammals. Ecol Lett 22: 365–376

Sih A, Ferrari MCO, Harris DJ (2011) Evolution and behavioural responses to human-induced rapid environmental change. Evol Appl 4:367–387

Simon RN, Fortin D (2020) Crop raiders in an ecological trap: optimal foraging individual-based modeling quantifies the effect of alternate crops. Ecol Appl 30:e02111

Sorokin PA, Zimmerman CC (1929) Principles of rural-urban sociology. Henry Holt, New York

Srinivasaiah NM (2019) Millennial male elephants of the Eastern Ghats. Hornbill (2):30–37

Srinivasaiah NM, Anand VD, Vaidyanathan S, Sinha A (2012) Usual populations, unusual individuals: insights into the behavior and management of Asian elephants in fragmented landscapes. PLoS One 7:e42571

Srinivasaiah NM, Vaidyanathan S, Sukumar R, Sinha A (2016) Landscape-level management of the Asian elephant through population- and individual-level monitoring in human-dominated landscapes of the eastern Ghats. Report submitted to the Karnataka Forest Department. National Institute of Advanced Studies, Bangalore

Srinivasaiah N, Kumar V, Vaidyanathan S, Sukumar R, Sinha A (2019) All-male groups in Asian elephants: a novel, adaptive social strategy in increasingly anthropogenic landscapes of southern India. Sci Rep 9:8678

Srinivasaiah NM, Vaidyanathan S, Sinha A (in press) Elephas flexibilis: behavioural plasticity in Asian elephants in a human-dominated landscape in southern India. In: Kumar A, Vasudev D (eds) Ecological and anthropological studies in India: the science of conserving wildlife. Springer India, New Delhi

State of Indian Agriculture (2016) State of Indian agriculture 2015–2016. Ministry of Agriculture and Farmers Welfare, Department of Agriculture, Cooperation and Farmers Welfare, Government of India, New Delhi

Stephens DW, Brown JS, Ydenberg RC (2007) Foraging: behavior and ecology. University of Chicago Press, Chicago

Sukumar R (1989) The Asian elephant: ecology and management. Cambridge University Press, Cambridge, UK

Sukumar R (2003) The living elephants: evolutionary ecology, behavior and conservation. Oxford University Press, New York

Sukumar R (2006) A brief review of the status, distribution and biology of wild Asian elephants *Elephas maximus*. Int Zoo Yearb 40:1–8

Sukumar R, Gadgil M (1988) Male-female differences in foraging on crops by Asian elephants. Anim Behav 36:1233–1235

Tuomainen U, Candolin U (2011) Behavioural responses to human-induced environmental change. Biol Rev 86:640–657

Venter O, Sanderson E, Magrach A, Allan JR, Beher J, Jones KR, Possingham HP, Laurance WF, Wood P, Fekete BM, Levy MA, Watson JEM (2016) Sixteen years of change in the global terrestrial human footprint and implications for biodiversity conservation. Nat Commun 7: 12558

Western D, Russell S, Cuthill I (2009) The status of wildlife in protected areas compared to non-protected areas of Kenya. PLoS One 4:e6140

Wong BBM, Candolin U (2015) Behavioral responses to changing environments. Behav Ecol 26: 665–673

Young KD, Ferreira SM, van Aarde RJ (2009) The influence of increasing population size and vegetation productivity on elephant distribution in the Kruger National Park. Austral Ecol 34: 329–342

**Nishant Srinivasaiah** currently based at the Foundation for Ecological Research, Advocacy and Learning in Bangalore, India, is interested in understanding behavioral decision-making in elephants. He has a doctoral degree, having studied the behavioral ecology of male Asian elephants in human-dominated landscapes, from the National Institute of Advanced Studies, Bangalore. He has also established and leads the Frontier Elephant Programme, a multi-institutional collaborative effort, designed to understand the lives and times of the Asian elephant in the Anthropocene.

**Srinivas Vaidyanathan** is a Wildlife Biologist with particular interest in understanding changes in landscape-level processes and structure, and how the same affect large mammal populations and distributions. He is Senior Fellow at the Foundation for Ecological Research Advocacy and Learning in Bangalore, India. He is also a guest faculty at the Master's course in Wildlife Biology and Conservation at the National Centre for Biological Sciences, Bangalore, where he has been co-guiding students for their Master's dissertations. Srinivas's expertise is varied, ranging from using a spatial approach to understanding various ecological and environmental questions to finding innovative and practical solutions to conservation problems.

**Raman Sukumar** (PhD, 1985) is professor of ecology at the Indian Institute of Science, Bangalore, India. His work originally focused on the behavior and conservation biology of elephants, expanding later to tropical forest ecology and climate change. Sukumar is Fellow of the Indian Academy of Sciences, the Indian National Science Academy, and the National Academy of Sciences, India. He has held visiting positions at Princeton University, Columbia University, and Kyoto University. He has contributed extensively to Indian government policy on nature conservation and has been a member of the Indian Board for Wildlife and the Project Elephant Steering Committee. He was Chair, IUCN/SSC Asian Elephant Specialist Group (1997–2004). He is the author of four books and over 200 scientific papers, and the recipient of several awards.

**Anindya Sinha** is based at the National Institute of Advanced Studies in Bangalore, India, and is an adjunct faculty at nine other institutions in India and the UK. Although he earned a doctorate in molecular biology in 1993 from the Tata Institute of Fundamental Research in Mumbai, India, his current research interests are primarily in the areas of animal behavioral ecology, cognitive ethology, evolutionary biology, population and behavioral genetics, urban ecologies, animal studies—particularly of primates, elephants, and other mammalian species, philosophy of biology, performance studies, and the intangible as well as tangible cultural heritage of India.

# NFUA and Biodiversity: Current State of the Research and Potential Opportunities 17

Harpreet Kaur

**Abstract**

The relationship between cities and biodiversity is not well understood in theory or in practice. Nonetheless, cities influence biodiversity by influencing the flora and fauna living within urban centres, of surrounding areas, and of far-away places directly or indirectly. With more than 50% of global population living in cities, it is imperative to understand how biodiversity is conserved in urban areas and how it influences the provision of ecosystems services. City landscapes are constrained in terms of land availability, and maintaining green or vegetative areas for provision of ecosystem services and biodiversity conservation is challenging. Urban agriculture systems exist in many forms and vary from low tech traditional systems such as allotment gardens to high tech rooftop hydroponic systems. Because of the wide variation in vegetation cover, diversity of species, management, and structure, urban agriculture (UA) can exhibit high levels of biodiversity. This chapter documents the motivations behind conserving biodiversity in urban centres and cites examples from various cities. Next, the chapter moves to understand the relation between various forms of UA and biodiversity. The examples of various forms of UA from different cities described here enrich the discourse on the role of these systems in food security within urban spaces, in biodiversity management and conservation, and provision of ecosystem services.

**Keywords**

New forms of urban agriculture · Biodiversity conservation in cities · Biodiversity and urban agriculture · Ecosystem services

H. Kaur (✉)
Agroparistech, Paris, France

 
J. A. Diehl, H. Kaur (eds.), *New Forms of Urban Agriculture: An Urban Ecology Perspective*, https://doi.org/10.1007/978-981-16-3738-4_17

## 17.1 Introduction

Cities are global nodes of consumption, and with the expansion of existing cities and emergence of new ones, material consumption is predicted to grow even faster. An urban area can be called a 'city proper' depending upon its administrative boundaries, an 'urban agglomeration' which is the extent of the contiguous urban area, or a 'metropolitan area' where boundaries are defined by the degree of economic and social interconnectedness to nearby areas. The proportion of people living in cities (city proper, urban agglomerations, or metropolitan area) is increasing all over the world especially in Southeast Asia, Africa, and Middle East. Around 55% of the world's population—4.2 billion people—live in cities today and this number is projected to double its current size by 2050 (DESA 2016). This increasing number represents both an increase in the migration of people to cities and increase in the urban area. Consequently, a substantial increase in agricultural production is needed to feed the increasing city population. This demographic shift has many profound implications on the food security of people living in urban as well as rural areas and on the structure and functioning of food production systems. Morgan and Sonnino (2010) estimate that the increase in GDP and incomes especially in the global south will change the rate of resource consumption. For example, the increase in food consumption, e.g., meat and dairy, for 9 billion people may be equivalent to that of 12 billion. Food security has officially become a matter of national security. Growing conflicts for natural resources and climate change further exacerbate this issue.

There is ample scientific evidence that human activities are destabilizing the earth's global climate and ecosystems. The exponential growth of human activities constitutes a dominant driver of change to the earth's system such that scientists believe that earth has entered a new epoch: the Anthropocene. Urbanization and conventional input-intensive agricultural practices to support the urban systems are primary drivers of this change. This raises concern as further pressure on the Earth System could destabilize key biophysical systems and trigger irreversible changes. Regrettably, the predominant model of social and economic development remains largely unaware of the environmental risks induced by human on continental to global scales. Rockström et al. (2009) proposed the concept of 'planetary boundaries' (PBs)—a safe operating space for humanity with respect to the functioning of the Earth System. The authors deliberate the non-negotiable planetary conditions that we need to recognize and regard to avoid the risks of catastrophic environmental change. Nine PBs are defined: climate change, ocean acidification, stratospheric ozone, global nitrogen and phosphorous cycles, atmospheric aerosol loading, freshwater use, land use change, biodiversity loss, and chemical pollution. Almost half of the world's total GDP, i.e., $44 trillion is potentially at risk because most of the businesses are dependent on nature and ecosystem services. Loss of biodiversity and ecosystem function are among the top five threats to humankind in the next decade.

The concerns related to impacts of climate change on food production and volatility of food prices on food accessibility are growing among national and international political and humanitarian institutions. Whether it is sustainable intensification to increase food production or development of alternate forms of farming based on organic and agroecological principles, new paradigms are emerging to meet the food requirements in environmentally friendly ways. The driving principles of these paradigms are to build diverse supplies of food that are close to population centres and improve local management of food systems. As much of the urban food requirement is supplied by rural farms, the need to develop urban and peri-urban food systems to complement the rural food supply is being recognized in urban development policies. There is a growing recognition that urban agriculture (UA) has the potential to suffice for the inadequate food access in cities. The benefits of maintaining traditional forms and developing new forms of UA in an urban setting are many—food accessibility, reduction in food miles, contribution to household income, creation of jobs, biodiversity conservation, etc. Agriculture has existed in urban areas for a long time but increasing competition for land and development of supply chains for food availability has replaced farmed land with built-up area. With the acknowledgement of ecosystem services provided by green spaces and UA structures such as allotment gardens, community supported agriculture, and rooftop gardens, city development plans aim to incorporate various forms of UA in cities.[1]

This chapter provides an overview of the advantages of maintaining various forms of food and non-food UA systems. Section 17.1 focuses on why biodiversity in an urban area should be conserved and common constraints for its conservation. The section also presents examples from various cities on the new approaches undertaken to maintain and develop green spaces for improving floral and faunal biodiversity. Section 17.2 summarizes traditional and new forms of UA and their role in conservation of agri- and related biodiversity. In general, examples of various forms of UA from different cities are described to enrich the discourse on the role of these systems in food security within urban spaces, in biodiversity management and conservation, and provision of ecosystem services.

### 17.1.1 Constraints and Opportunities for Biodiversity in the Context of Urbanization

Cities should be looked at as places with a high potential for biodiversity conservation (de Oliveira et al. 2011). They are complex hotspots and melting points where a unique biological gradient occurs such that even rare species can be found in them. There are three main reasons for high biodiversity in cities: (1) many urban areas were originally established at ecosystem junctions such as on rich flood plains where agriculture was possible, (2) cities often include relicts of natural habitats such as forests or meadows, and (3) cities provide habitats that are distinctive and dynamic, providing exclusive physical and ecological conditions for a combination of local and exotic species (Given 2000). However, owing to urban growth and urbanization, cities are also the places where biodiversity faces the greatest challenges.

[1] About—C40 Cities, https://www.c40.org

United Nations Population Division (UNDP) makes a clear distinction between urbanization and urban growth. Urban growth is characterized by an increase in the size of the urban areas in terms of population and infrastructure influenced by the internal dynamics, making it a one-way process. On the other hand, urbanization is a two-way process characterized by the population living in an urban area. It is the movement of people from rural to urban areas along with the changes in the perceptions of migrants in terms of attitude, beliefs, and behaviour patterns within the urban boundaries. Either way, urbanization and urban growth are 'demanding' in terms of utilization of natural resources and are almost always associated with having a negative influence on the ecology within the urban bounds. The impacts arising from cities also vary with their economic status. For example, Hoornweg et al. (2016), while developing a methodology to assess a city's impact on biodiversity, estimated that cities whose residents consume products grown on lands cleared in the ecologically sensitive parts of the world have a higher negative biodiversity index. In effect, the impacts on biodiversity are reflected by the larger global purchasing power of cities. This and similar studies provide important scientific evidence that urban areas not only influence local habitats but also ecosystems that are situated elsewhere.

Urban areas in their very distinct structural and ecological make-up provide unique opportunities for biodiversity conservation as well. There is mounting evidence that urban and sub-urban areas can contain high levels of biodiversity (Alvey 2006). Forms of green spaces such as vegetation along roads, urban parks, woodlots, abandoned sites, along with various forms of UA positively impact the biodiversity in city space including conservation opportunities for rare species. In case of undisturbed or natural ecosystems, the definitions of biodiversity are well-defined and straight-forward. Convention on Biological Diversity (CBD 1994) defines biodiversity as, '*the variability among living organisms from all sources including, inter alia, terrestrial, marine and other aquatic ecosystems and the ecological complexes of which they are part; this includes diversity within species, between species, and of ecosystem*'. However, defining biodiversity in human-dominated systems can often be difficult and controversial. For example—the term 'agriculture biodiversity' became widely used only after the 1992 Convention on Biological Diversity, but it was only after CBD Conference of Parties 3[2] that a solid foundation was built for the concept of agricultural biodiversity. Nonetheless, the understanding of agro-ecosystems and agro-biodiversity as we know today developed over three decades. The present definition of agro-biodiversity not only takes into account the genetic and the species-level biodiversity but also the cultural diversity influenced by human interactions. It has spatial, temporal, and scale dimensions (FAO Netherlands Conference Paper). Food and Agriculture Organization (FAO 1999) defines agro-biodiversity as, '*The variety and variability of animals, plants and micro-organisms that are used directly or indirectly for food*

[2]UNEP/CBD/COP/3/38 Report of the Third Meeting of the Conference of the Parties to the Convention on Biological Diversity.

*and agriculture, including crops, livestock, forestry and fisheries. It comprises the diversity of genetic resources (varieties, breeds) and species used for food, fodder, fibre, fuel and pharmaceuticals. It also includes the diversity of non-harvested species that support production (soil micro-organisms, predators, pollinators), and those in the wider environment that support agro-ecosystems (agricultural, pastoral, forest and aquatic) as well as the diversity of the agro-ecosystems*'.

Similarly, defining biodiversity in an urban context is challenging especially taking into consideration the presence of exotic species introduced by humans, which often dominate and thrive well in urban spaces. Studies report that sub-urban and peri-urban areas often have a higher species diversity than the native ecosystems they replace because of the increase in resource availability, heterogeneity, and management (Alvey 2006). Such an increase in species diversity can be considered valuable to maximize the exposure of urban residents to biodiversity, but this perception may be less appreciated if the intention is to conserve native and local species. Keeping these conservation dilemmas in mind, I discuss motivations that often drive conservation of biodiversity in an urban area in the following section.

## 17.2 Motivations for Biodiversity Conservation in Urban Areas

### 17.2.1 Providing Environmental Education by Connecting People to Nature

With more than half the global population residing in cities, the importance of natural areas within the urban space cannot be ignored. City residents have most of their contact with nature in the urban green spaces in the form of parks, urban farms, conserved relict patches of forests, green belts around the city boundaries, etc. Such areas provide an opportunity to impart knowledge of local ecology and conservation to a large number of people who lack the means or motivation to travel to nature reserves for learning about wildlife. There is a growing realization that education and outreach programs are a cornerstone towards wildlife and habitat conservation (Dearborn and Kark 2010). However, conservation within urban landscapes where the human demography is least exposed to natural ecosystems is still not well developed. Nonetheless, several cities have created or are in the process of creating conservation areas for native vegetation and wildlife. Many cities across the globe have initiated restoration and stewardship programs that also include significant community engagement. Research on the role of such programs in increasing the understanding of city dwellers about native biodiversity and ecosystem services they provide is still in initial stages. However, the existing literature shows a positive impact of urban environmental education on the attitudes of people towards their natural surroundings. Most of the programs on environmental education projects are local in nature, their outreach can be leveraged by developing global partnerships with non-governmental organizations, businesses, and UN Education for Sustainable Development. Box 17.1 presents an example of development of biodiversity parks in Delhi along the floodplain of Yamuna river—a project brought to fruition by collaboration between city's development authority and research universities. The Biodiversity Park is not only an important site for research studies on ecological rehabilitation within a city boundary but also for environmental education of local residents and children and ecotourism.

**Box 17.1 Yamuna Biodiversity Park, Delhi, India**

Yamuna Biodiversity Park (YBP) was established in 2002 to restore the dwindling native biodiversity of Delhi. Delhi Development Authority (DDA) in collaboration with University of Delhi developed the park to restore and conserve the natural habitat of local flora and fauna and provide recreational and educational services to Delhi residents. The park is a combination of wetlands and forest communities located on the West bank of Yamuna River spread over 457 acres of land surrounded by dense human settlements. YBP as of today is home to some 2000 species of plants and animals. The development of the park was in two phases—Phase I covered an area of 157 acres on the inactive floodplain of Yamuna River and Phase II covered an area of 300 acres on the active floodplain. Both zones are connected by a 200-m-long corridor to allow animal mobilization for foraging during hot and monsoon season, which often culminates in flooding. The park has emerged as an important educational and research Centre and is the capital's most visited park. YBP consists of a nature interpretation Centre with an amphitheatre and auditorium for events, Bambusetum, conservatory for fruit yielding plants, wetlands for migratory birds, herbal garden, sacred grove, butterfly conservatory, and a nature reserve.

The butterfly conservatory is a major ecotourism destination. The plantation of native plant species has attracted numerous butterflies and 36 species can be sighted here. The sacred grove is home to 30 species of Ficus—a tree of religious importance in northern India. There are two wetlands—one is long and shallow simulating the river and the other is deep and wide. The presence of wetlands improved the biodiversity of YBP and provide important ecosystem services of water purification, storage of rainwater and ground water recharge, and are home to hundreds of native and migratory birds (more than 5000 migratory ducks from Siberia, Central Asia, and Europe visit YBP wetlands each year in winter). Along with this, the wetlands help to preserve aquatic fauna such as invertebrates, turtles, fish, and plankton. The nature reserve zone is reserved for conservation purposes and not open to visitors. Within an area of 130 acres, 20 different biotic communities with three trophic levels have been developed in YBP using massive plantation programmes. A number of animals such as monitor lizard, nilgai, jungle cats, and civets, which disappeared from this region due to rapid urbanization, now inhabit YBP. The park which is in various stages of development provides an ideal location to learn about local biodiversity and redevelopment of ecosystems using bioremediation techniques.

### 17.2.2 Preserving Important Local Biodiversity

Urban landscapes are becoming increasingly large. World Bank, 2015 reports that more than 60% of land is projected to be developed by 2030. Ninety percent of this growth will happen in the developing countries such as Asia and Sub-Saharan Africa and almost one million housing units will be constructed by the year 2060 to accommodate the projected urban population. Many cities were originally established along river floodplains, ecological transition zones or locations rich in endemic species; their importance in conservation of regional and global biodiversity cannot be overlooked (Luck 2007). As the land use and land cover change modifies wilderness areas to agricultural fields or asphalted urban landscapes, cities are becoming an important refuge for wild life. A number of studies report that population of many species are recurring in urban spaces—they not only include native species or urban exploiters but also migratory birds, insect species that rely on green patches in the cities to navigate through monocropped agricultural patches (LaPoint et al. 2013; Tam and Bonebrake 2016; Lewis et al. 2019). It is well established that cities need natural capital to function—whether it is located far away from cities or within built-up area to provide the critical ecosystem services for a balanced, sustainable lifestyle. However, whether such a dependence exists for nature or biodiversity to thrive—is an important question. There is a growing evidence from the conservation studies that urban areas play an important role in maintaining and conserving biodiversity and that the conservation efforts in cities are essential. Stressing on the importance of conservation in urban areas, The Intergovernmental Science-Policy Platform on Biodiversity and Ecosystem Services (IPBES et al. 2019) report on global assessment of biodiversity highlights that we have a small but critical window to create functional landscapes that benefit both people and biodiversity. Boxes 17.2 and 17.3 represent two examples of biodiversity conservation efforts within densely populated cities.

**Box 17.2 The City of Cape Town, South Africa**

Cape Town is globally known for its high beta and gamma diversity,[3] endemic species, and plant species richness. It is an important part of Cape Floristic Region (CFR) which is a biodiversity hotspot and a World Heritage Site. CFR is a member of the Mediterranean Biome and has the second highest population density and growth in the Biome. CFR faces high levels of biodiversity loss and is a global priority for conservation. Seventy percent of the population

(continued)

[3] Biodiversity can be measured at the three levels—Alpha, Beta, and Gamma diversity. Alpha diversity is *within habitat or intra-community* diversity. It has two components—species richness and evenness. Beta diversity is *between habitat and inter-community* diversity. It measures change in species composition along a gradient. Gamma diversity encompasses diversity at the *landscape level* (Whittaker 1972).

**Box 17.2** (continued)

of Cape Town Province lives in Cape Town, and the city economic activities contribute to 11% of the GDP of South Africa. Biodiversity conservation within the city is a challenging task. Although agriculture in the city has important negative impacts on biodiversity such as competition for land, fragmentation of habitats, and presence of alien species, the presence of farming has contributed to less dense urban centres. The Municipal Spatial Framework recognizes the positive impact of retention of green spaces and agricultural fields in reducing highly dense urban structures. However, the presence of small fragmented sites in an urban area often poses challenges in conserving animals with large home ranges. This is typically the case of Cape Town as well where numerous reserves are required to conserve highly localized species. Presence of agricultural land or open, unused spaces ensures the connectedness of such reserved sites. Nonetheless, conservation of wildlife within city boundaries often can lead to human wildlife conflict, which is the case in Cape Town. Partnership programs such as Dassenberg Coastal Catchment Partnership are landscape scale programs involving a number of stakeholders (state, communal, and private) for conservation of native biodiversity within the urban bounds. It is important to realize that not all biodiversity conservation occurs within such protected areas. Many species are not confined to protected areas and the presence of green spaces, kitchen and home gardens, green spaces along roads provides microhabitats for their survival and conservation. For example, in the city of Cape Town, the endangered Leopard Toad spends a significant proportion of its life in urban green spaces.

**Box 17.3 Green Roofs on Bus Stops in Utrecht**

Undoubtedly, the traditional forms (as described in Boxes 17.1 and 17.2) of biodiversity conservation can be adapted to save, maintain, or increase wildlife in urban areas. However, provisioning space for such structures within the already built-up areas may not always be possible. Pockets of green spaces can be installed in innovative ways to support invertebrates and birds within concretized landscapes. The city of Utrecht, Netherlands focuses on establishing green roofs not only for capturing fine particulates of air pollution and collect rainwater but also providing islands of foliage for small animals. Residents who want to install green roofs on their houses can avail grants to cover 50% of the costs. The idea is to create a vertical forest wherein apartments of varying sizes can be nestled among the plants. It is often difficult to access green spaces in city centres where buildings are packed together and housing structures usually replace parks and gardens. Taking it a step further,

(continued)

**Box 17.3** (continued)
the city of Utrecht gave 316 bus stops a green makeover! Bus stop roofs are covered with sedum and other flower species to act as local bee sanctuaries. Use of sedum flowers in creating green spaces has gained popularity because of their low maintenance.

### 17.2.3 Understanding How Species Respond to Rapidly Changing Environments

Defining biodiversity of a natural ecosystem is relatively easy compared to describing it in an urban context. In human-dominated systems, exotic species thrive well because of their management, absence of competitors, predators, and pests. On the other hand, many native species succumb to the dramatic changes in the urban environment and ecology and only a subset can cope up with such shifts (Kark et al. 2007; McDonnell and Hahs 2008). A number of studies have explained this phenomenon with the change in bird diversity in urban environments. Birds like butterflies serve as indicators of environmental change; they are vulnerable to land use changes such as habitat fragmentation, conversion of green spaces to built-up areas, and changes at the lower trophic levels (bioaccumulation of pesticides, changes in prey population), and are easy to monitor for population level changes. In general, it has been shown that the bird diversity declines along an increasing urbanization gradient (Faeth et al. 2005; McKinney 2006). A gradient exists from relatively undisturbed areas outside urban agglomeration to the city core. In those areas with relatively high native vegetation, 'urban avoiders' dominate the landscape—these are mostly native species. In peri-urban environments with increasing levels of urbanization, native as well as non-native species live and are referred to as 'sub-urban adapters'; these areas have a higher species richness.

Blair (1996) in his study on the distribution of birds across an urban gradient in California, categorized birds into three types: urban avoiders, sub-urban adapters, and urban exploiters. Areas with an intermediate level of urbanization have been reported to show richness peaks in several studies (Crooks et al. 2004). Lastly, 'urban exploiters' are the species that exist in the most urbanized area—the city core with built area. These are mostly non-native species adapted to urban environments, also referred to as 'urban commensals' by Shochat et al. (2006) because of their dependency on urban resources. The local native species are few in number because the built-up and paved spaces disrupt the native habitat. Similarly, studies conducted on plant communities across an urban gradient show that land under urban development had the greatest number of non-native species. Construction of houses, roads, and trails increases the prevalence of non-native species by acting as corridors for their dispersal.

In a broader sense, it can be said that urban ecosystems can be used as models to understand the impacts of environmental change on the behaviour of species.

Moreover, the presence of natural habitats in urban areas may provide insights to conservationists to anticipate and mitigate climate change impacts. There is a growing body of evidence that traditional approaches to wildlife and biodiversity conservation have limited success in urban areas. Rosenzweig (2003) argued that human-dominated landscapes that are completely different from their natural counterparts can also be ecologically valuable without being pristine or wild. The conventional wilderness based conservation ideology is entrenched in the idea that biodiversity and heavy human presence are incompatible. On the other hand, the principles of reconciliation ecology that see a merging of conservation ideas within human-dominated spaces are largely necessary to guide the conservation practices in urban agglomerates. It is true that it may be impossible to completely protect or recreate an ecosystem that looks and functions like the native ecosystem that the city replaced. In an urban system, where the majority of human population lives and works, the primary objectives of biodiversity conservation differ from, say, a forested area. The more realistic conservation aspects would be to decide what biodiversity and ecosystem functions are needed, what purposes they serve, and which species assemblages can achieve that. It is true that such an approach can lead to 'engineered urban ecosystems' dedicated to one or a set of ecosystem services or it can also lead to emergence of novel ecosystems at equilibrium. Either way, urban systems present unique logistical hurdles that require creativity in setting conservation goals and in the methods used to attain them (Hobbs et al. 2006).

### 17.2.4 Ecosystem Services

Millennium Ecosystem Assessment (MEA) (2005, p. 27) defines an ecosystem as, 'a dynamic complex of plant, animal and micro-organism communities and their non-living environment interacting as a functional unit'. And ecosystem services (ES) are 'the benefits that humans obtain from ecosystems'. The ES approach includes all the benefits: ones perceived by humans as 'required or beneficial' and others that are not perceived or quantified. Nonetheless, the definitions and evaluations of ESs are anthropocentric. The interest in the quantification of ES has grown in the past few decades partly because of the recognition of the impacts of human growth on the ability of natural systems to sustain it and the rate at which natural ecosystems are being modified or even lost to human intervention. MEA (2005) classifies ESs into four main categories: provisioning services (provision of all the food and raw material required for human existence), regulatory services (climate regulation, erosion control), supporting services (maintenance of biogeochemical cycles), and cultural services (recreation, aesthetics, spiritual, and religious experience).

Cities, conventionally viewed as regions of economic growth with limited space available for developing areas for protecting biodiversity and ES, are largely dependent on natural hinterlands existing elsewhere to take care of the inputs required to run the city functions and manage its outputs. However, provision of ES locally provides solutions to many problems that face cities such as urban heat island effect, noise reduction, provision of clean air. ES can be generated within cities or outside cities. For example, development of urban parks provides regulatory services such as noise and air pollution regulation, regulating temperatures, provision of spaces for recreational and education activities. Trees and vegetation lower air temperatures by providing shade and evapotranspiration—the location and structure of vegetation are, however, important. Shade and evapotranspiration have been shown to reduce peak summer temperatures by 1–5 °C. Planting and marinating vegetation incurs costs which includes initial planting, pest control, pruning, etc. A study of urban forestry programs in US cities shows that the annual expenditure on maintaining a single urban tree was roughly US$15–65. The net benefits obtained from the trees in the form of ESs ranged from US$30 to 90 per tree. But benefits vary considerably by tree species and the biodiversity they support (Bolund and Hunhammar 1999; McPherson et al. 2005).

Regrettably, the value of ESs is not adequately quantified in monetary terms. Therefore, benefits arising by adopting ES or losses incurred from their disappearance are often not incorporated in urban development. However, the scenario is changing as many cities now invest in green infrastructure which focuses on use of sustainable material and augmentation with green roofs or rooftop gardens to enable the provision of some ESs. The Economics of Ecosystems and Biodiversity (Berghöfer et al. 2011, TEEB[4]) calls for an urgent need to improve or develop tools for assessment of economic and non-economic benefits of ES for incorporation in city planning. With increasing urban expansion and densification, mechanisms for improving biodiversity and ES need to be identified and incorporated in land use planning with active participation form local governments and regional and national policies.

## 17.3 Urban Agriculture and Biodiversity

UA has many benefits; this book and plenty of literature available highlight the ways UA provides a sustainable sustenance within an urban setting. But how does it contribute to adding biodiversity and/or conserving the species already present in an urban ecosystem? The following section focuses on the types of urban farming systems, the crops grown, and other floral and faunal biodiversity supported by them. Various types of production systems exist in UA among which growing crops (with a focus on horticulture crops) are more common compared to rearing animals. A wide range of crops are grown depending upon the season such as leafy greens (spinach, pak choy, lettuce), cauliflower, cabbage, okra, eggplant, beans, cassava leaves, pea, potato, squash, and many more. Non-seasonal vegetables are also grown in hydroponic and aeroponic systems. In addition to food crops, farmers also grow aromatic and flowering plants because of their high demands in cities. Fruits such as

[4]TEEB Manual for cities (2011).

banana, strawberries, melon, oranges, peaches, strawberries are also grown but most of them are usually used for self-consumption.

New forms of urban agriculture (NFUA) along with traditional UA systems are capable of producing enough to support the needs of vegetables for the city dwellers. For example, the cities of Melbourne, Adelaide, and Sydney are able to meet the requirement of some of the city's vegetables and fruits needs through UA (Mok et al. 2014). Up to 98% of cauliflower in Adelaide comes from permaculture community farms and commercial peri-urban facilities. This is also the case for strawberry production in Melbourne—urban and peri-urban production of strawberries fulfils 97% of strawberry requirement in the city. Farming in cities is not a new phenomenon, a diversity of production systems exists. For simplicity, I divide the types of cultivation systems into traditional systems that have exist in and around cities and NFUA that use innovative farming systems. They have been categorized depending on the methods used, crops grown, and optimization of crop production in limited spaces.

## 17.3.1 Traditional Urban Farming Systems

### 17.3.1.1 Cultivated Open Spaces-Limited Acreage Urban Farms

Soil based urban farms belong to this category. Such farms can be a relic of traditional bigger farms, orchards, undeveloped floodplains under agriculture, community gardens, backyard or home gardens, farming on lands temporarily dedicated to horticulture activities, etc. The crops grown in such farms are season-dependent ranging from leafy greens to various types of beans, cucurbits, solanaceae, and tubers. Cereals such as wheat, barley, corn, rice are also common in such farming systems and fruit orchards (mango, guava, papaya, avocado, coconut, banana) also exist. Depending on the region, seasonal ornamental and flowering plants are also grown. Such systems are typically very diverse and the management practices can vary depending upon the family structure, economic status, and size of the farms. In general, intercropping is frequent to increase the number and amount of product harvested per unit area and open pollinated cultivars are grown. Most of the open space farms exist on river floodplains in developing areas.

#### Allotment Gardens and Community Gardens

Allotment Gardens (AG) and Community Gardens (CG) are important features of the urban landscape. These urban green spaces provide numerous benefits that are not restricted to those of food production only. Besides performing regulatory functions such as local climate modulation, carbon sequestration, AGs and CGs have a high cultural significance. AGs can be considered a subtype of CG which has a specific spatial features and social and economic organization. In contemporary terms, the underlying value of AGs is recreation, their origin was primarily to fulfil some economic purpose such as feeding the poor or a form of social welfare system. In fact, AGs date back to Elizabethan times when the term first appeared. At that time, AGs were small parcels of land given as a compensation to farmers who had lost their land to wealthy land lords and had no means of earning a livelihood (Allotment.org.uk 2012). Their role assumed importance during WW I and WW II, especially during WW II AGs were used to increase food production and lessen the demand of commercial food supplies and transportation facilities (Garnett 2000). However, with increasing urbanization and competition for infrastructure development, many AGs have been replaced by residential buildings, industries, etc.

AGs are best known for creation of microhabitats within the urban built-up space. Some of the most common forms of microhabitats are ornamental or vegetable plant beds, ponds, areas dedicated to fruit trees and bushes. Urban spaces in general have a lower faunal biodiversity, only a few taxonomic groups of birds and insects can be found in AGs. On the other hand, floral biodiversity including that of plant species enlisted in IUCN Red List as well as other vascular plants and agro-biodiversity in terms of in situ conservation is quite high. See Müller et al. (2013), Gilbert (2013), and Borysiak et al. (2017) for detailed floral biodiversity assessment in AGs. Different utilitarian forms of AGs exist such as vegetable cultivated, vegetable cultivated and ornamental, only ornamental cultivated, and abandoned or fallow. In general, they are 'wild-life friendly' labour dependent gardening systems where crops are fertilized mostly with compost and fertilizers are used in very limited amount for ornamental plants. Use of pesticides and herbicides is only occasional. According to conservationists, AGs should be considered hotspots for native species as such green infrastructure elements limit biotic homogenization (Bell et al. 2016).

Most of the studies focussing on ES provided by AGs have been done in Europe. In 0.75 ha of land under allotment gardens with low to high levels of management intensity, 290 species of edible and nonedible plants were found. Among these were 150 spontaneous species[5] and none of them was ornamental plants. Over 2000 crops and ornamental species of plants have been recorded in AGs in Germany (Speak et al. 2015). Both native and non-native species are present in AGs, the abundance of these species is however dependent on the management intensity of the garden (Smith et al. 2006; Speak et al. 2015; Borysiak et al. 2017). Not only are such studies few in number, the AGs studied have variable plot sizes, management practices, and cultural contexts. These factors make it difficult to ascertain scale of biodiversity and crop productivity. Nonetheless, AGs along with CGs are important green spaces within an urban landscape which nurture biodiversity related to agriculture—a feature usually ignored in conservation in cities.

CGs are large plots of land owned by the municipality, local institutions, a land trust, or a community. They are divided into subplots for cultivation by individual households. CGs or Community Supported Agriculture (CSA) are a well-known form of farming within urban boundaries in the USA and Canada. CGs have become an important way to combat food insecurity not only resulting from unaffordability due to poverty but also inaccessibility of quality food, i.e., presence of food deserts. In Australia the first CG was developed in 1977 in the city of Melbourne. CSAs though quite successful in the USA and Canada are not very popular in Australia. In

[5] Spontaneous species are those plant species that reproduce spontaneously in the gardens.

total there are around 220 CGs in the country and most of them are linked with permaculture projects. Permaculture differs from CSA in design—it aims at mimicking the ecosystem structures by building guilds that follow stages of plant succession. The primary focus of permaculture is to increase biodiversity and ES. CSA, permaculture, and commercial peri-urban agriculture generate about 25% of Australia's total agriculture production. Interestingly, the total area under these forms of agriculture is less than 3% of the land under conventional agriculture in five mainland states of Australia. The food production is particularly significant in Adelaide, Melbourne, and Sydney. Almost all of Sydney's Asian vegetable requirements are sourced from UA and peri-urban agriculture. Strawberries, peaches, mushrooms, lettuce, Asian vegetables, cauliflower, cabbage are some of the produce sourced mainly from farming within and around city boundaries for the states of Victoria, New South Wales, Queensland, South Australia, and Western Australia (Mok et al. 2014). The farmers' market is a relatively new phenomenon in Australia and there are 165 such markets across the country (Australian Farmers' Market Association 2012).

Most of the research literature on AGs and CGs comes from the USA, Europe, and Australia and the few studies that document their productivities report highly variable results. For example, Vitiello et al. (2009) estimated that gardeners were able to produce 6.84 kg/m$^2$ of tomatoes—extremely high yield given that tomatoes are a highly productive vertical crop. On the other hand, National Gardening Association (2009) reported a yield of 2.44 kg/m$^2$ in New York City similar to yields in Oakland CGs of around 2 kg/m$^2$ (McClintock and Cooper 2010). Algert et al. (2014) found the average yield in CGs in the city of San Jose to be 3.66 kg/m$^2$, which is closer to bio-intensive farming practices[6] than conventional agriculture. Yields for vertical crops such as tomatoes and green beans per square land area are higher. As there are multiple crops in each bed in case of AGs and CGs, reporting the weight produced per plant provides more information than per square land area and is a useful method to compare vegetable production between plots.

## Home Gardens

Home gardens (HGs) are green spaces around houses used for cultivation of ornamental plants, crops, fruit trees and are managed by the family. In an urban context, they are considered homologous to AGs in relation to the biodiversity and ES provided. HGs are considered as important in situ conservation sites that provide highly diverse ecological niches to floral and faunal species. In a traditional sense, HGs are shaped by the interaction between ecological and cultural entities. Galluzzi et al. (2010, p. 3637) refer to HGs as 'crucial reservoirs of agro-biodiversity, both at

[6]Bio-intensive agriculture is a sustainable organic farming system. Yields are optimized based on working with the basic elements needed for plant growth—soil, water, air, and sun and special focus is placed on increasing biodiversity and soil fertility. This form of farming is considered very suitable for small-scale farmers. This is because with minimal financial input farmers can harvest great amounts of produce from a small piece of land.

inter- and intra-specific taxonomic level resulting from an amalgamation of environmental and cultural contexts'.

HGs are usually present in areas with low population density and are demarcated from their surroundings by barriers such as hedges and wires. In rural, agricultural dominated systems, HGs act as a refuge to micro- and macro-fauna for pollination by providing them specialized and safe micro-climate. In this way they aid in maintaining the sustainability of the larger agricultural system surrounding them. Unlike rural systems, HGs in urban areas are not surrounded by agricultural fields, they interrupt the built-up continuum by providing pockets of greenery that contribute to improving air quality, regulate temperatures, absorb carbon dioxide emissions, and have a recreational value. The sizes of HGs vary depending upon the competition for land, proximity to city centre, etc.

Traditionally, HGs resemble an agro-forestry system with their multi-layered structure. Different plant species exist in temporal and spatial succession and this stratified structure makes HGs a resilient system (Smith et al. 2006). Moreover, a minimal use of pesticides protects this natural micro-habitat for wild or native floral and faunal species. Soil microbial biodiversity is also high. Food, fodder, medicinal, ornamental, fuel, plants with a religious value can be found in a typical HG. Some researchers also report the presence of 'relic' crops in HGs that are no longer cultivated in commercial agriculture such as lima beans in Cuban and sponge gourd in Nepal (Hodgkin 2002). See Galluzzi et al. (2010) for the variety of local, wild, or weedy relic plant diversity in HGs from different countries. The genetic diversity of HGs is often a reflection of the socio-economic status of the household and this correlation has been proven for HGs in China. Nonetheless, the nature of this relationship is highly variable and many studies indicate that biodiversity is influenced by people who tend HGs. For example, HGs with local varieties were mostly nurtured by elderly household members or women (Brush 2000; Sordi et al. 2008).

#### 17.3.1.2 Intensive Horticulture Systems

Intensive horticulture systems are cultivated open spaces that have been categorized differently from other UA types because of the intensive farming practices used to grow crops. These types of farming systems are quite common in and around cities of developing nations. Farming is conducted on land owned or rented by the producers. For example, in case of Delhi, 954 ha of area along the Yamuna river is under cultivation (National Green Tribunal, India 2020). In a study conducted by Cook et al. (2015), majority of the surveyed farmers had been practising agriculture for more than 20 years. Some cross-generational farm plots also exist but most of the farmers are migrants paying rent to private owners or occupying governmental land. Farmers here grow year round, harvesting two to three crops per year. The crops are grown for selling purposes but are also consumed at home. The commonly grown crops are seasonal vegetable crops: summer varieties of gourd, cucumber, aubergine, pumpkins, and autumn/winter varieties of cauliflower, cabbage, carrot, spinach, mustard (leaves), okra, and tomato. Farmers also grow flowers such as roses and marigolds and maintain orchards of mango and guava; but, rearing of orchards is not a common practice among migrant farmers.

In African countries as well, UA is a long standing indigenous tradition with around 20 million people widely practicing different forms of UA (Drechsel et al. 2006). As in the case of Delhi, open spaces in Dar es Salam are cultivated by farmers growing crops that have a high market demand—dominant crops are vegetables, nonetheless, cereals and oil seed crops are also cultivated. Proximity to the market allows farmers to make choices depending upon demand and supply, prices of the crops, etc. such that the same farmer may or may not grow the same crops in the subsequent year. As is true with most of the open farming spaces, tenure security, absence of a proper status of the immigrant farmer population, lack of subsidies, and high land competition offer a peculiar set of challenges. When irrigated, these horticulture systems are highly productive. For example, it is possible to obtain nine harvests of leafy vegetables like lettuce interrupted by one cabbage crop per year from the same plot of land. With no time to leave the land fallow, high agrichemical inputs are required to replenish the exported nutrients from the soil each year. Besides chemical fertilizers, farmers also prefer to use poultry manure as it releases the nutrients sufficiently fast and is suitable for fast growing vegetable crops. Livestock production is also done in these farms but it is mostly for self-consumption. The farmers use some level of mechanization but mostly rely on cheap labour available in the city.

A very interesting aspect of UA in Dar es Salem is that the overall open area under cultivation has remained same for the past two decades, their locations have changed considerably. Drechsel and Dongus (2010) report that of the total area under cultivation in Dar es Salam, only 23% has remained stationary since 2005 most of which is located in the floodplains. As these low lying areas are prone to flooding, they are not suitable for infrastructural development. The authors also report that most of the area that has been recently put under cultivation also exists around the river or railway lines where construction activities are prohibited. The income generated from these farms in various cities of West and East Africa largely depends on the irrigation facilities and availability of labour. To exemplify, the monthly revenue generated from lettuce-only farms in Dakar can be between US$213 and 236 per month. UA here can be an important source of income to alleviate poverty (people under poverty line have earnings less than US$1 per day) (Balde et al. 2005). In general, the average income generated in various African cities range from US$140 to 800 depending upon the proximity of farms to the urban centres and markets, types of crops grown, availability of irrigation facilities, and season. Some vegetables such as spring onions, lettuce, and cabbage that can be grown year round and with high demand in the urban centres generate high monthly incomes (Danso et al. 2002; Drechsel and Dongus 2010).

#### 17.3.1.3 Mono-cropping Systems

Farms using mono-cropping systems are usually found in the city peripheral regions. Examples of this system are described by Lemeilleur et al. (2003) and Bhatta and Doppler (2010). Mono-cropping systems are less input intensive with little or no irrigation and little augmentation by organic or inorganic fertilizers. Because of this reason the crops grown are locally adapted but with high demands in the surrounding area such as cassava, yam, sweet potatoes, corn, or rice. The land used for such cultivation systems is mostly unsuitable for housing. The crops produced are consumed at home and sold in the nearby markets, taking advantage of reduced transportation costs due to proximity to the cities. These systems, however, owing to their constraints associated with theft, labour availability, and competition with small-scale production within the city are losing relevance. But, monocropped systems can still be found in Cameroon and Nepal.

## 17.3.2 New Forms of Urban Agriculture

In response to little or no availability of land within city, low soil fertility, low water availability, pollution in urban areas and to minimize the impacts of conventional farming systems and use resources available locally, NFUA bank upon technological advances and support from local governments. To improve the profitability, these systems also need to be integrated into local farming and food systems. Mostly being soil less in nature, natural substrates such as coconut coir or other kinds of composts derived from urban wastes are used. NFUA are intensive in nature and their location and sizes vary on the availability of space. Organoponics, hydroponics, rooftop farming are some of the most common forms of this type of crop production.

### 17.3.2.1 Organoponics

Widely adopted in Cuba and Venezuela, organoponics is a form of organic farming where crop production occurs in raised platforms or beds filled with organic matter and soil. These structures are mainly located in areas where soil quality is poor. In Cuba, organoponic systems along with intensive gardens (located in areas with good drainage, adequate water supply, and fertile soil) were promoted and supported by the local government and Ministry of Agriculture in the 1980s. There were 1613 organoponic farms in Cuba in the year 1996 covering an area of 259 ha. These farms are also highly productive with an average yield of 16 kg/m$^2$. In the city of Havana in 1997, 15,092 ha of land was under some form of UA with around 8000 gardens with mixed ownership (Novo and Murphy 2000).

Production is based on agroecological principles and the farmers are encouraged to use locally available resources as agricultural inputs. Not only are soils augmented by bio-fertilizers, crop rotation and intercropping are used to reduce pest incidences. In case of pest infestations, biological controls in the form of entomopathogens, beneficial insects, and antagonists are employed. Botanical pesticides such as extracts of Neem (*Azadirachta indica*) and Melia (*Melia azedarach*) are also used along with pheromone traps. Four principle components of choosing varieties of seasonal crops (both local and hybrid), augmenting soils with compost and bio-fertilizers, using integrated pest management, and capacity building are used to make the NFUA system highly productive and sustainable such that today it meets

50% of the total needs of fruits and vegetables in Havana (Altieri et al. 1999; Orsini et al. 2013).

FAO supports the sustainable intensification of commercial market gardens in peri-urban areas. In the densely populated areas, the complementary strategy is to help low-income countries to build micro-gardens for a wide range of vegetables, roots, herbs, and tubers. Micro-gardening is an intensive cultivation of horticultural crops in small spaces such as patios, rooftops, and balconies using wooden crates, tyres, and custom build containers. The idea is to integrate the practice of vegetable and fruit production in city life using environmentally friendly techniques such as rain water harvesting and household waste management. A micro-garden can be as small as 1 $m^2$ and it is fairly easy to keep it productive. The substrate for growing crops is usually a mix of soil and other local materials such as coconut fibre, rice husks, laterite, etc. In the absence of availability of a substrate, water enriched in soluble fertilizers can be used. Like in the case of organoponic farms, the pests are controlled by non-chemical ways, for example, using insect proof nets or coloured sticky traps. Growing herbs such as basil, mint, parsley is also advised as they not only act as insect repellents, but they can also be used in the kitchen. Such micro-gardening projects have been successfully launched in a number of Central and South American countries. For example, a programme in Caracas helped 10,000 low-income families to grow a variety of vegetables. A similar intervention in Dakar helped more than 4000 urban residents, mostly women, to establish micro-gardens in backyards, patios, and terraces. Such interventions have also been promoted in Namibia, Niger, Rwanda, Senegal, and Gabon.

#### 17.3.2.2 Hydroponics

Hydroponics has existed from ancient times. The Hanging Gardens of Babylon and Floating Gardens of Aztecs used the principles of hydroponics to grow plants and crops. Hydroponics in cities in today's context has two important roles: hydroponics for greener and cleaner cities and hydroponics for food production. Hydroponic systems in cities often go up vertically because of limited space availability. They capture particulate matter and pollutants from the air through the plant foliage. Installation of hydroponics often leads to an immediate improvement in the temperature and air quality of the surroundings. The captive action is highest during the vegetative growth phase of the plants. Urban hydroponics plays an important role in food production in cities as well. It provides a sustainable and scalable solution to farming in urban area in the developing and less developed countries that often face shortage of water, high air pollution, competition for land, etc. Hydroponic systems range from highly sophisticated, capital-intensive that require highly skilled labour technologies to simple structures for the low-income urban population. Forms of hydroponics systems have been discussed in detail in Chap. 9.

Plants in a hydroponic system are grown without soil but use an inert material for anchoring roots such as clay, sand, rock wool, or coconut coir. The trays holding the inert material are mounted at an angle to allow the flow of nutrient solution back to the holding tank. The crops grown in hydroponic systems need to be monitored closely and pruned at regular intervals to maintain high yields. Almost all

**Table 17.1** Yields of Crops grown in a hydroponic system. Source: Orsini et al. (2014)

| Crops | Yield ($kg/m^2/day$) |
|---|---|
| Cucumber | 2.5 |
| Leeks | 0.5 |
| Green bean | 1.3 |
| Lettuce | 2.7 |
| Onion | 0.6 |
| Peapod | 1.5 |
| Salad greens | 2.3 |
| Tomatoes | 1.4 |

horticultural crops such as leafy greens, tomatoes, beans, squash, egg plants, peppers, cucumbers, leeks can be grown. However, cereal crops that need a high acreage to have notable yields such as wheat and maize are not suited for these systems. Underground vegetables that require deep growing beds are also not suitable for hydroponics. Yields of horticulture crops grown in hydroponics have very high yields—almost 100 times higher than conventionally grown crops. The produce is cleaner and contains less contaminants or pesticides. Table 17.1 shows the yields of some commonly grown horticulture crops using hydroponic techniques.

#### 17.3.2.3 Rooftop Gardens

To regulate and mitigate the plethora of environmental problems that arise in a city, development of green infrastructure has received a lot of attention. Urban green spaces, rooftop farms for food and non-food purposes, parks, allotment gardens, CSA, etc. all form a continuum of green spaces that have a variety of benefits for the city dwellers. Rooftop gardens (RTGs) are man-made green structures installed on the roof spaces of industrial, commercial, and residential buildings. RTGs can be used to grow food, provide shade and shelter, for storm water management, amelioration of air pollution, and thermal and energy benefits. However, limited research exists on the potential of RTGs in food production. In terms of biodiversity conservation in cities, RTGs create important green corridors to improve the dispersive capacity of species that are less mobile. This greatly limits the impacts of habitat fragmentation—common in built-up areas. High installation costs, significant maintenance requirements, and building weight restrictions are some of the factors that make implementing intensive horticultural production systems on rooftops challenging.

RTGs, if exploited and managed properly, can be utilized for producing large volumes of vegetables. Peck (2003) estimated that 4.7 million kg of vegetable crops could be produced from 650,000 $km^2$ of greened rooftops in Toronto. Hydroponic RTGs can produce around 20 $kg/m^2/year$ of food crops compared to conventional urban farms with average productivity of 1.5 $kg/m^2/year$ (Grewal and Grewal 2012). Besides, the usual crops of potato, lettuce, eggplants, peppers, and chillies, and fruits such as watermelons and cantaloupes can also be grown on RTGs. The amount of crop harvested varies widely with the type of cultivated medium Nutrient Film Technique—NFT, Floating and Substrate Systems. Seasonal production peak variations for crops for substrate and floating systems are higher compared to NFT. A combination of different techniques can be used to optimize production on RTGs because some crops grow well only in certain media. For example, fruit crops need substrate systems, whereas leafy greens do better with floating systems and NFT. If all the 3500 flat roofs of the city Bologna, Italy, with a total surface area of 82 ha, were converted to RTGs with estimated yields around 42 g/m$^2$ of vegetable crops per day, the entire rooftop surface area could yield 12,000 tonnes of vegetables per year. This implies that 77% of the city's vegetable requirement could be achieved by RTGs (Orsini et al. 2014).

In terms of increasing biodiversity, RTGs can be used as 'hotspots' in the network of green infrastructure. A network of RTGs can be designed such that they connect rooftops within 500-m distance and can be provided with wild flowers for pollen and nectar and shelters for insects. Studies indicate that this distance is appropriate for wild bees that have a foraging distance of 1 km (Osborne et al. 2008). Coccinellidae insects such as ladybirds are beneficial predators with longer flying distances that can rely on the alternative food sources provided by RTGs (Lundgren 2009). Such green roofs and RTGs can connect the bigger biodiversity centres of cities such as reserve forest area or parks constituting long flying routes that would ensure long term persistence and resilience of floral and faunal biodiversity in a city space. RTGs, while still in early phases of development, are listed as one of the viable options for 'novel opportunities for local food production'. Cities such as Toronto and Brooklyn are planning to install large scale rooftop food production systems not just for their provisioning ecosystem service but also for the many other environmental benefits they offer.[7,8]

## 17.4 Conclusion

This chapter explores the various forms of UA and NFUA systems that exist in cities and the variation of biodiversity that inhabit these systems depending on location, function, socio-economic importance as well as influences from policy frameworks. The traditional forms of UA described in this chapter provide many ecosystem services other than provisioning services. For example, allotment gardens, community gardens, and home gardens can have high to low productivities in terms of how they are managed, but their role is not just confined to food production. They act as important reservoirs of agri-biodiversity, native and spontaneous species, as well as ornamental plants. On the other hand, NFUA may have more defined functions such as food production, regulation of temperatures, improvement of pollination by providing green corridors, etc.

[7]https://www.toronto.ca/city-government/planning-development/official-plan-guidelines/green-roofs/

[8]https://brooklyneagle.com/articles/2019/04/18/green-roofs-will-soon-be-required-on-all-new-buildings-in-new-york-city/

In general, the reader will notice that the examples or case studies cited in this chapter do not provide a complete picture of how biodiversity is integrated in an urban ecosystem of one particular city. This is because the research studies on the various forms of UA in relation to provision of ES as well as conservation of biodiversity are quite scattered in terms of their study sites, scale, and types. Though most forms of UA are present in urban areas, their role in maintaining urban systems is not well documented. The chapter presents mosaic of such studies form various parts of the world. The review emphasizes a need to consider these systems in totality to understand their function in ES within city boundaries. No single form of UA can provide all ES; therefore, there is a need for better city planning to incorporate various forms. In general, this chapter points to a lack of research on the presence and functioning of these systems in developing and developed nations.

## References


Algert SJ, Baameur A, Renvall MJ (2014) Vegetable output and cost savings of community gardens in San Jose, California. J Acad Nutr Diet 114(7):1072–1076

Altieri MA, Companioni N, Cañizares K, Murphy C, Rosset P, Bourque M, Nicholls CI (1999) The greening of the "barrios": urban agriculture for food security in Cuba. Agric Hum Values 16 (2):131–140

Alvey AA (2006) Promoting and preserving biodiversity in the urban forest. Urban For Urban Green 5(4):195–201

Australian Farmers' Market Association (2012) https://farmersmarkets.org.au/

Balde D, Chaudhuri N, Dieng Y, Fall A, Gueye-Girardet A (2005) Providing the city with fresh vegetables from urban and peri-urban spaces: social and economical benefits and constraints, impacts on public health; the case of Patte d'Oie (Commune of Dakar, Senegal), juin 2004–juin 2005

Bell S, Fox-Kämper R, Keshavarz N, Benson M, Caputo S, Noori S, Voigt A (eds) (2016) Urban allotment gardens in Europe. Routledge, New York

Berghöfer A, Mader A, Patrickson S, Calcaterra E, Smit J, Blignaut J et al (2011) TEEB manual for cities: ecosystem services in urban management. The economics of ecosystems and biodiversity, Suiza

Bhatta GD, Doppler W (2010) Socio-economic and environmental aspects of farming practices in the peri-urban hinterlands of Nepal. J Agric Environ 11:26–39

Blair RB (1996) Land use and avian species diversity along an urban gradient. Ecol Appl 6 (2):506–519

Bolund P, Hunhammar S (1999) Ecosystem services in urban areas. Ecol Econ 29(2):293–301

Borysiak J, Mizgajski A, Speak A (2017) Floral biodiversity of allotment gardens and its contribution to urban green infrastructure. Urban Ecosyst 20(2):323–335

Brush S (2000) Ethnoecology, biodiversity and modernization in Andean potato agriculture. In: Minnis P (ed) Ethnobotany a reader. University of Oklahoma Press, Oklahoma, pp 283–306

CBD (1994) https://www.cbd.int/convention/articles/?a=cbd-02

Cook J, Oviatt K, Main DS, Kaur H, Brett J (2015) Re-conceptualizing urban agriculture: an exploration of farming along the banks of the Yamuna River in Delhi, India. Agric Hum Values 32(2):265–279

Crooks KR, Suarez AV, Bolger DT (2004) Avian assemblages along a gradient of urbanization in a highly fragmented landscape. Biol Conserv 115(3):451–462

Danso G, Drechsel P, Wiafe-Antwi T, Gyiele L (2002) Income of farming systems around Kumasi. Urban Agric Mag 7:5–6

de Oliveira JP, Balaban O, Doll CN, Moreno-Peñaranda R, Gasparatos A, Iossifova D, Suwa A (2011) Cities and biodiversity: perspectives and governance challenges for implementing the convention on biological diversity (CBD) at the city level. Biol Conserv 144(5):1302–1313
Dearborn DC, Kark S (2010) Motivations for conserving urban biodiversity. Conserv Biol 24 (2):432–440
Drechsel P, Dongus S (2010) Dynamics and sustainability of urban agriculture: examples from sub-Saharan Africa. Sustain Sci 5(1):69
DESA U (2016) The world's cities in 2016: Data booklet. Statistical Papers-United Nations (Ser. A), Population and Vital Statistics Report. United Nations, New York. https://doi.org/10.18356/8519891f-en
Drechsel P, Graefe S, Sonou M, Cofie OO (2006) Informal irrigation in urban West Africa: an overview, vol 102. IWMI, Colombo
Faeth SH, Warren PS, Shochat E, Marussich WA (2005) Trophic dynamics in urban communities. Bioscience 55(5):399–407
FAO (1999) https://www.fao.org/3/y5609e/y5609e01.htm
Galluzzi G, Eyzaguirre P, Negri V (2010) Home gardens: neglected hotspots of agro-biodiversity and cultural diversity. Biodivers Conserv 19(13):3635–3654
Garnett T (2000) Urban agriculture in London: rethinking our food economy. Growing cities, growing food. German Foundation for International Development, Feldafing, Germany, pp 477–500
Gilbert PR (2013) Deskilling, agrodiversity, and the seed trade: a view from contemporary British allotments. Agric Hum Values 30(1):101–114
Given DR (2000) Biodiversity of the urban environment: the importance of indigenous species and the role urban environments can play in their preservation. In: Urban biodiversity and ecology as a basis for holistic planning and design: proceedings of a workshop held at Lincoln University, pp 22–33
Grewal SS, Grewal PS (2012) Can cities become self-reliant in food? Cities 29(1):1–11
Hobbs RJ, Arico S, Aronson J, Baron JS, Bridgewater P, Cramer VA et al (2006) Novel ecosystems: theoretical and management aspects of the new ecological world order. Glob Ecol Biogeogr 15(1):1–7
Hodgkin T (2002) Home gardens and the maintenance of genetic diversity. In: Home gardens and in situ conservation of plant genetic resources in farming systems. Proceedings of the second international home gardens workshop, Witzenhausen, Federal Republic of Germany, 17–19 July, 2001. International Plant Genetic Resources Institute (IPGRI), Rome, Italy, pp 14–18
Hoornweg D, Hosseini M, Kennedy C, Behdadi A (2016) An urban approach to planetary boundaries. Ambio 45(5):567–580
IPBES S, Díaz J, Settele ES, Brondízio ES, Hien N, Ngo M et al (2019) The global assessment report on biodiversity and ecosystem services—Summary for policymakers. Intergovernmental Science-Policy Platform on Biodiversity and Ecosystem Services, Bonn
Kark S, Iwaniuk A, Schalimtzek A, Banker E (2007) Living in the city: can anyone become an 'urban exploiter'? J Biogeogr 34(4):638–651
LaPoint S, Gallery P, Wikelski M, Kays R (2013) Animal behavior, cost-based corridor models, and real corridors. Landsc Ecol 28(8):1615–1630
Lemeilleur S, Temple L, Kwa M (2003) Identification des systèmes de production du bananier dans l'agriculture urbaine et périurbaine de Yaoundé
Lewis AD, Bouman MJ, Winter A, Hasle E, Stotz D, Johnston MK, Czarnecki C (2019) Does nature need cities? Pollinators reveal a role for cities in wildlife conservation. Front Ecol Evol 7: 220
Luck GW (2007) A review of the relationships between human population density and biodiversity. Biol Rev 82(4):607–645
Lundgren JG (2009) Nutritional aspects of non-prey foods in the life histories of predaceous Coccinellidae. Biol Control 51(2):294–305
McClintock N, Cooper J (2010) Cultivating the commons an assessment of the potential for urban agriculture on Oakland's public land

McDonnell MJ, Hahs AK (2008) The use of gradient analysis studies in advancing our understanding of the ecology of urbanizing landscapes: current status and future directions. Landsc Ecol 23 (10):1143–1155
McKinney ML (2006) Urbanization as a major cause of biotic homogenization. Biol Conserv 127 (3):247–260
McPherson G, Simpson JR, Peper PJ, Maco SE, Xiao Q (2005) Municipal forest benefits and costs in five US cities. J For 103(8):411–416
MEA (2005) Ecosystems and human well-being, vol 5. Island Press, Washington, DC, p 563
Mok HF, Williamson VG, Grove JR, Burry K, Barker SF, Hamilton AJ (2014) Strawberry fields forever? Urban agriculture in developed countries: a review. Agron Sustain Dev 34(1):21–43
Morgan K, Sonnino R (2010) The urban foodscape: world cities and the new food equation. Camb J Reg Econ Soc 3(2):209–224
Müller N, Ignatieva M, Nilon CH, Werner P, Zipperer WC (2013) Patterns and trends in urban biodiversity and landscape design. In: Urbanization, biodiversity and ecosystem services: challenges and opportunities. Springer, Dordrecht, pp 123–174
National Gardening Association (2009) The impact of home and community gardening. National Gardening Association, Burlington, VT
National Green Tribunal India (2020). https://www.outlookindia.com/newsscroll/954-hectares-of-land-along-yamuna-floodplain-still-under-cultivation-ngt-panel/1886622
Novo MG, Murphy C (2000) Urban agriculture in the city of Havana: a popular response to a crisis. In: Bakker N, Dubbeling M, Gündel S, Sabel-Koshella U, de Zeeuw H (eds) Growing cities, growing food. Urban agriculture on the policy agenda. Zentralstelle für Ernährung und Landwirtschaft (ZEL), Feldafing, Germany, pp 329–346
Orsini F, Kahane R, Nono-Womdim R, Gianquinto G (2013) Urban agriculture in the developing world: a review. Agron Sustain Dev 33(4):695–720
Orsini F, Gasperi D, Marchetti L, Piovene C, Draghetti S, Ramazzotti S et al (2014) Exploring the production capacity of rooftop gardens (RTGs) in urban agriculture: the potential impact on food and nutrition security, biodiversity and other ecosystem services in the city of Bologna. Food Secur 6(6):781–792
Osborne JL, Martin AP, Carreck NL, Swain JL, Knight ME, Goulson D et al (2008) Bumblebee flight distances in relation to the forage landscape. J Anim Ecol 77(2):406–415
Peck S (2003) Towards an integrated green roof infrastructure evaluation for Toronto. Green Roof Infrastruct Monitor 5(5):4–5
Rockström J, Steffen W, Noone K, Persson Å, Chapin FS III, Lambin E et al (2009) Planetary boundaries: exploring the safe operating space for humanity. Ecol Soc 14(2):32
Rosenzweig ML (2003) Reconciliation ecology and the future of species diversity. Oryx 37 (2):194–205
Shochat E, Warren PS, Faeth SH, McIntyre NE, Hope D (2006) From patterns to emerging processes in mechanistic urban ecology. Trends Ecol Evol 21(4):186–191
Smith RM, Thompson K, Hodgson JG, Warren PH, Gaston KJ (2006) Urban domestic gardens (IX): composition and richness of the vascular plant flora, and implications for native biodiversity. Biol Conserv 129(3):312–322
Sordi M, Polegri L, Negri V (2008) Biodiversitadi interesse agrario nel comprensorio del Lago Trasimeno. University of Perugia/Trasimeno Lake Regional Park, Italy, pp 978–988
Speak AF, Mizgajski A, Borysiak J (2015) Allotment gardens and parks: provision of ecosystem services with an emphasis on biodiversity. Urban For Urban Green 14(4):772–781
Tam KC, Bonebrake TC (2016) Lepidoptera-habitat relationships in urban parks. Urban Ecosyst 19:721
Vitiello D, Nairn M, Planning P (2009) Community gardening in Philadelphia: 2008 harvest report, Penn planning and urban studies. University of Pennsylvania, Philadelphia, p 68
Whittaker RH (1972) Evolution and measurement of species diversity. Taxon 21(2–3):213–251

# 18 The Future of New Forms of Urban Agriculture: A Call for Evidence

Harpreet Kaur and Jessica Ann Diehl

The chapters included in this volume present an overview of frameworks, technical innovations, and case studies for the analysis of urban and peri-urban agriculture from all over the world. The book spotlights new and emerging forms of farming in a city and its interaction with the four pillars of food production—land, labour, water, and biodiversity. The contributing authors have used exploratory, descriptive, and experimental methodologies to examine the transformation of urban agricultural practices in cities with different ecologies, development pressures and priorities, and policy engagement. Chapters feature management of land for protecting and supporting agriculture in cities, use of city waste for enabling production, the changing face of an urban farmer, and impacts on biodiversity. In this volume, we have compiled diverse research on the intrinsic relationships and dependency of urban agriculture in its many manifestations on basic resources of land and water as well as humans and biota. The purpose of this book was to address what we saw as the absence of reliable empirical data on the scale and impact of urban resources on NFUA. We attempt to fill a gap in our understanding of the viability and sustainability of UA and ease urban planner's reluctance to embrace the concept. In this final chapter, we summarize some of the key findings and attempt to elaborate on synergies and ways forward.

H. Kaur
Agroparistech, Paris, France

J. A. Diehl (✉)
Department of Architecture, National University of Singapore, Singapore, Singapore
e-mail: akijac@nus.edu.sg



J. A. Diehl, H. Kaur (eds.), *New Forms of Urban Agriculture: An Urban Ecology Perspective*, https://doi.org/10.1007/978-981-16-3738-4_18

## 18.1 Space Required. . .Land and Soil Optional

A fundamental question for urban agriculture is land availability—due to scarcity, politics, and value fluctuation. Part I addresses the issue of where to locate NFUA. Growing food requires a dedicated space for cultivation—and while open, undeveloped land is the default, NFUA can occupy smaller, underutilized spaces and structures, e.g., vacant rooftops, across the urban gradient. But, what the research in this section makes evident is that policies, data, incomes, and contextual conditions influence land availability and spatial requirements.

Urban planning practice is rather young worldwide and far from being well established and commonly accepted. There are only a few countries where urban agriculture is added as a mandate to city development plans. Demographic and economic expansion of cities tends to be accompanied by spatial expansion, resulting in encroachments of green spaces generating land use conflicts. Importance of creating green spaces within the city boundaries though is gaining importance in recent times, urban planning focusing on integration of such spaces is nascent and not commonly accepted at the policy level. Only a few cities worldwide currently focus on this including different forms of urban green infrastructure in long-term development plans. In Chap. 2, Weichold highlights the challenges and potentials of managing land use in an urban set-up in a developed country. To achieve this, the author uses a contemporary case study addressing the future of spatial development in the Luxembourg region. Using landscape suitability analysis through composite mapping—Analytical Hierarchy Process (AHP)—she investigates how Luxembourg can develop land without abandoning its fertile, productive agricultural land. Weichold uses a combination of parameters primarily used in terms of agricultural production along with AHP which had not previously been applied to the case of Luxembourg. The study adds valuable insights into current planning discussions in Luxembourg by providing spatial planning guidelines—at an urban and regional scale—where to develop land while respecting the ecological qualities of its land, in this case the soil quality. She highlights peculiar issues such as land speculation on undeveloped but buildable land are prevalent and have increased the pressure on undevelopable land recently. Agricultural land-use suitability using GIS and the AHP technique for planning areas can be marked for UA. The AHP, however, is highly dependent on the availability of accurate datasets; the methodology cannot be used for developing areas where data sets are incomplete or most of the time not reliable and where rapid rural-urban transformation is happening.

Alarmingly, it is in places where rapid urbanization is occurring that land suitability and spatial planning for NFUA are needed most. By 2030, the number of megacities (cities having >10 million inhabitants) is expected to rise to 43 from the current total of 33, and incipient megacities (5–10 million inhabitants) will rise to 66. Most of these emerging cities, clustered in African and Asian countries, are often haphazardly planned. India presently has five such megacities: Delhi, Mumbai, Kolkata, Bengaluru, and Chennai. In Chap. 3, Jain estimates the loss of soil-based agriculture in rapidly urbanizing Delhi by the year 2030 using remote sensing. She further distinguishes UA into food and non-food forms and discusses its role in mitigating the by-products of urbanization, viz. land use land cover change and urban heat islands. The author draws attention to the fact that although the city is witnessing a continuous shrinking in traditional agricultural lands that support crop production and animal husbandry, adoption of several NFUA presents a promising picture. But there is a greater emphasis on non-food UA practices, which is not helpful in ensuring food security of urban poor; however, the other benefits provided by these systems cannot be side-lined. Nonetheless, the author highlights that 'no single UA strategy fits best' especially in case of Delhi owing to its sheer size, population density, and variability in the economic status of Delhi's residents. Noting the paucity of research on NFUA, she emphasizes that more qualitative and quantitative research is required based on city-specific financial, socio-economic, environmental, and ecological cost-benefits for the implementation and planning of UA strategies.

Echoing Jain's call for research, Kumar et al. (Chap. 4) contribute a descriptive case study summary of different models of NFUA across different cities in India. Their sample frame includes diverse cities such as New Delhi, Chennai, Mumbai, Thiruvananthapuram, Pune, Bangalore, Coimbatore, Chandigarh, and Jaipur. They compare the per capita availability of land in these cities acting as a proxy to the land development pressure and the constraints it poses to allocate dedicated land parcels for urban agriculture. Dwelling further, the authors discuss drivers and institutional arrangements facilitating NFUA models. To enrich their discourse, the team also conducted interviews with experts and practitioners selected from the key case study cities.

Among the variety of NFUA options discussed in the literature and this edited book that range from personal use kitchen gardens to high-tech vertical farming structures, roof top farming (RTF) is one of the few that does not compete with the existing land use. Su and Ow (Chap. 5) argue that despite the theoretical arguments in the favour of RTF, very little activity of this type of NFUA has been observed. Such farms are mostly established for non-profit purposes such as social, educational, or experimental, and even at times to enhance the quality of life. Regardless of scenario, RTF set-ups are usually small scale with low production capacity. The authors highlight the fact that the potential earnings from roof farming are still not particularly attractive compared to an average worker's income. In an attempt to bridge this informational gap, Su and Ow calculated the potential profits generated from vegetable production with a reflector-assisted double-layer system in 150 $m^2$ screen house in Guangzhou, China. Using such innovations, the authors estimate that the incomes would be equivalent to 84% of average monthly compensation of a local worker. This study presents a compelling example of upscaling production using low-tech innovation to achieve competent incomes. The team, nevertheless, heeds the need for governmental initiatives to make this practice part of urban development, in the absence of which RTF will likely remain an interesting academic exercise and a missed opportunity.

Cities are pressed for land availability all around the world, however, this case scenario is rather complicated for Small Island Developing States like Vanuatu (Chap. 6) because of their isolated location, small land areas separated by vast oceans, and the costs of providing basic services. Vanuatu is one of the most vulnerable countries in the world to climate change and is predicted to experience sea level rise and saltwater inundation, rising average temperatures, as well as days of extreme heat with greater variation in rainfall including more intense droughts and rain. Increasing and securing peri-urban food production are therefore important for the city's long-term food and nutrition security. Urban spread, however, threatens prime agricultural land on the peri-urban fringe. In a case study of the historic and current planning context of Vanuatu, James concludes that strong political will, transparency in decision-making, and economic incentives to maintain peri-urban farms are essential to ensuring farmland is not only available (i.e., not developed) but that it is also accessible to and utilized by those who want to farm.

## 18.2 Reframing Water Scarcity

Part II of the book includes four chapters that focus on water—as an input and output of urban agriculture; as a necessity and a by-product. Agriculture consumes up to 70% of fresh water available globally. Water stress has become a global denominator in wet and dry climates alike, particularly in urban settings. At the same time, urban centres all over the world are experiencing accelerating rates of failure in water treatment and distribution infrastructure. In fact, municipalities on almost every continent consume the smallest share of water—the water resources used to produce their food have footprints elsewhere. Water for food (as it is a basic need) is almost incomparable to other beneficial uses, with urban agriculture perhaps even more so because it tends to be smaller scale, more accessible to the poor, and socially more equitable than imported food from large farms. Increasing food production using less water (SDG 6) and reversing land degradation and increasing the sustainable management of soils (SDG 15) can be addressed in urban agriculture just as it can in large scale farms, where SDGs 6 and 15 are primarily targeted.

In Chap. 7, Fisher emphasizes an engineering perspective on water use for urban agriculture. He discusses the role of effective and informed policy decisions and technology innovation as a panacea for improving the water (especially wastewater) consumption efficiency in urban agriculture. He adds and presents case studies of countries where a combination of innovation in policy and technology has resulted in tremendous growth in urban agriculture. The author analyses and compares the amount of water required to grow fresh vegetables in the large commercial farms in the peri-urban areas and small-scale gardens in an urban area based on studies from the USA, Western Europe, and Africa. He cites two specific cases—one from Holland and the other from Kigali, Rwanda—that exemplify the role of region-specific, low-tech advancements in the impacts of engineering perspectives to upscale farming in an urban set-up. In the case of Holland, establishing partnerships with research institutions, using wastewater, waste heat, and energy from power plants and factories, efficient lighting technologies, hydroponics, and green houses, tremendously increased crop productivity and resource use efficiency. Strikingly, most horticulture vegetables grown in such facilities occupy only 0.2% of the country's land mass—yet Holland is the world's second largest exporter in terms of value of horticultural vegetables. Fisher highlights the role of Life Cycle Analysis (LCA) to evaluate controlled-environment farming impacts especially on energy consumption. Comparatively, in Kigali, Rwanda, he highlights how the development of region-specific engineering innovations can provide an impetus to urban agriculture. There, planners and designers created a watershed land use pro-forma known as an Environmental Treatment Zone (ETZ) to manage and collect stormwater and wastewater and to compost other organic wastes in the proximity. Wastewater treatment, biogas production, and composting would make resources usable for farms co-located in the ETZ.

Investigating further on the use of wastewater in farming in urban set-ups, Miller-Robbie and Ramaswami (Chap. 8) describe the complex interactions of food-energy-water-health (FEW-Health) nexus in a city. They assess location-specific impacts in a site study in Hyderabad, India. The authors highlight the fact that in many low- and middle-income cities, a large proportion of domestic wastewater is not treated, and nearby rivers receive the contaminated water. Often in such cities, the surrounding agricultural land and irrigation water is forced downstream of urban riverine/wastewater discharges. The reuse of wastewater in UA is not new or rare—an estimated 200 million farmers irrigate at least 20 million hectares with raw or partially treated wastewater. The question, therefore, is no longer if wastewater should be used, but how it can be made more sustainable and safer for irrigation. The authors quantify the FEW-Health nexus impacts of wastewater treatment including GHG emissions (energy and non-energy related), economic benefits to farmers (food production), water reuse (water savings), monetary cost (infrastructure), and health benefits to society (pathogen risk reduction in food). They evaluate trade-offs for three farm sites irrigated by different sources of water: groundwater, treated effluent water, and untreated surface water representative of the sewage-contaminated riverine system. The results indicate that wastewater treatment improved water quality as expected. However, crop quality especially for spinach did not improve significantly as measured by the presence of *E. coli* and nematode ova. They found both behavioural and other environmental causes to contribute to such contamination. The study also reported a reduction in system wide GHG emissions as a result of investments in constructing, operating, and maintaining wastewater treatment facilities. The authors conclude that more field studies that measure pathogens on crops in different locations, climates, and seasons are required to make informed decisions on the use of wastewater for irrigation purposes in cities.

Chapter 9 discusses the solutions Frontier Agriculture (FA) provided to bridge disparities in food security status for refugee populations living in the Middle East and North African Region (MENA). FA comprises of climate-smart, water saving technologies both low (such as Wick system, Deep Water Culture—Kratky Method, Ebb and Flow method) or high tech (Nutrient Film Technique, Hydroponics, Aeroponics, Aquaponics). Verner et al. estimate the suitability of such systems to improve the livelihoods and well-being of refugee as well as host communities. The authors point out that MENA region faces two main challenges: water crisis (already a water scarce region, MENA uses 85% of its water in agriculture) and a recent escalation of refugees (with 18 million adult and youth living under refugee-like conditions). They discuss the applicability and potential of FA using FEW-nexus to address these challenges. The authors further argue that water and agriculture are key to stabilization and ultimately to peacebuilding processes. These practices will help produce, process, and sell food items without long transportation needs, generate income and employment, and rebuild household-level food security. Most importantly, this will help reconstruct social integration and cohesion following a bottom-up approach. This argument is supported by policy simulations which show that the typical development policies that only invest in education, skills, and employability are unlikely to succeed in improving welfare unless accompanied by measures that parallel create adequate economic opportunities.

Chapter 10 shifts the focus back on utilization of waste materials for farming purposes in and around Durban, South Africa. Like most developing countries, South Africa is witnessing migration of people from rural areas to cities like Durban. Such migration has burdened the health and food security scenario in the city. In response to the growing demand for fresh and local produce, many smallholder gardeners and entrepreneurs have established urban farms in and around the city. However, these farms face challenges such as availability of safe water, organic waste, protection of crops from theft, and destruction by animals. Leech reports that this food growing practice was taking place despite regulations issued by the Provincial Town and Regional Planning Commission of the KwaZulu-Natal Land Use Management Scheme, which prevented smallholder farmers from practising agriculture in built environments in KwaZulu-Natal, including Durban metro. It also meant that the food produced within the urban built environment bypassed the checks and regulations otherwise imposed on food articles produced by conventional rural agriculture. Nonetheless, the Land Use Management Act, 2013 allowed farming in urban areas in the interest of supporting disadvantaged communities in cities. Under this scheme, several food security projects were launched to encourage and provide support to the local farming community in selected areas for cultivation of cost-effective crops mostly vegetables. Participants in these projects formed Agricultural Hubs according to their geographical proximities. Leech worked with one such Agricultural Hub formed by a local farming enterprise in Mariannridge and surroundings in western Durban. He describes the services, guidance, and trainings provided to the identified food gardeners on sustainable and organic farming practices for treating and using waste for fertilization, collecting and using rain water for irrigation in autumn and winter months, and integrated pest control techniques. The author however raises an important issue that although the National Environmental Health Norms and Standards, 2015 highlights nine important check points for food and nutritional health of which five are applicable to food production such water quality monitoring, the urban agriculturists are made responsible to attend to these checks themselves which raises concerns over the quality of food produced.

## 18.3 The People Who Grow and Harvest the Crops

Part III of this edited book focuses on labour—the people who engage in urban agriculture and its associated activities. While humans are not a requirement of plants in an ecological system, they are a necessary part of a cultivated system. Globally, agriculture provides a livelihood for roughly 2 billion people globally and about 800 million in urban contexts. Different types of NFUA require different knowledge base and skills—ranging from manual labourers cultivating traditional fields to highly skilled operators of high-tech cropping systems, people are needed to grow and harvest crops. With NFUA occupying diverse spaces and with multifunctional uses and activities, there are a variety of people involved across diverse social groups in formal and informal capacities. In some cases, agriculture is a livelihood pursuit, for others it meets food security needs. The practitioners have a direct impact on what is grown, how it is grown, and where it is sold—in effect, labour is a critical resource influencing how NFUA integrates or separates from the social, ecological, and economic systems of the city.

Oviatt (Chap. 11) demonstrates how the personal characteristics of producers influence the way urban agriculture manifests and the benefits associated with it. The author stresses on the importance of taking producer traits into consideration when studying urban agriculture. Through a case study of farmers associated with AGRUPAR (Agricultura Urbana Participativa: Participatory Urban Agriculture) in Quito, she found agricultural practice to be influenced by three producer characteristics: migration history, age, and gender. AGRUPAR is a long-standing program through the municipal government of Quito that supports thousands of producers throughout the city, ranging from hobbyist gardeners to larger-scale producers who depend on intensive production for their livelihood. The program targets low-income communities, using urban agriculture as a tool for addressing poverty and food insecurity, but membership is open to anyone interested in joining. Her findings demonstrate the differences that exist because of the producer characteristics. For example, the women who participated in AGRUPAR on average had less economic returns from their garden sales compared to men, nonetheless they benefitted more in terms of developing relationships and engaging with others in the community. The author with this work provides a unique perspective on the benefits of urban agriculture. Because producers are situated differentially in the social context, Oviatt suggests that the practice of urban production and the benefits associated with it do not accrue uniformly among all producers. Thus, it is important to reflect critically on how NFUA interact with other factors in a producer's life to lead to differential outcomes.

Shifting to the other side of the globe and to a high-income city context, Chap. 12 describes the changing profile of farmers in Singapore. With less than 1% of farmland dedicated to agricultural purposes, the city-state is heavily dependent upon imports to meet the nation's food requirements. With limited area dedicated to farming practices, Singapore offers high entry barriers to set-up farms. Farming enthusiasts thus started using underutilized spaces such as rooftops and indoor spaces for growing crops. The farming systems in these vary from soil based to vertical farming systems which are tech-oriented. As a result, fewer and skilled labour is required to maintain the farms. The transition from conventional to high-tech farming methods has changed the demography of labour involved. In this chapter Sia and Diehl compare how the labour force in the food production system is changing in conventional and urban-integrated farms. They dive deeper into labour characteristics—education background, skills, wages earned, and motivations to join the food industry. The labour force in conventional farms is constituted by immigrants from neighbouring developing countries such Bangladesh and Myanmar who were farm workers in their country of origin, either as commercial farmers or involved in subsistence farming. As a result, most of them had experience in farming, with specific knowledge in traditional farming in contrast to the urban Singapore farmers who turned to farming as a career choice with no prior farming experience except short-term volunteering at a farm. In terms of wages earned, workers on conventional farms earned less compared to urban farmers. In both cases, however, the average incomes were far below the median pay in the country even if the urban farmers were degree holders. While majority of the urban farmers turned to farming to be part of the food system most commercial farm owners also acknowledged the difficulties of sustaining farming in Singapore, and would rather not continue farming if they had the financial resources to exit the industry.

Chapter 14 brings this section together by conceptualizing labour. Diehl describes labour as a fundamental resource and a social network as part of a larger goal of creating sustainable farming livelihoods. She uses case studies from four cities to study diverse farmer-labour social networks: Delhi, India; Jakarta, Indonesia; Singapore; and Sydney, Australia. Drawing on her research in these cities presenting very different takes on urban farming, she raises an important question—who is a farmer? What skills, qualifications, licenses, land titles, formal contracts must the person possess to undertake the role of a farmer? Not surprisingly, these characteristics are poles apart while defining a famer in a developed versus developing country. Mostly farming in urban set-ups is primarily driven as a livelihood making pursuit by farmers and then as an effort to increase food security at city level. Therefore, the day-to-day and operational incentives and barriers that farmers face play a primary role in farmer behaviour and decision-making. Using Sustainable Livelihoods Framework, Diehl focuses specifically on how influence and access to livelihood assets impact livelihood strategies in order to achieve livelihood outcomes. Farmer-labour associations vary widely among the studied cities. In Delhi, where the farmers hired farmers to assist in farming activities in exchange for their own labour, urban farms in Jakarta were tended by the farmers' family with members helping during key times such as for planting or harvesting. The labour force in urban farms in Singapore mostly consisted of immigrants. However, language barriers limited interactions between farmers and workers and a distinct hierarchy in farmer-worker social networks existed which was practically absent in urban farmers and labours in Delhi. Sydney presented the most diverse farmer demographic among the four cities. The urban farmers ranged from multi-generational pioneer families to recent migrants with diverse origins. The four case cities demonstrate that even within one type of urban agriculture, namely commercial urban farms, labour requirements and practices can be diverse. Diehl concludes that the farmer-labour social network profoundly impacts the feasibility and sustainability of operating a farm.

## 18.4 From Pollinators to Pachyderms: NFUA and Urban Habitats

The final section of this book, Part IV puts the focus on biodiversity. Urban development converts natural environments into built-up, man-made environments, destroying and/or displacing the original ecological habitats. Biodiversity is threatened globally by urbanization, exacerbated by pollution and fragmented ecosystems. But what do we know about preservation and restoration potential of NFUA? And how do changing urban ecosystems impact the potential for growing food in the city? That this section is the shortest in this edited volume is a reflection of the gap in research—and an important area for future studies.

The section begins with the issue of pollinators. There is a decline in bee populations around the world—a critical threat to the food security. The primary causes are use of pesticides, parasites, and absence of floral resources. Although wild bees are impacted by these stressors, much of the focus on declining bee populations has been on honey bees. Smith et al. (Chap. 15) explore questions about urban apiculturists' perceptions and knowledge of wild bees, as well as the impact of urban apiculture on wild bees in community gardens. While agricultural intensification (e.g., transitions to monocultures or innovations in systemic pesticides) makes rural areas increasingly inhospitable to bees, cities can be important refuges that allow diverse bee communities to persist in spite of changes in surrounding natural and rural areas. Authors found that bees—both wild and bumble bees—were present regardless of whether the gardens kept beehives. The wild bee groups were almost as abundant as the bumble bee groups. They also found that in case of California there was no strong evidence that honey bees depressed wild bee communities or populations; they co-existed. The authors report that there was no consistent negative relationship between the abundance of honey and bumble bees—in most gardens, where there were more resources, there were more bees of all kinds. Wild bee abundance was positively related to weed density and that of honey bees to density of woody vegetation, which suggests that some degree of resource partitioning exists. However, such resource partitioning could be a problem for wild bees if the nectar and pollen in weeds were of lower nutritional quality or if foraging on weeds took more time or energy. To understand whether urban beekeeping provided any benefits to wild bees, the authors report that this was more dependent on the gardeners. In general, in the gardens where people kept honey bees, there was a significantly higher number of flowering plant species and weed species than in gardens without honey beehives.

In Chap. 16, Srinivasan et al. shifts our focus to human wildlife conflict in peri-urban areas, in this case: elephants. They call this unique process *synurbisation*, adaptation in a nonhuman species to human-induced change. A change in the land use, shifting from single- to double- or multiple cropping every year, and an overall increase in built-up areas, has transformed the foraging behaviours of elephants in southern India. Animals that can adapt increase their movement rates in human-dominated areas or completely move out of high human-density sites to settle in alternate areas. The Asian elephant *Elephas maximus* has home ranges varying from 250 to 1000 $km^2$ and their ranging behaviour is largely influenced by ecological as well as anthropogenic factors. In this chapter the authors assess the impact of changing land use and management practices on the distribution of elephants in a human-dominated landscape over a period of 16 years. Crop fields accounted for approximately 18.33% of their study area, a significant proportion of this is peri-urban in nature, adjacent to several important urban centres such as Bangalore and Hosur of southern Karnataka and western Tamil Nadu. These human-dominated areas harbour elephants as well, even though elephant populations may be in significantly lower densities. The elephants in the study landscape typically exhibited important behavioural adaptations at the level of the population, manifest principally in age-, gender-, or group size-based differences in seasonal habitat use, but also at the individual level through the display of idiosyncratic behaviours. For example, female elephant groups with young ones adopted a more risk-averse strategy compared to their male counterparts. Their study revealed that rurban male elephants not only frequented the intensely human-dominated agricultural regions but also became resident in those areas. There has been a 750-fold increase in crop-loss claims consequent to elephant depredation in the past one and a half decades. Encounters between the two species often become critical, with approximately 400 people and 150 elephants succumbing to human–elephant conflict across the country annually. This chapter presents a unique example of human wildlife conflict involving large mammals. Urban sprawl changes the agriculture patterns around densely populated areas, which often tend to encroach the natural habitats where wildlife conservation laws are not strict—mainly in developing nations. This presents a stark difference between peri-urban agriculture characteristics between the developed and developing nations. Human wildlife conflict remains a poorly researched area in terms of its interaction with peri-urban farming practices.

In the final Chap. 17, Kaur provides an overview of the various types of food and non-food UA systems. The first section of the chapter mainly focuses on the 'why' and 'how' of biodiversity conservation in cities. Cities are found in all types of natural systems; therefore, 'which biodiversity needs to be conserved' remains a complex question. She cites various examples from different cities to exemplify how the objectives and forms of biodiversity conversation vary. It is not always necessary that biodiversity in a dense urban agglomeration can accommodate native biodiversity; the way a city fosters biodiversity varies with its individual context. For example, the city of Cape Town is an urban hotspot within the global biodiversity hotspot Cape Floristic Region with a high proportion of endemic species that are threatened. Because it is important for this region to save its endemic species, the arrangements made (such as leaving corridors) differ from facilities made by other cities where conservation of a particular set of biodiversity is not a priority. In general, the motivations of biodiversity conservation within a city space vary considerably. Part II of the chapter focuses specifically on the forms of UA both traditional and new and the role they play in maintaining agri-biodiversity and provision of ecosystem services. Traditional UA forms such as allotment gardens and home gardens act as important reservoirs of native and introduced species. The ecosystem services provided by them are not restricted to food production, they provide many regulatory as well as cultural services. On the other hand, the new forms of UA usually have a specified function such as regulation of temperature, food production, etc. The author concludes that no single form of UA can provide a variety of ecosystem services and that a combination of such forms is necessary for improving the sustainability of a city.

## 18.5 NFUA: A Call for More Evidence

Globalisation, climate change, and urbanization increasingly have significant and deep impacts on local and global food systems. With growing urban populations' rising demand for food coupled with the inherent risks of relying on the global food system, many city governments are developing planning strategies for implementing urban agriculture at different scales. The explicit framing of urban agriculture as embedded in the urban 'system' challenges cities to address agriculture not as a discrete activity, but as an important component of economic, social, and ecological systems. Forms of urban agriculture that provide benefits in addition to food production—ecosystem services such as those mentioned throughout this volume—have been termed New Forms of Urban Agriculture (NFUA). While NFUA have the potential to provide diverse benefits to humans, we shift the focus to investigate the potential impacts of urban resources on NFUA.

Using an urban ecology lens, we wanted to know more about how urban resources of land, water/waste, labour, and biodiversity impact NFUA. As a concept, urban ecology is complex and dynamic, comprised of energy flows and feedbacks among the components—boundaries are not fixed. And, investigation requires diverse disciplines, expertise, and geographic contexts—which was precisely our starting point as editors. We invited academics and practitioners embedded in every continent and across developing and developed contexts. Initially, we thought the majority of research would be originating in the science disciplines: environmental sciences, ecology, engineering, etc. However, urban planners, anthropologists, public health and behavioural scientists, and economists, among others were found entrenched in this small but active space. Thus, the authors included in this edited book represent a diverse range of technical, scientific, planning, horticultural, and social expertise.

NFUA research is in its infancy, and yet it is imperative to understand the interactions between the resources needed to make it feasible as well as the tangible and intangible urban conditions that make it sustainable. We also need to understand how NFUA impact urban *to rural* systems. We know how to grow food in cities without land—rooftop, vertical, and indoor systems are found in nearly every major city globally. We know how to do it (feasibility), but is it sustainable?

We suggest a few areas of research that could prove useful. At the city-scale, we need to optimize diverse and varied land use types contextualized by current/future actual use. We suggest more research on how UA can be integrated across landscape typologies including but not limited to degraded forests, urban parks and open green spaces, common spaces in residential areas, unused and underutilized spaces, commercial areas, institutional zones, streetscapes, underutilized built structures, and water and drainage networks. What are the forms and functions of agriculture that are most suitable for these different contexts, conditions, and timeframes, e.g., fallow land that will be developed? Rather than approaching NFUA as a potential conflicting use, we should consider it an asset—a way to add value to existing urban spaces. And, we should consider temporality and short-term and flexible systems that can be adapted as needs change. This will push the boundaries of the multifunctional potential of NFUA.

Agriculture in the urban context is fundamentally different from rural agriculture precisely because it is embedded in urban economic, social, environmental, *and* infrastructure systems. Thus, research should be driven by theoretical frameworks and models that reflect the complex and dynamic city system—even if only to contextualize study of a singular attribute. It is not enough to know how, where, or when it happens, we also need to know why. In terms of technology, more research on open versus closed loop systems is needed. How do we leverage urban infrastructure, e.g., utilities including water and energy, while also making use of natural resources of solar and hydrology? Site suitability models can be designed to consider data on urban heat island effects, wind flows, and stormwater runoff. And, Life Cycle Assessments (LCAs) that incorporate social and economic measures in addition to energy flows should be a standard for evaluating NFUA across global cities. Without data, we cannot know if or how it works.

NUFA are increasingly discussed and promoted by academics, practitioners, and governments and the social and policy research is trending up; however, more empirical data is needed to bolster the conversation. Going forward, we recommend more research overlapping themes that emerged throughout this volume including land availability versus utilization; technology and economics; human–animal relationships; water scarcity and human health; food security and income; and comparisons across developed and developing contexts—what can be learned? There is important research being conducted in all these areas, but it is like fragmented forest patches requiring corridors to establish linkages. We hope this book offers that.

The manufacturer's authorised representative in the EU is Springer Nature Customer Service Centre GmbH, Europaplatz 3, 69115 Heidelberg, Germany. If you have any concerns regarding our products, please contact ProductSafety@springernature.com

Printed and bound by CPI Group (UK) Ltd, Croydon, CR0 4YY
30/09/2026
02225333-0011